Advances in Mineral Processing and Hydrometallurgy—3rd Edition

Advances in Mineral Processing and Hydrometallurgy—3rd Edition

Guest Editor

Corby G. Anderson

MDPI

Basel • Beijing • Wuhan • Barcelona • Belgrade • Novi Sad • Cluj • Manchester

Guest Editor
Corby G. Anderson
Kroll Institute for Extractive
Metallurgy
Colorado School of Mines
Golden, CO
USA

Editorial Office
MDPI AG
Grosspeteranlage 5
4052 Basel, Switzerland

This is a reprint of the Special Issue, published open access by the journal *Metals* (ISSN 2075-4701), freely accessible at: https://www.mdpi.com/journal/metals/special_issues/9259176SUL.

For citation purposes, cite each article independently as indicated on the article page online and as indicated below:

Lastname, A.A.; Lastname, B.B. Article Title. *Journal Name* **Year**, *Volume Number*, Page Range.

ISBN 978-3-7258-6442-3 (Hbk)
ISBN 978-3-7258-6443-0 (PDF)
https://doi.org/10.3390/books978-3-7258-6443-0

Contents

About the Editor

Corby G. Anderson

Corby G. Anderson is a Licensed Professional Chemical Engineer with over 45 years of global experience in industrial operations, corporate-level management, engineering, design, consulting, teaching, research, and professional service. He is a native of Butte, America. His career includes positions with Thiokol Chemical Corporation, Key Tronic Corporation, Sunshine Mining and Refining Company, H. A. Simons Ltd., and at CAMP-Montana Tech. He holds a BSc in Chemical Engineering from Montana State University, an MSc from Montana Tech in Metallurgical Engineering, and a PhD from the University of Idaho in Mining Engineering—Metallurgy. He is a Fellow of both the Institution of Chemical Engineers and the Institute of Materials, Minerals, and Mining. He has directed or co-directed over 40 graduate students. He shares 19 international patents and a new patent application covering several innovative technologies, 6 of which were successfully reduced to industrial practice with 6 licensed. He currently directs the Kroll Institute for Extractive Metallurgy, and serves as the Harrison Western as part of the Mining Engineering at the Colorado School of Mines. He is also the CSM Director for the Center for Resource Recovery and Recycling. In 2009, he was honored by the Society for Mining Metallurgy and Exploration with the Milton E. Wadsworth Extractive Metallurgy Award for his contributions in hydrometallurgical research. In 2015, he was awarded the International Precious Metals Institute's Tanaka Distinguished Achievement Award. In 2016, he received the Distinguished Member Award from the Society for Mining, Metallurgy, and Exploration, and became a Distinguished Member of the U of Idaho Academy of Engineering. In 2017, he received the EPD Distinguished Lecturer Award from The Minerals, Metals, and Materials Society. In 2019, he was named as a Henry Krumb Distinguished SME Lecturer. In 2019, he was also appointed and now serves as a Visiting Faculty within the Minerals Engineering Department of Central South University in China, the largest program of Mineral Processing and Extractive Metallurgy in the world. In both 2016 and in 2021, he received an Outstanding Faculty Award from the Colorado School of Mines. He was also elected in 2021 to the Sigma Xi Scientific Research Honor Society. In 2022, he received the TMS EPD Distinguished Service Award for his career contributions. Recently, Stanford University listed him in their Top 2% Scientists Worldwide University Rankings. Finally, in 2022, he and his co-authors received the SME Taggart Award for a notable contribution to the science of mineral processing.

metals MDPI

Editorial

Advances in Mineral Processing and Hydrometallurgy—3rd Edition

Corby G. Anderson

Kroll Institute for Extractive Metallurgy, Mining Engineering Department, Colorado School of Mines, Golden, CO 80401, USA; cganders@mines.edu

1. Introduction and Scope

After the success of the two prior editions, a third edition of a Special Issue of *Metals* has been commissioned, specifically devoted to aspects of mineral processing and hydrometallurgy. This is a unique, focused Special Issue, as the specific combination of hydrometallurgy and mineral processing is not common. Both hydrometallurgy and mineral processing will be relied upon to produce the critical minerals and metals needed for renewable energy.

As Free [1] states, water makes aqueous processing of metals possible. Water is an unusual substance from a chemical perspective. Most substances with similar molecular weights, such as methane and ammonia, are gases at room temperature. However, water is a liquid at room temperature. Water's unusual properties are primarily due to hydrogen bonding effects. These are related to the tendency of hydrogen to donate electrons and the tendency of oxygen to accept electrons. As such, hydrometallurgy offers superior separations, as well as the capability to treat low-grade materials to produce fungible compounds and metals.

Wills and Napier Munn [2] recently published a seminal distinct book on mineral processing. They stated "As-mined" or "run-of-mine" ore consists of valuable minerals and gangue. Mineral processing—sometimes called ore dressing, mineral dressing, or milling—follows mining and helps prepare the ore for extraction of the valuable metal in the case of metallic ores, and it produces a commercial product of products, such as iron ore and coal. Apart from regulating the size of the ore, it is a process of physically separating the grains of valuable minerals from the gangue minerals to produce an enriched portion, or concentrate, containing most of the valuable minerals, as well as a discard, or tailing, containing predominantly the gangue minerals. The importance of mineral processing is taken for granted today, but it is interesting to reflect upon the fact that less than a century ago, ore concentration was often a crude operation, involving simple gravity and hand-sorting techniques performed by the mining engineers. The twentieth century saw the development of mineral processing as a serious and important professional discipline, and without physical separation, the concentration of many ores, and particularly the metalliferous ores, would be hopelessly uneconomic.

Fuerstenau and Han [3] published a book devoted exclusively to the combination of mineral processing and hydrometallurgy. In this unique publication, they stated that to advance technology in the production of material resources, nations look towards practicing and future engineers. Current and future mineral processing engineers must obtain sound and rigorous training in the sciences and technologies that are essential for effective resource development. Many industrial and academic leaders have recognized the need for more textbooks and references in this critical area. This was the driving force for writing a comprehensive reference book that covers mineral processing and hydrometallurgical extraction.

Hence, this third-edition Special Issue was designed to highlight research papers on characterization, recycling and waste minimization, mineralogy, geometallurgy, thermodynamics, kinetics, comminution, classification, physical separation, liquid–solid separation, leaching, solvent extraction, ion exchange, activated carbon, precipitation, reduction, process economics, and process control. Specific metals of interest include gold, silver, PGMs, aluminum, copper, zinc, lead, nickel, and titanium. Critical *Metals* articles on topics such as lithium, antimony, tellurium, gallium, germanium, cobalt, graphite, indium, and rare earth metals were welcomed. Eleven high-quality, peer-reviewed articles from around the globe were selected.

2. Overview of the Published Articles

Wang et al. [Contribution 1] provided an article on the in situ leaching of ionic rare earths. The ion exchange reaction between rare earth ions and leaching agent ions was carried out, allowing the rare earth ions to be leached from the ore body as the leaching solution flows through the pores. This indicates that the leaching process of rare earth ions is closely related to the seepage field, ion exchange field, and ion migration process of the leaching solution. In this study, an ionic rare earth mine located in Longnan of Jiangxi Province was taken as the research object. By conducting nuclear magnetic resonance scanning on the ore samples of this mine and vectorizing the nuclear magnetic resonance images, a two-dimensional geometric model of pores was obtained. Then, COMSOL Multiphysics software was used to establish a coupled numerical model of seepage–exchange–migration of the ionic rare earth mine during the leaching process at the pore scale to study the seepage situation of leaching solution with different injection strengths and concentrations, as well as the exchange and migration process. The results show that increasing the concentration of magnesium ions can increase the difference in ion diffusion concentration, accelerate the forward exchange rate of ions, promote the forward exchange reaction, and improve the concentration gradient of rare earth ions in the leaching solution. The more significant the diffusion effect, the higher the ion migration rate, simultaneously inhibiting the reverse adsorption of rare earth ions and accelerating the leaching efficiency of rare earth ions. In addition, increasing the strength of the injection solution allows rare earth ions to leach out of the ore body earlier; shortens the leaching cycle; and, thus, reduces the peak concentration of leached rare earth ions. By analyzing the effects of the strength of the injection solution and leaching concentration on ionic rare earth leaching, the influence of those two factors on engineering economy can be briefly evaluated, which can be provided as a reference for the optimization of ionic rare earth mining technology

Qin et al. [Contribution 2] investigated the effects of nickel content in nickel-bearing pyrite on photocatalytic properties, light absorption properties, and oxidative decomposition of thiosulfate were studied. The leaching experiments show that the consumption of thiosulfate in the Cu^{2+}-ethylenediamine (en)-S_2O_{32-} system increases with an increase in nickel content in nickel-bearing pyrite. The consumption of $Cu(en)_{22+}$ initially increases and then decreases with an increase in leaching time. There is a clear correlation between the change trend in its consumption and the doping amount of nickel in pyrite. The XPS results show that, in the Cu^{2+}-ethylenediamine (en)-S_2O_{32-} gold leaching system (temperature 25 °C, time 35 h, solution: 0.1 mol/L S_2O_{32-}, 5 mmol/L $Cu(en)_{22+}$, 200 mL solution), the nickel of pyrite-containing nickel can be transferred to the leaching solution and becomes nickel ion. In this leaching system, Cu(II), which was originally complexed with en, is reduced to Cu(I) in a brief time. The consumption of $Cu(en)_{22+}$ increased rapidly in the 5 h period and then decreased gradually after 5 h. The results showed that the presence of free Ni^{2+} in the solution facilitated the conversion of bivalent copper ions to monovalent copper ions. Free Ni^{2+} ions can compete with Cu^{2+} ions for en ligands. When ethylenediamine complexes with Ni^{2+}, the decomposition of $Cu(en)_2^{2+}$ into Cu(en)+ and en occurs more rapidly. And the en, which was originally to be oxidized with Cu(en)+ to form $Cu(en)_2^{2+}$, forms $Ni(en)_2^{2+}$. As a result, the concentration of $Cu(en)_2^{2+}$ continues to decrease in a brief period

Alguacil [Contribution 3] reported that though indium has been removed from the fifth list (2023) of critical raw materials for the European Union list of critical metals, its recovery is still of paramount importance due to its wide use in a series of high-tech industries. As its recovery is strongly associated with zinc mining, the recycling of In-bearing wastes is also of interest, for both profitable and environmental reasons. With unit operations (in hydrometallurgy and pyrometallurgy or extractive metallurgy) playing a key role in the recycling of indium, the present work reviewed the most recent innovations (2024) regarding the use of these operations in the recovery from this valuable metal from different solid or liquid wastes

Wang et al. [Contribution 4] provided another article noting that the vast seabed holds tremendous resource potential that can provide necessary materials for future human societal development. This study focuses on the mineralogy of seafloor manganese nodules off the coast of China in the Western Pacific and the primary techniques for extracting valuable metal elements from manganese nodules. The research indicates that the main valuable metal elements in the manganese nodules from this region include Cu, Co, Ni, Mn, Fe, etc. The key to extracting these valuable metals lies in reducing Mn(IV) to Mn(II) to disrupt the structure of the nodules, thereby releasing the valuable elements. The extraction processes for the main valuable metal elements of manganese nodules are mainly divided into two categories: pyrometallurgical–hydrometallurgical and solely hydrometallurgical. To cope with the challenges of environmental change and improve utilization efficiency, bioleaching, hydrogen metallurgy, and co-extraction are gaining increasing attention. For promoting commercialization, the future development of manganese nodule resources can refer to the technical route of efficient short-process extraction technology, the comprehensive recovery of associated resources, and tail-free utilization.

Menad et al. [Contribution 5] noted that currently Electric Arc Furnace Slag (EAFS) is undervalued and only used in road construction, while blast furnace slag (BFS) is used as an interesting alternative in construction materials to replace natural aggregates in the manufacture of concrete. Steel slag (SS) represents a promising secondary resource due to its high content of critical metals, such as chromium (Cr) and vanadium (V). These metals are essential for various strategic industries, making it crucial to consider slag as a resource rather than waste. However, the primary challenge lies in selectively recovering these valuable metals. In this work, we explore the development of a hydrometallurgical process aimed at efficiently extracting Cr and V from Electric Arc Furnace Slag (EAFS). The characterization of the investigated EAFS shows that the main crystalline phases contained in this heterogeneous material are srebrodolskite, larnite, hematite, and spinel (magnesio-chromite). The targeted metals seem to be dispersed in various mineral species contained in the SS. An innovative hydrometallurgical method has been explored, involving physical preparation by co-grinding slag with alkaline reagents followed by treatment in a microwave furnace to modify the metal-bearing species to facilitate metal processing dissolution. The results obtained showed that the leaching rates of Cr and V were, respectively, 100% and 65% after 15 min of treatment in the microwave furnace, while after 2 h of conventional heat treatment, as explored in a previous study, 98% and 63% of the Cr and V were leached, respectively.

Redrovan [Contribution 6] noted that the thiosulfate–glycine–copper system has emerged as a promising alternative for gold recovery, offering significant advantages over cyanidation and ammoniacal thiosulfate leaching. Recognizing the limitations of thiosulfate degradation in ammoniacal systems, this study focused on optimizing the thiosulfate–glycine–copper system for gold recovery using an auriferous ore with (10 g t^{-1}) of Au. The ore was associated with aluminosilicates such as grossular (64%) and clinochlore (12%). The leaching conditions were systematically varied, including thiosulfate (0.5–1 M); glycine (0.3–1.75 M); copper sulfate (2–10 mM); pH (9.3–10.5); temperature (20–60 °C), 6 h; and potassium permanganate concentrations (0.004–0.04 M). Dosing intervals were also optimized. Thus, the best conditions were thiosulfate (0.7 M), glycine (1.75 M), copper sulfate (5 mM), pH 9.3, 60 °C, and permanganate addition every 2 h. This system achieved 89.3% gold recovery in just 6 h, comparable to cyanidation (89.8% in 24 h) and ammoniacal thiosulfate (58% in 6 h), but without generating toxic effluents, such as in the cyanidation process. Additionally, a gold dissolution mechanism was proposed, highlighting glycine's role in stabilizing cupric ions and enhancing thiosulfate efficiency. This study underscores the thiosulfate–glycine–copper system as a sustainable and effective method for gold recovery.

Kenzhaliyev et al. [Contribution 7] studied an innovative approach to processing refractory zinc-bearing clinker through the synergistic application of microwave thermal treatment and ultrasonic-assisted leaching. Microwave irradiation induces phase transformations in the clinker, improving its reactivity and facilitating subsequent zinc dissolution, while ultrasonic cavitation enhances mass transfer by disrupting passivation layers. Key process parameters, including acid concentration, temperature, pulp density, and leaching time, were systematically investigated using response surface methodology (RSM) and central composite design (CCD). The results demonstrate that the optimized process conditions led to a significant increase in zinc recovery from refractory materials.

Hirata-Miyasaki and Anderson [Contribution 8] noted that demand and prices for antimony have increased over the last few years. Recycling supplied 15% of domestic consumption in the US, while the remaining 85% was imported. Hydrometallurgical processes have long used alkaline sulfide systems and hydrochloric acid, closing doors on innovative approaches. Bromine compounds have been recently used to recover PGMs and REEs successfully; thus, antimony leaching with bromine compounds is theoretically feasible. This research was conducted to develop a viable technology for hydrobromic acid between 50 °C and 70 °C as a leaching reagent on dross through single- and two-stage leaching using design of experiment (DoE) and adding sustainability to current industrial processes while minimizing waste products in recycling processes. The preliminary results showed that bromine, specifically hydrobromic acid, can be used as a leaching reagent for antimony dissolution. By decreasing the lead content in the solids and increasing the concentration, temperature, and reaction time, antimony leaching from the dross increased from 20% to 50%. The findings, coupled with acid regeneration, can be implemented as an alternative to other reagents in industrial plants.

Safiulina et al. [Contribution 9] stated that N,O-donor hybrid heterocyclic extractants have immense potential for separation of actinides from lanthanides in spent nuclear fuel reprocessing processes. We demonstrate that this type of reagent can be used for primary concentration of actinides contained in eudialyte, a promising mineral containing a heavy group of lanthanides. With respect to lanthanide ions, the efficiency of their extraction decreases in the series L3 >> L1 > L2, and the extraction of actinides decreases in the series L1 $\approx$ L3 >> L2. For the extractant, L2 based on 2,2′-bipyridine-6,6′-dicarboxylic acid diamide, the efficiency of lanthanide purification from U, Th exceeds 50. The structure and stereochemical features of the ligands do not have a significant effect on the composition of the formed complexes. The solvation numbers are close to 1 for all range f-elements studied, except for thorium, which indicates the predominant formation of complexes with the composition ratio of 1:1. Solvation numbers 1.4–1.5 are observed for thorium(IV), and the established values indicate the formation of a mixture of complexes with the composition ratios of 1:1 and 2:1.

Seo et al. [Contribution 10] addressed the recycling of spent lithium ion batteries that generates Na_2SO_4-containing wastewater, resulting in environmental problems and resource losses. This study investigates a treatment method using bipolar membrane electrodialysis (BMED) to recover H_2SO_4 and NaOH from Na_2SO_4 solutions. The acid and/or base recovery efficiency, energy consumption, operational stability, and economic feasibility of two BMED configurations, i.e., two- and three-compartment systems, were systematically compared. The current density, initial concentrations, and initial concentrations and volumes of the acid and base were optimized under constant current conditions. The three-compartment system exhibited higher acid purity and stability, whereas the two-compartment system offered lower energy consumption and membrane degradation. Under optimal conditions, both systems successfully recovered H_2SO_4 and NaOH from Na_2SO_4-containing wastewater. A techno-economic analysis based on a lab-scale process demonstrated the cost advantages of the two-compartment system versus the long-term operational stability of the three-compartment system. These findings suggest that BMED is a viable and effective solution for the treatment of Na_2SO_4-containing wastewater from battery recycling processes.

Jovanović et al. [Contribution 11] examined the leaching behavior of copper and iron from a sphalerite concentrate in sulfuric acid utilizing an ensemble MnO_2–KI oxidation system. Temperature was shown to significantly influence leaching kinetics, with efficiency notably improving between 40 °C and 80 °C. The introduction of KI affected the balance between sulfur passivation and oxidant availability, facilitating increased leaching efficiencies. At 3 wt.% KI, maximum recoveries of 82.1% Cu and 85.3% Fe were achieved, indicating a notable decrease in surface passivation. Kinetic study analysis revealed low activation energies of 28.90 kJ mol^{-1} for copper and 18.94 kJ mol^{-1} for iron, indicating that both processes proceed readily at moderate temperature regimes. Despite being diffusion-controlled, the mechanisms of dissolution are different—iron leaching is more complicated, involving pyrite oxidation, sulfur layer formation, transformation to marcasite, and, ultimately, iron(III) release, whereas copper leaching involves direct interaction of chalcopyrite with the oxidants, like the behavior of sphalerite. The MnO_2–KI-H_2SO_4 system's promise for more sustainable polymetallic concentrate processing is demonstrated by its reduced energy requirements and enhanced leaching efficiency.

3. Summary and Outlook

This third Special Issue of *Metals* was again well supported by diverse submissions and a final book publication of eleven high-quality peer-reviewed articles. Due to this continued success, another Special Issue ("Advances in Mineral Processing and Hydrometallurgy—4th Edition" https://www.mdpi.com/journal/metals/special_issues/M92897U5GK (accessed on 7 August 2025)) has been commissioned as a follow-up to highlight contributions from the global hydrometallurgy and mineral processing community.

Conflicts of Interest: The author declares no conflicts of interest.

List of Contributions:

1. Wang, D.; Wu, F.; Rao, Y.; Zhao, Z.; Xu, W.; Han, M. Microscopic Simulation of RE3+ Migration in Ion-Type Rare Earth Ores Based on Navier–Stokes Equation—Exchange Reaction—Ion Migration Coupling. *Metals* **2024**, *14*, 1130. https://doi.org/10.3390/met14101130.
2. Qin, X.; Zhang, T.; Qin, W.; Zhang, H. Effect of Nickel Impurities in Pyrite on Catalytic Degradation of Thiosulfate. *Metals* **2024**, *14*, 1256. https://doi.org/10.3390/met14111256.

3. Alguacil, F.J. Recent Advances in Indium Recovery. *Metals* **2024**, *14*, 1282. https://doi.org/10.3390/met14111282.
4. Wang, X.; Qin, W.; Li, M.; Liu, X.; Cheng, Y.; Chen, S.; Yang, C. Mineralogy of Deep-Sea Manganese Nodules and Advances in Extraction Technology of Valuable Elements from Manganese Nodules. *Metals* **2024**, *14*, 1359. https://doi.org/10.3390/met14121359.
5. Menad, N.-E.; Seron, A.; Bensamdi, S. Innovative Process for Strategic Metal Recovery from Electric Arc Furnace Slag by Alkaline Leaching. *Metals* **2024**, *14*, 1364. https://doi.org/10.3390/met14121364.
6. Redrovan, A.S.; Torre, E.d.l.; Aragón-Tobar, C.F., Gold Leaching from an Auriferous Ore by Alkaline Thiosulfate–Glycine–Copper Solution. *Metals* **2025**, *15*, 204. https://doi.org/10.3390/met15020204.
7. Kenzhaliyev, B.; Surkova, T.; Berkinbayeva, A.; Baltabekova, Z.; Smailov, K.; Abikak, Y.; Saulebekkyzy, S.; Tolegenova, N.; Omirbek, T.; Dosymbaeva, Z. Innovative Methods for Intensifying the Processing of Zinc Clinker: Synergy of Microwave Treatment and Ultrasonic Leaching. *Metals* **2025**, *15*, 246. https://doi.org/10.3390/met15030246.
8. Hirata-Miyasaki, A.; Anderson, C.G. Effect of Lead in Antimony and Tin Dissolution from Recycled Lead–Acid Battery Dross in Hydrobromic Acid Solution. *Metals* **2025**, *15*, 356. https://doi.org/10.3390/met15040356.
9. Safiulina, A.M.; Lizunov, A.V.; Ivanov, A.V.; Borisova, N.E.; Matveev, P.I.; Aksenov, S.M.; Ivanets, D.V. Uranium(VI), Thorium(IV), and Lanthanides(III) Extraction from the Eudialyte Concentrate Using the N,O-Hybrid Heterocyclic Reagents. *Metals* **2025**, *15*, 494. https://doi.org/10.3390/met15050494.
10. Jovanovi'c, A.; Andic, D.; Bugarcic, M.; Jelic, I.; Vujovic, N.; Anderson, C.; Sokic, M. Recovery of Cu and Fe from a Sphalerite Concentrate by the MnO_2–KI Leaching Oxidation System. *Metals* **2025**, *15*, 1039. https://doi.org/10.3390/met15091039.
11. Seo, M.; Lee, Y.; Kim, J.; Chang, J.; Cho, Y.; Ahn, J. Treatment of Na_2SO_4-Containing Wastewater Generated During the Recycling of Spent Lithium-Ion Batteries: Comparative Study on the Operating Modes of Bipolar Membrane Electrodialysis. *Metals* **2025**, *15*, 1067. https://doi.org/10.3390/met15101067.

References

1. Free, M.L. *Hydrometallurgy Fundamentals and Applications*, 2nd ed.; The Minerals, Metals and Materials Society; Springer: Berlin/Heidelberg, Germany, 2022; 458p, ISSN 2367-1182.
2. Wills, B.A.; Napier Munn, T.J. *Mineral Processing Technology*; Elsevier Science & Technology Books: Amsterdam, The Netherlands, 2006; 220p, ISBN 0750644508.
3. Fuerstenau, M.C.; Han, K.N. *Principles of Mineral Processing*; Society for Mining, Metallurgy and Exploration: Englewood, CO, USA, 2003; 573p, ISBN 0-87335-167-3.

metals

MDPI

Article

Treatment of Na_2SO_4-Containing Wastewater Generated During the Recycling of Spent Lithium-Ion Batteries: Comparative Study on the Operating Modes of Bipolar Membrane Electro-Dialysis

Minhyuk Seo [1], Youngjae Lee [1], Junhee Kim [1], Jaehyuk Chang [1], Yeonchul Cho [2] and Jaewoo Ahn [1,*]

1 Department of Advanced Materials Sci. & Eng., Daejin University, Pocheon 11159, Republic of Korea; ab224008@daejin.ac.kr (M.S.); ab224009@daejin.ac.kr (Y.L.); ab225025@daejin.ac.kr (J.K.); ab225026@daejin.ac.kr (J.C.)

2 SungEel HiTech Co., Ltd., Gunsan 54002, Republic of Korea; ycjo@sungeelht.com

* Correspondence: jwahn@daejin.ac.kr; Tel.: +82-010-8644-1982

Abstract: The recycling of spent lithium-ion batteries generates Na_2SO_4-containing wastewater, resulting in environmental problems and resource losses. This study investigates a treatment method employing bipolar membrane electrodialysis (BMED) to recover H_2SO_4 and NaOH from such wastewater. The acid and base recovery efficiencies, energy consumption, operational stability, and economic feasibility of two BMED configurations, i.e., two- and three-compartment systems, were systematically compared. The current density, initial concentrations of the feed, and initial concentrations and volumes of the acid and base were optimized under constant current conditions. The three-compartment system exhibited higher acid purity and stability, whereas the other system exhibited lower energy consumption and membrane degradation. Under optimal conditions, both systems successfully recovered H_2SO_4 and NaOH from the Na_2SO_4-containing wastewater. A techno-economic analysis based on a lab-scale process revealed that the two-compartment system exhibited cost effectiveness while the three-compartment system showed long-term operational stability. These findings suggest that BMED is a viable and effective solution for the treatment of Na_2SO_4-containing wastewater generated from battery recycling processes.

Keywords: lithium-ion battery recycling; wastewater; bipolar membrane electrodialysis

1. Introduction

The current recycling processes for spent lithium-ion batteries (LIBs) involve using H_2SO_4 during leaching and solvent extraction stages to recover valuable metals, such as Li, Co, Ni, and Mn, and using NaOH during precipitation and solvent extraction stages [1–3]. Figure 1 illustrates a representative LIB recycling process, wherein approximately 0.81 tons of Na_2SO_4 is generated per ton of treated NCM622-derived black mass. Approximately 0.41 tons of Na_2SO_4 is generated during the Li recovery step, wherein Li is precipitated as Li_2CO_3 using Na_2CO_3 after the recovery of Co, Ni, and Mn. This results in a total Na_2SO_4-containing wastewater mass of approximately 1.22 tons [4,5]. This wastewater is also a major byproduct of cathode precursor synthesis, wherein co-precipitation reactions occur between sulfate-based metal salts and NaOH [6,7].

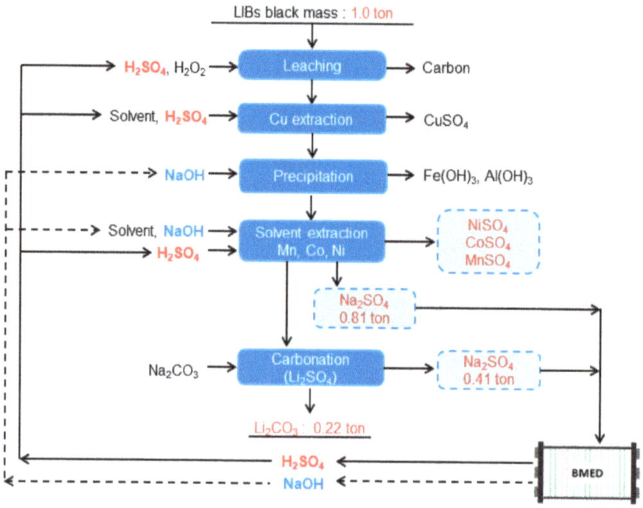

Figure 1. Schematic of spent LIB recycling process.

Currently, there are no regulations for the direct discharge of Na_2SO_4-containing wastewater. Na_2SO_4 is chemically stable and nontoxic, posing little immediate environmental threat. However, elevated concentrations of Na_2SO_4 in water systems can increase the water ionic strength, disturb aquatic ecosystems, and cause issues such as scaling and corrosion in industrial systems [7–11]. For treatment, evaporation and concentration processes, e.g., mechanical vapor recompression, are primarily used to convert wastewater into a solid form suitable for sale. However, these methods require high-temperature or vacuum conditions, which increases their capital and energy costs [12].

Recently, bipolar membrane (BPM) electrodialysis (BMED), which is an electro-separation technology, has gained attention as a regeneration method for NaOH and H_2SO_4 from Na_2SO_4-containing wastewater [13,14]. This process involves water dissociation within the BPM to generate H^+ and OH^-, which convert the salt into an acid and a base. There are two main BMED operating modes, i.e., two- and three-compartment systems [15,16].

Figure 2a shows the membrane structure of a two-compartment BMED system, consisting of a BPM and a cation exchange membrane (CEM). Under applied voltage, sodium ions (Na^+) pass through the cation exchange membranes and react with OH^- to form NaOH [13–17]. S. Xue investigated the effects of operating mode and process parameters on the treatment of CH_3COONa using a two-compartment BMED system [18], and Y. Lee reported that NaOH (2.96 M) and H_2SO_4 (1.24 M) were recovered from Na_2SO_4-containing wastewater; however, the recovered acid solution contained Na_2SO_4 (0.58 M) as an impurity [19].

Figure 2b illustrates the structure of a three-compartment BMED system, which employs CEM, an anion exchange membrane (AEM) and BPM. In this configuration, Na^+ and SO_4^{2-} migrate through the CEM and AEM, respectively, forming NaOH and H_2SO_4 [13–17]. J. Kroupa successfully recovered NaOH (4.00 wt%) and H_2SO_4 (4.44 wt%) with purities of 99% and 98%, respectively, using this system [20]. Wang et al. recovered 1.22 M H_2SO_4 from 15 wt% Na_2SO_4 treatment (energy consumption = 2.59 kWh/kg) [21].

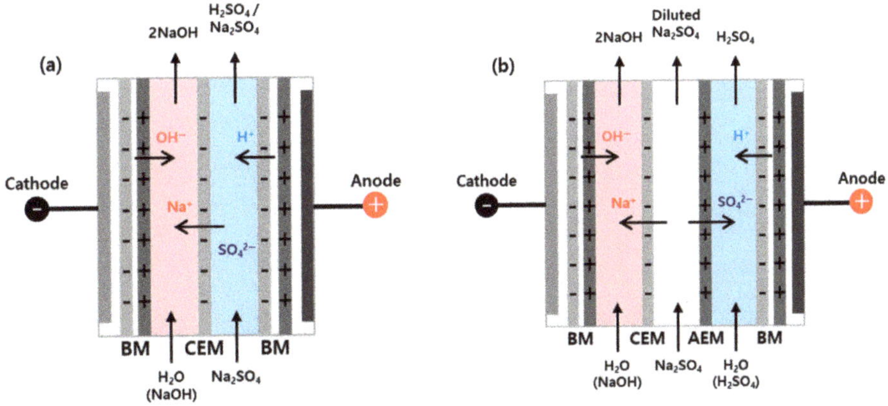

Figure 2. Schematic of the two- and three-compartment BMED process.

The two-compartment BMED system exhibits a simpler structure with fewer membranes, leading to easy construction and operation. However, only an acid or a base can be recovered with high purity, and the ion-exchange membrane is exposed to acidic and basic environments, reducing its durability [21–23]. In contrast, the three-compartment BMED system allows high-purity recovery of acids and bases and minimizes back-diffusion and cross-contamination effects. However, it has a more complex structure and involves a greater number of membranes, increasing the maintenance costs [21–23].

This study compared the operational results of a two-compartment BMED system (using a CEM and BPM) and a three-compartment BMED system (using a CEM, AEM, and BPM). Experimental analyses under various operating conditions and different process parameters were conducted to evaluate the Na_2SO_4 desalination performance as well as acid and base recovery characteristics, focusing on the recovery rate of NaOH, concentrations of recovered NaOH and H_2SO_4, current efficiency (CE), and energy consumption. The optimal operating conditions were determined for each system and economic feasibility and operational stability were validated.

Previous studies did not conduct experiments that directly compared two-compartment and three-compartment BMED systems under identical operating conditions. In contrast, this study provides a direct comparison of these two configurations, specifically focusing on Na_2SO_4-containing wastewater from LIB recycling. It offers an integrated evaluation of recovery efficiency, energy consumption, membrane durability, and cost factors, aspects that have rarely been investigated simultaneously in the existing literature.

2. Materials and Methods

Herein, two types of BMED systems were designed for the recovery of NaOH and H_2SO_4 from Na_2SO_4 solutions and compared. As shown in Figure 3, the performance of the two- and three-compartment systems was evaluated under identical current conditions and similar solution compositions.

The experimental setup consisted of feed, base recovery, and acid recovery compartments as well as a membrane stack, circulation pumps, a power supply, conductivity and pH meters, and a control panel. Both systems employed Neosepta-series membranes (ASTOM Co., Ltd., Minato, Japan).

The membrane stack in the three-compartment system was composed of alternately arranged membranes: 11 BPMs, 10 CEMs, and 10 AEMs (effective membrane area = 0.055 m^2).

The structure of the two-compartment system was similar to that of the three-compartment system but without the AEMs.

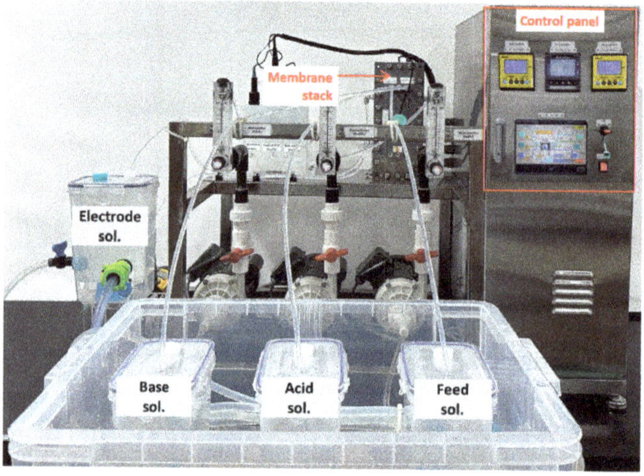

Figure 3. Photograph of the used lab-scale BMED.

The feed solution was a simulated solution of Na_2SO_4 (1.30 M) prepared to resemble the composition of Na_2SO_4-containing wastewater generated from the recycling of spent LIBs. Analytical-grade anhydrous Na_2SO_4 (Daejung Chem.) was dissolved in ultrapure water. In the base and acid recovery compartments, NaOH (0.1 M) and H_2SO_4 (0.1 M) solutions were used, respectively. The NaOH and H_2SO_4 solutions were prepared by diluting a NaOH solution (40 wt%) and a H_2SO_4 solution (95 wt%), respectively, with ultrapure water. A Na_2SO_4 solution (5 wt%) was used as the electrolyte to maintain conductivity between the membrane stack and electrodes.

Both systems were operated under galvanostatic (constant current) conditions. Depending on the experimental conditions, the initial concentrations (0.05–0.50 M) and volumes (750–1500 mL) of the acid and base, feed solution concentration (0.43–1.30 M), and current density (250–450 A/m^2) were adjusted. In the two-compartment system, the experiment was terminated after 480 min, whereas the experiment in the three-compartment system was terminated when the cell voltage exceeded 3 V because the resistance increase due to the decreasing conductivity during operation may considerably affect membrane stability at voltages higher than this threshold.

Samples were collected at predetermined time intervals for determining the degree of desalination of the feed solution and confirm the generation of H_2SO_4 and NaOH. The concentrations of H_2SO_4 and NaOH in the collected samples were quantitatively analyzed via acid–base titration. Titration was performed using standard solutions (0.1 N) of NaOH and H_2SO_4 as well as methyl orange and phenolphthalein as indicators for the base and acid samples, respectively.

To precisely determine the Na^+ and $SO_4{}^{2-}$ concentrations in each compartment, inductively coupled plasma–atomic emission spectroscopy was employed. To assess water transport characteristics, changes in feed, base and acid solution volume before and after the reaction were measured based on scale markings on solution chambers.

Next, obtained experimental data were used to quantitatively evaluate the water migration rate (W_{mg}; %), recovery yield of the base and acid (%), energy consumption (E_{weight}; kWh/kg), $C.E$ (%), and ion flux (J; $mol/m^2{\cdot}h$) using Equations (1)–(5).

W_{mg} was calculated using Equation (1):

$$W_{mg}(\%) = \frac{V_f - V_i}{V_i} \tag{1}$$

where V_i is the initial volume of the solution and V_f is the volume of the solution at the end of the experiment.

The recovery of the base and acid (%) was calculated using Equation (2):

$$Recovery(\%) = \frac{m_f}{m_i} \times 100 \tag{2}$$

where m_f is the number of moles of Na^+ or SO_4^{2-} in the base or acid recovery compartment and m_i is the number of moles of Na^+ or SO_4^{2-} initially present in the feed solution.

$C.E$ was calculated using Equation (3):

$$C.E(\%) = \frac{m_{actual}}{m_{theortical}} \tag{3}$$

where $m_{theortical}$ and m_{actual} are the theoretical and actual amounts of the generated product, respectively.

E_{weight} was calculated using Equation (4):

$$E_{weight} = \frac{I \times \int_0^t V\,dt}{W} \tag{4}$$

where V is the applied voltage (V), I is the applied current (A), t is the operation time (h), and W is the weight of initially loaded Na_2SO_4 (kg).

J was calculated using Equation (5):

$$J = \frac{m_f}{At} \tag{5}$$

where m_f is the number of moles of Na^+ or SO_4^{2-} transferred to the base or acid recovery compartment, A is the effective membrane area of the ion-exchange membrane (0.055 m^2), and t is the operation time.

3. Results and Discussion

3.1. Comparison Between the Acid and Base Recovery Efficiencies of the Two- and Three-Compartment Systems

Experiments were conducted under identical operating conditions to compare the Na_2SO_4 desalination efficiencies as well as acid and base recovery efficiencies of the two- and three-compartment systems. The feed solution consisted of 1.5 L of Na_2SO_4 (1.30 M) and the acid and base recovery compartments were filled with 1.5 L of H_2SO_4 (0.1 M) and NaOH (0.1 M), respectively. The experiments were conducted at a current density of 360 A/m^2 and terminated after 480 min (Figure 4 and Table 1).

As shown in Figure 4a, the Na^+ flux in the two-compartment system was >2.5 mol/m^2·h higher than that in the three-compartment system in the initial stage. However, the decrease in the Na^+ flux was more pronounced in the two-compartment system as time progressed. After 240 min, the three-compartment system exhibited a higher Na^+ flux. In both systems, the Na^+ flux gradually decreased over time. This phenomenon was attributed to a reduction in the Na^+ concentration in the feed compartment owing to ion transport to the base compartment. This resulted in concentration polarization [24,25] as well as the reverse diffusion of H^+ and OH^-, hindering ion migration [26].

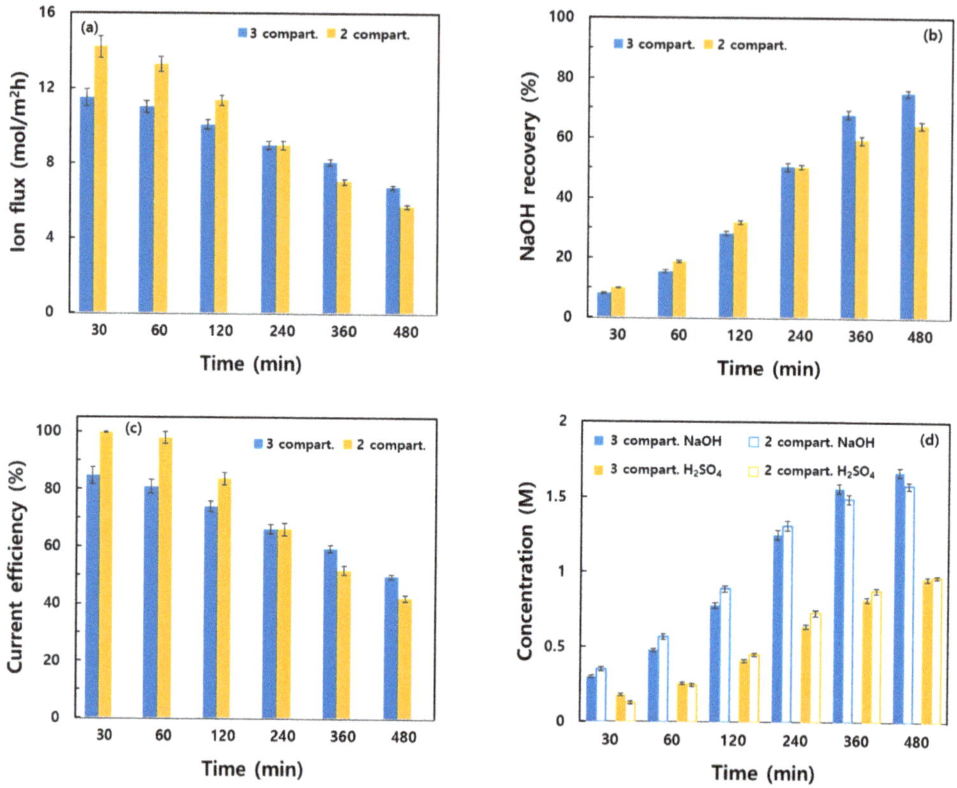

Figure 4. Effect of the operating mode on (**a**) Na^+ flux, (**b**) NaOH recovery rate, (**c**) current efficiency, and (**d**) NaOH and H_2SO_4 concentrations.

Table 1. Effect of the operating mode on system performance.

Time (min)	Operating Mode	Na^+ Flux (mol/m^{2}h)	Recovery (%)		Recovered Concentration (M)		C.E (%)	Energy Consumption (kWh/kg) *
			NaOH	H_2SO_4	NaOH	H_2SO_4		
30	2 compart.	14.19	9.97	9.55	0.35	0.13	99.99	-
	3 compart.	11.50	8.08	6.72	0.30	0.18	84.74	-
60	2 compart.	13.28	18.67	18.02	0.57	0.25	97.88	-
	3 compart.	10.99	15.44	13.15	0.48	0.26	80.96	-
120	2 compart.	11.35	31.91	31.96	0.89	0.45	83.67	-
	3 compart.	10.05	28.25	25.43	0.78	0.41	74.07	-
240	2 compart.	8.95	50.32	49.90	1.31	0.73	65.97	-
	3 compart.	8.96	50.34	45.95	1.25	0.64	66.00	-
360	2 compart.	7.02	59.22	57.15	1.49	0.88	51.76	-
	3 compart.	8.05	67.85	63.88	1.56	0.82	59.30	-
480	2 compart.	5.72	64.29	59.88	1.58	0.97	42.14	1.52
	3 compart.	6.74	75.18	77.35	1.67	0.96	49.69	1.43

* Energy consumption for the processing of 1 kg of Na_2SO_4.

Moreover, in the two-compartment system, the formation of H_2SO_4 within the feed compartment reduced the electrical resistance and consequently the operating voltage. This decreased the driving force for Na^+ migration, which further accelerated the decrease in the Na^+ flux. Similar changes were observed in the NaOH recovery rate Figure 4b. In the first 240 min, the two-compartment system exhibited a NaOH recovery rate that was 2–5%

higher than that exhibited by the three-compartment system. However, after 240 min, the three-compartment system exhibited a recovery rate that was >8% higher than that of the two-compartment system.

Figure 4c shows the concentrations of recovered NaOH and H_2SO_4. The NaOH concentration exhibited fluctuations similar to those of the Na^+ flux and NaOH recovery rate. Till 240 min, the two-compartment system exhibited a 0.06-M higher concentration (1.31 M) than the three-compartment system. However, after 360 min, the three-compartment system exhibited a NaOH concentration that was ~0.6 M higher than that of the two-compartment system.

The H_2SO_4 recovery rate showed fluctuations similar to those of the NaOH recovery rate, but the concentration of recovered H_2SO_4 was higher in the two-compartment system. This was attributed to the migration of Na^+ and SO_4^{2-} in hydrated forms through the ion-exchange membranes, increasing the volume of the acid and base recovery compartments and decreasing the volume of the feed compartment [27]. This volume change reduced the concentrations of recovered NaOH and H_2SO_4 in the three-compartment system. In the two-compartment system, NaOH was also diluted but H_2SO_4 directly formed in the feed compartment became concentrated, resulting in a higher final concentration than that in the three-compartment system.

The purity of H_2SO_4 recovered from the three-compartment system exceeded 99%. Meanwhile, in the two-compartment system, recovered H_2SO_4 was mixed with residual Na_2SO_4, resulting in lower purity (58.2%).

The CE values of the two systems are shown in Figure 4d. In the two-compartment system, CE exceeded 80% within 30–120 min, which was >10% higher than that in the three-compartment system during the same period. However, after 240 min, CE exhibited a substantial decline, reaching 42.14% after 480 min. The three-compartment system also showed high CE (>70%) before 120 min. However, CE declined afterward, reaching 49.69% after 480 min. This decline was attributed to the reduction in the Na^+ flux, which decreased the number of target ions contributing to the current.

Similar performances have been reported in previous studies. For example, León reported a decrease in CE from 79.20% to 52.70% over time during the recovery of NaOH from NaCl using BMED. The researcher attributed this to reverse ion diffusion and water transport [28].

The energy consumption values of the two systems after 480 min were also compared. The energy consumption of the two-compartment system was 1.52 kWh/kg, which was 0.09 kWh/kg higher than that of the three-compartment system. This difference was attributed to the lower recovery efficiency of the two-compartment system, which required additional energy for treating 1 kg of Na_2SO_4.

3.2. *Effect of Process Parameters on the Recovery Characteristics of the Two Systems*

The effects of process parameters, i.e., current density, feed solution concentration, and the concentrations and volumes of the acid and base, on the Na_2SO_4 desalination characteristics as well as NaOH and H_2SO_4 recovery characteristics of the two- and three-compartment systems were studied and compared.

The effect of current density on the desalination and recovery performances of the two systems were investigated via experiments conducted at current densities of 250–450 A/m^2 under identical conditions. The feed, acid recovery, and base recovery compartments were each filled with 1.5 L of Na_2SO_4 (1.30 M), 1.5 L of H_2SO_4 (0.1 M), and 1.5 L of NaOH (0.1 M), respectively. Each experiment was terminated after 240 min (Figure 5 and Table 2).

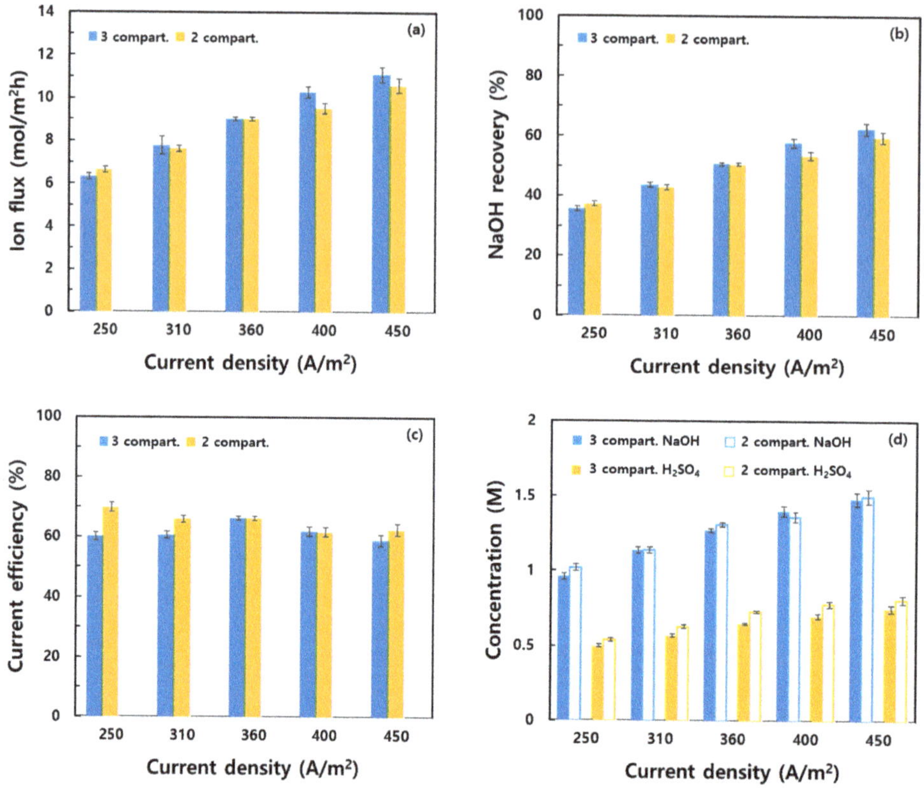

Figure 5. Effect of current density on the performance of the two- and three-compartment BMED systems: (**a**) Na^+ flux, (**b**) NaOH recovery, (**c**) current efficiency, and (**d**) energy consumption.

Table 2. Effect of current density on the performance of the two- and three-compartment systems.

Current Density (A/m^2)	Operating Mode	Na^+ Flux (mol/m^2h)	Recovery (%)		Recovered Concentration (M)		C.E (%)	Energy Consumption (kWh/kg) *
			NaOH	H_2SO_4	NaOH	H_2SO_4		
250	2 compart.	6.64	37.35	37.21	1.02	0.54	69.94	0.87
	3 compart.	6.32	35.65	32.80	0.96	0.50	60.09	0.97
310	2 compart.	7.62	42.87	42.52	1.14	0.63	66.11	0.96
	3 compart.	7.78	43.72	40.02	1.14	0.57	60.69	1.03
360	2 compart.	9.00	50.57	49.66	1.31	0.73	66.30	1.00
	3 compart.	9.01	50.66	46.78	1.27	0.65	66.42	1.11
400	2 compart.	9.51	53.46	50.67	1.36	0.78	61.73	1.09
	3 compart.	10.27	57.72	51.18	1.40	0.70	61.91	1.11
450	2 compart.	10.61	59.63	51.64	1.50	0.81	62.54	1.13
	3 compart.	11.11	62.44	57.25	1.48	0.75	58.93	1.25

* Energy consumption for the processing of 1 kg of Na_2SO_4.

Figure 5a shows the Na^+ flux. With increasing current density, the Na^+ flux increased. At 250 A/m^2, the Na^+ flux was 6.64 and 6.32 $mol/m^2{\cdot}h$ for the two- and three-compartment systems, respectively. At 450 A/m^2, the flux increased to 10.61 and 11.11 $mol/m^2{\cdot}h$, respectively. Except at 250 A/m^2, the three-compartment system consistently exhibited higher Na^+ flux. This was likely due to the structural advantage of the three-compartment system: the feed compartment was located between the acid and base compartments, enabling physical separation of ion flows and minimization of ion interference.

The NaOH recovery rate increased with increasing current density Figure 5b. While the recovery rate of the three-compartment system was approximately 1.5% lower than that of the two-compartment system at 250 A/m^2, it was 1–3% higher at current densities of >310 A/m^2.

As shown in Figure 5c, the CE of the two-compartment system decreased with increasing current density, whereas that of the three-compartment system increased up to 360 A/m^2 and then decreased. The decrease in the CE of the two-compartment system was likely due to heat accumulation because of its simpler structure and fewer compartments, which resulted in a higher temperature (2°C) compared to the three-compartment system. This may have caused side reactions, such as water dissociation, at the electrodes or within the BPM, reducing CE [29].

In the three-compartment system, CE decreased at current densities of >400 A/m^2, likely owing to ion depletion at the membrane surface due to over-limiting current. This led to concentration polarization, wherein ion transport no longer proportionally increased with current [25,30].

Figure 5d shows the energy consumption of both systems. In both systems, energy consumption increased with increasing current density, which was primarily due to an increase in the average voltage. Although the two-compartment system exhibited lower energy consumption than the three-compartment system, it may require additional thermal management equipment such as heat exchangers in large-scale applications since the process temperature needs to be reduced to ensure membrane stability.

The effect of the Na_2SO_4 concentration in the feed compartment on Na_2SO_4 desalination and acid or base recovery was investigated via experiments conducted at different initial Na_2SO_4 concentrations. For the three-compartment system, the optimal current density was found to be 360 A/m^2 because higher current densities substantially increased cell voltage, potentially affecting membrane stability, CE, and energy consumption. For the two-compartment system, the optimal current density was determined to be 450 A/m^2 based on the recovery performance and process duration.

The feed compartment was filled with 1.5 L of Na_2SO_4 solutions with concentrations of 0.43–1.30 M. All other conditions were identical to those set in the current density experiments. The three-compartment system was operated until the cell voltage reached 3 V, while the two-compartment system was operated for a fixed duration (480 min). The results are shown in Figure 6 and Table 3. Figure 6a–d show the Na^+ flux, NaOH recovery rate, concentrations of recovered NaOH and H_2SO_4, and energy consumption, respectively.

In the three-compartment system, moderately high Na^+ fluxes were observed at low Na_2SO_4 concentrations, i.e., 10.80 and 9.15 $mol/m^2{\cdot}h$ at 0.43 and 0.65 M, respectively. At high Na_2SO_4 concentrations ($\geq$0.87 M), the Na^+ flux decreased to 8.65–7.07 $mol/m^2{\cdot}h$. This reduction was attributed to the decrease in the average voltage due to the lower resistance of the feed solution at higher Na_2SO_4 concentrations, resulting in a low driving force for Na^+ migration through ion-exchange membranes [31].

The NaOH recovery rate was >75% under all conditions. However, at concentrations of 0.43 and 0.65 M, the recovery rates were moderate because of short process times (120 and 199 min, respectively) despite high the Na^+ flux values.

Low Na_2SO_4 concentrations yielded NaOH and H_2SO_4 concentrations of up to 1.0 and 0.6 M, respectively. In contrast, high Na_2SO_4 concentrations enabled the recovery of up to 1.67 M NaOH and 0.96 M H_2SO_4. The energy consumption under all conditions (except for a Na_2SO_4 concentration of 1.30 M) ranged from 1.23 to 1.27 kWh/kg.

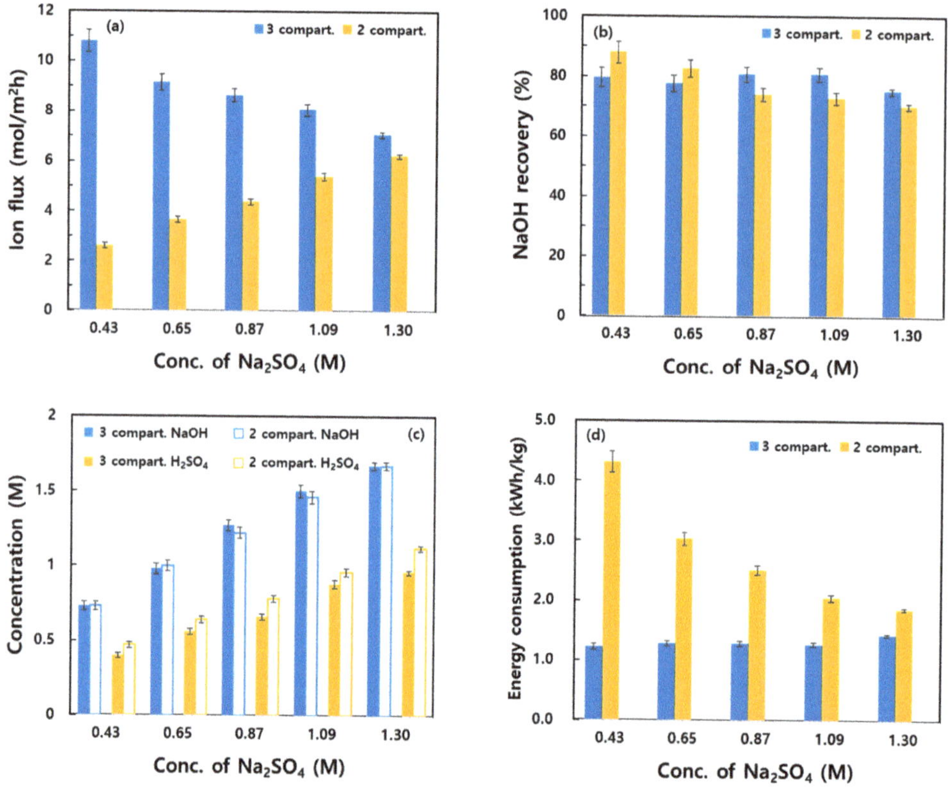

Figure 6. Effect of Na_2SO_4 concentration on the performance of the two- and three-compartment BMED systems: (**a**) Na^+ flux, (**b**) NaOH recovery, (**c**) NaOH and H_2SO_4 concentrations, and (**d**) energy consumption.

Table 3. Effect of Na_2SO_4 concentration on the performance of the two- and three-compartment systems.

Conc. of Na_2SO_4 (M)	Operating Mode	Na^+ Flux (mol/m^2h)	Recovery (%)		Recovered Concentration (M)		C.E (%)	Energy Consumption (kWh/kg) *
			NaOH	H_2SO_4	NaOH	H_2SO_4		
0.43	2 compart.	2.61	87.98	79.08	0.73	0.47	15.38	4.31
	3 compart.	10.80	79.62	76.51	0.73	0.40	69.58	1.23
0.65	2 compart.	3.67	82.61	72.52	1.00	0.64	21.66	3.03
	3 compart.	9.15	77.62	78.82	0.98	0.56	61.36	1.29
0.87	2 compart.	4.38	73.91	70.20	1.22	0.78	25.84	2.51
	3 compart.	8.65	80.82	73.64	1.27	0.66	59.69	1.29
1.09	2 compart.	5.39	72.73	68.60	1.46	0.96	31.78	2.05
	3 compart.	8.06	80.85	83.75	1.50	0.88	56.38	1.27
1.30	2 compart.	6.23	70.00	63.78	1.67	1.12	36.39	1.86
	3 compart.	7.07	75.18	77.35	1.67	0.96	49.69	1.43

* Energy consumption for the processing of 1 kg of Na_2SO_4.

In the two-compartment system, Na^+ fluxes at low Na_2SO_4 concentrations were 2.61 and 3.67 $mol/m^2 \cdot h$, which were 1–3 $mol/m^2 \cdot h$ lower than those at high Na_2SO_4 concentrations. In the three-compartment system, Na^+ and $SO_4{}^{2-}$ migrated through membranes, whereas in the two-compartment system, only Na^+ migration occurred. Therefore, the effect of concentration polarization relaxation at high Na_2SO_4 concentrations played a more dominant role than the reduced driving force due to the low voltage, which increased ion flux [32].

A NaOH recovery rate of >82% was achieved at Na_2SO_4 concentrations of <0.65 M, whereas moderate recovery rates of 70–73% were observed at higher concentrations (≥0.87 M). However, similar to the three-compartment system, the concentrations of recovered NaOH and H_2SO_4 increased with increasing Na_2SO_4 concentration: up to 1.0 M NaOH and 0.65 M H_2SO_4 at low concentrations and up to 1.67 M NaOH and 1.12 M H_2SO_4 at high concentrations.

The CE value also increased from 15.38% to 36.39% with increasing Na_2SO_4 concentration, which was in good agreement with changes in the Na^+ flux. Unlike that of the three-compartment system, the energy consumption of the two-compartment system decreased with increasing Na_2SO_4 concentration owing to the lower recovery rates at lower Na_2SO_4 concentrations, which increased the energy requirements for processing per kilogram of Na_2SO_4.

To increase the concentrations of recovered NaOH and H_2SO_4 and investigate their effects on NaOH and H_2SO_4 recoveries and Na_2SO_4 desalination, experiments were conducted by varying the initial concentrations of NaOH and H_2SO_4 in their respective recovery compartments. The current densities were set to 360 and 450 A/m^2 for the three- and two-compartment systems, respectively, based on the conditions set in the above-mentioned experimental investigations regarding the Na_2SO_4 concentration.

The feed compartment was filled with 1.5 L of a Na_2SO_4 solution (1.30 M), and the base and acid recovery compartments were each filled with 1.5 L of NaOH and H_2SO_4 solutions, respectively, with concentrations of 0.05–0.50 M. The experimental results are presented in Figure 7 and Table 4.

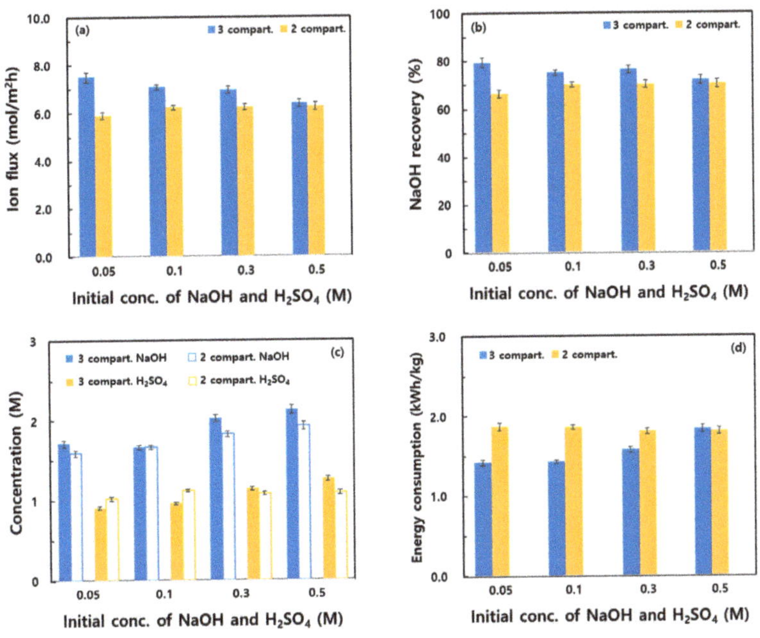

Figure 7. Effect of the initial concentrations of NaOH and H_2SO_4 on the performance of the two- and three-compartment BMED systems: (**a**) Na^+ flux, (**b**) NaOH recovery, (**c**) NaOH and H_2SO_4 concentrations, and (**d**) energy consumption.

Table 4. Effect of the initial concentrations of NaOH and H_2SO_4 on the performance of the two- and systems.

Initial Conc. of NaOH and H_2SO_4 (M)	Operating Mode	Na^+ Flux (mol/m²h)	Recovery (%)		Recovered Concentration (M)		Water Migration (%)		
			NaOH	H_2SO_4	NaOH	H_2SO_4	Base	Acid	Feed
0.05	2 compart.	5.90	66.39	63.42	1.59	1.02	20.30	-	−20.30
	3 compart.	7.50	79.41	77.78	1.72	0.91	23.34	16.69	−40.04
0.10	2 compart.	6.23	70.00	63.78	1.67	1.12	17.13	-	−17.13
	3 compart.	7.07	75.18	77.35	1.67	0.96	27.40	16.65	−44.05
0.30	2 compart.	6.24	70.14	63.32	1.83	1.08	16.16	-	−16.16
	3 compart.	6.95	76.37	78.39	2.03	1.14	13.20	16.25	−29.45
0.50	2 compart.	6.25	70.33	63.38	1.93	1.09	15.68	-	−15.68
	3 compart.	6.39	71.91	74.62	2.13	1.26	12.98	16.69	−29.67

Figure 7a shows the Na^+ flux, revealing opposite performances of the two systems. In the three-compartment system, the Na^+ flux decreased from 7.50 to 6.39 mol/m^2·h with increasing initial concentrations of NaOH and H_2SO_4. In contrast, in the two-compartment system, the lowest flux (5.90 mol/m^2·h) was observed at 0.05 M. The Na^+ flux increased to 6.23 mol/m^2·h at 0.10 M and maintained a similar level up to 0.50 M.

In the three-compartment system, high NaOH and H_2SO_4 concentrations in the acid and base compartments may have increased the concentration gradient relative to the feed compartment. This may have caused ion accumulation on the membrane surface, hindering ion transport. According to a study by Zhu [33], reverse diffusion of Na^+ can occur with increasing NaOH concentration in the base compartment during operation, which was reported to reduce the NaOH recovery rate from 95.6% to 75.1%. These phenomena likely contributed to the observed decrease in the Na^+ flux.

In contrast, in the two-compartment system, the electrical resistance decreased with increasing NaOH concentration. In addition, since H_2SO_4 was generated in the feed compartment, the concentration gradient relative to the base compartment became smaller than that in the case of the three-compartment system, enhancing Na^+ transport.

Similar trend was observed for the NaOH recovery rate Figure 7b. In the three-compartment system, the NaOH recovery rate decreased from 79.41% at 0.05 M to 71.91% at 0.50 M. In the two-compartment system, the NaOH recovery rate increased from 66.39% at 0.05 M to 70.00% at 0.10 M and remained moderately stable up to 0.50 M.

Figure 7c shows the concentrations of recovered NaOH and H_2SO_4. Except for an initial NaOH concentration of 0.10 M, the concentration of recovered NaOH in the two-compartment system was ~0.20 M lower than that in the three-compartment system. The two-compartment system showed an approximately 0.15-M higher concentration of recovered H_2SO_4 at an initial H_2SO_4 concentration of $\leq$0.10 M. However, at concentrations of $\geq$0.30 M, the concentrations of recovered H_2SO_4 in the three-compartment system were >0.10 M higher than those in the two-compartment system.

These results were attributed to differences between the water migration and recovery rates of the two systems. At concentrations of <0.30 M, the three-compartment system exhibited 6–13% higher recovery rates than the two-compartment system. Moreover, at concentrations of $\geq$0.30 M, the rate of water migration into the base compartment was >3% lower in the three-compartment system than in the two-compartment system, which led to a high concentration of recovered NaOH in the former.

Figure 7d shows the energy consumption of the two systems. In the three-compartment system, the energy consumption substantially increased from 1.42 kWh/kg at 0.05 M to 1.84 kWh/kg at 0.50 M. In contrast, in the two-compartment system, the highest energy consumption was observed at 0.05 M (1.87 kWh/kg), which then gradually decreased to 1.81 kWh/kg at 0.50 M.

To increase the concentrations of recovered NaOH and H_2SO_4 and investigate their effects on Na_2SO_4 desalination and recovery performance, experiments were conducted using different initial volumes of NaOH and H_2SO_4 solutions loaded into the base and acid recovery compartments (Figure 8 and Table 5). Similarly to aforementioned experiments, the current densities were set to 360 and 450 A/m^2 for the three- and two-compartment systems, respectively. The feed compartment was filled with 1.5 L of a Na_2SO_4 solution (1.30 M). The base and acid recovery compartments were each filled with NaOH (0.1 M) and H_2SO_4 (0.1 M) solutions, with volumes of 0.75–1.5 L. Figure 8a shows the Na^+ flux. For both systems, the Na^+ flux increased with increasing solution volume. For solution volumes of <1000 mL, the two-compartment system exhibited a Na^+ flux that was 0.8–0.1 mol/m^2·h higher than that exhibited by the three-compartment system. However, for volumes of ≥1250 mL, the three-compartment system exhibited an approximately 0.8 mol/m^2·h higher Na^+ flux.

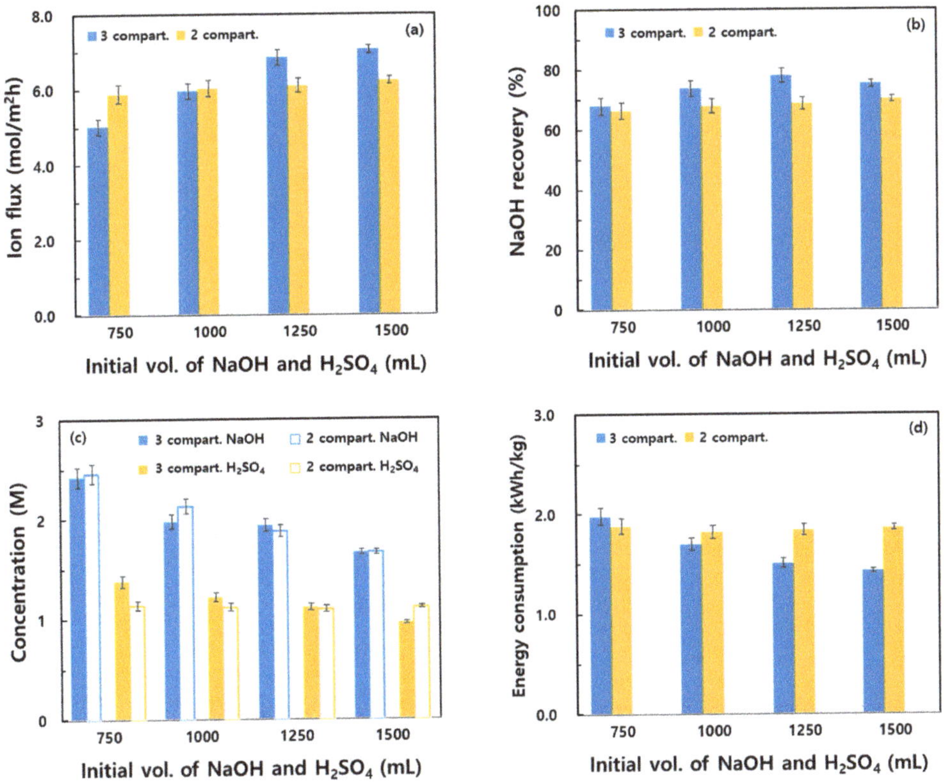

Figure 8. Effect of the initial volume of NaOH and H_2SO_4 on the performance of the two- and three-compartment BMED systems: (**a**) Na^+ flux, (**b**) NaOH recovery, (**c**) NaOH and H_2SO_4 concentrations, and (**d**) energy consumption.

Table 5. Effect of the initial volumes of NaOH and H_2SO_4 on the performance of the two- and three-compartment systems.

Initial Vol. of NaOH and H_2SO_4 (mL)	Operating Mode	Na^+ Flux (mol/m^2h)	Recovery (%)		Recovered Concentration (M)		Water Migration (%)		
			NaOH	H_2SO_4	NaOH	H_2SO_4	Base	Acid	Feed
750	2 compart.	5.89	66.32	62.01	2.46	1.14	44.47	-	−22.23
	3 compart.	5.02	67.90	66.46	2.42	1.38	50.55	31.97	−41.04
1000	2 compart.	6.03	67.85	63.55	2.13	1.12	28.80	-	−19.20
	3 compart.	5.96	73.75	73.01	1.98	1.22	50.11	25.16	−50.48
1250	2 compart.	6.10	68.60	63.65	1.88	1.10	20.30	-	−16.92
	3 compart.	6.85	77.93	76.44	1.94	1.12	35.89	19.76	−44.52
1500	2 compart.	6.23	70.00	63.78	1.67	1.12	15.68	-	−15.68
	3 compart.	7.07	75.18	77.35	1.67	0.96	27.40	16.65	−44.05

This result was attributed to the fact that larger volumes result in a more gradual increase in the acid and base concentrations in the respective compartments. This allowed the system to maintain concentration gradient in the feed compartment, reducing back diffusion and enhancing ion transport [34].

The NaOH recovery rate exhibited a similar trend (Figure 8b). It increased with increasing solution volume in both systems. In the two- and three-compartment systems, the NaOH recovery rate increased from 66.32% to 70.00% and from 67.90% to 75.18%, respectively. The increase in the Na^+ flux may have directly contributed to the increased recovery rate.

Figure 8c shows the concentrations of recovered NaOH and H_2SO_4. At low solution volumes, higher concentrations were observed. At a solution volume of 750 mL, the concentrations of recovered NaOH and H_2SO_4 in the three-compartment system were 2.42 and 1.38 M, respectively, whereas those in the two-compartment system were 2.46 and 1.14 M, respectively. At volumes of $\geq$1250 mL, the concentration of recovered NaOH in the three-compartment system was consistently 0.06 M higher than that in the two-compartment system. In contrast, at volumes of $\leq$1000 mL, the concentration of recovered NaOH in the two-compartment system was 0.04–0.15 M higher than that in the three-compartment system.

This difference at low volumes was attributed to the higher water migration rate (>6%) in the three-compartment system, which led to dilution in the base compartment. In the two-compartment system, H_2SO_4 was generated in the feed compartment. Owing to the lower recovery rates compared to those in the three-compartment system, lower final concentration of recovered H_2SO_4 was observed.

Figure 8d shows the energy consumption of the two systems. In the three-compartment system, the energy consumption decreased from 1.98 to 1.43 kWh/kg with increasing solution volume. This was attributed to the recovery rate increase observed at high solution volumes, which reduced the energy consumed per kilogram of Na_2SO_4 treated. Gao et al. reported a similar observation for a three-compartment BMED system, wherein increasing the initial volumes of H_2SO_4 and NaOH from 60 to 120 mL resulted in an increase in the energy consumption from 1.35 to 2.67 kWh/kg [35].

In contrast, the two-compartment system showed moderately stable energy consumption regardless of the solution volume (1.88–1.82 kWh/kg) under all tested conditions.

3.3. Techno-Economic Feasibility Analysis

The optimal operating conditions for each system were determined based on the results of the experiments discussed. Further, a techno-economic feasibility analysis was conducted.

The optimal conditions of the two-compartment system were a current density of 450 A/m^2, Na_2SO_4 (1.30 M) volume of 1.5 L in the feed compartment, and NaOH (0.1 M) volume of 0.75 L in the base compartment. The optimal conditions of the three-compartment system were a current density of 360 A/m^2, a Na_2SO_4 (1.30 M) volume of 1.5 L in the feed compartment, and NaOH (0.1 M) and H_2SO_4 (0.1 M) volumes of 0.75 L each in the base and acid recovery compartments, respectively.

The process flow diagrams for the two- and three-compartment systems under these conditions are shown in Figures 9 and 10, respectively. The results of the optimal process and corresponding economic analysis are summarized in Table 6.

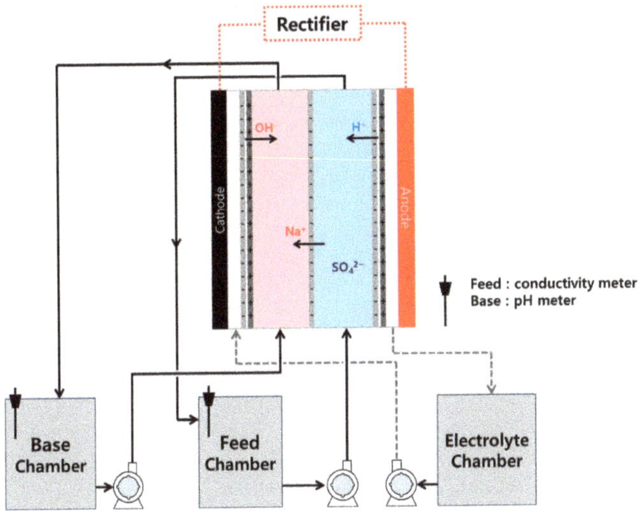

Figure 9. Schematic of the two-compartment BMED process.

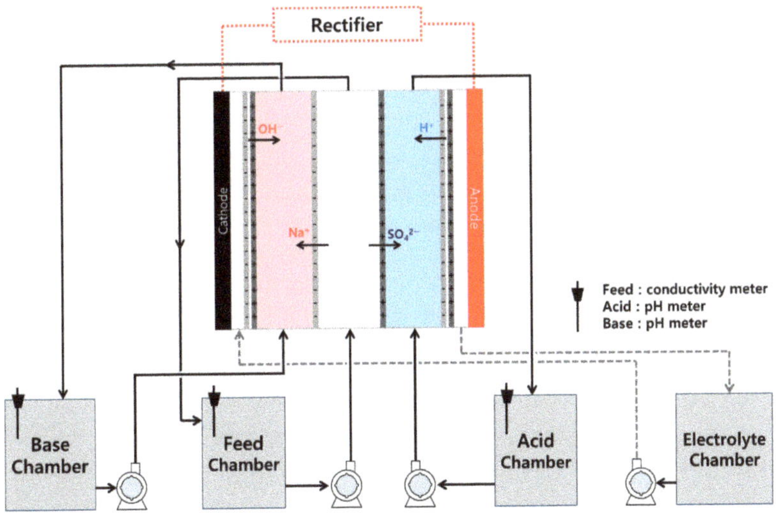

Figure 10. Schematic of the three-compartment BMED process.

Table 6. Comparison between the techno-economic factors of the two- and three-compartment systems.

Techno-Economic Factors	Two-Compartment System	Three-Compartment System
Membrane used	Cation exchange membrane and bipolar membrane	Cation exchange membrane, anion exchange membrane, and bipolar membrane
Bipolar membrane reaction	$H_2O \rightarrow H^+ + OH^-$	
Recovery of Na_2SO_4 to NaOH	66.32%	67.90%
Recovery of Na_2SO_4 to H_2SO_4	62.01%	66.46%
Concentration of recovered NaOH	2.46 M	2.42 M
Concentration of recovered H_2SO_4	1.14 M (H_2SO_4 + Na_2SO_4 mixed solution)	1.38 M
NaOH purity	Comparably to commercial NaOH obtained from membrane plants	
H_2SO_4 purity	63.3% (1.14 M H_2SO_4 + 0.66 M Na_2SO_4)	Comparable to commercial H_2SO_4 obtained from membrane plants
Operating temperature	27–30 °C	25–28 °C
Average voltage (V)	17.34	18.92
CE (%)	34.74	36.09
Process time (h)	9.00	9.87
Energy consumption (kWh/kg) #1	1.88	1.98
Process capacity (kg/year) #2,3	162.01	131.31
Total energy consumption (kWh/year)	305	260
Electricity charge (USD/kWh) #4	0.10	0.10
Total energy cost (USD/year)	30.5	26.0
Supplied Na_2SO_4 price (USD/year) #5	14.0	11.4
Produced NaOH price (USD/year) #6	27.88	21.08
Produced H_2SO_4 price (USD/year) #7	12.15	9.87
Total process cost (USD/year)	4.43	6.43

#1 kg = weight of Na_2SO_4, #2 1 year = 260 days, #3 1 day = 18 h, #4 0.10 USD/kWh, #5 Na_2SO_4 = 0.09 USD/kg, #6 NaOH = 0.35 USD/L, #7 H_2SO_4 = 0.31 USD/L.

The economic analysis was based on the process time and energy consumption per plant. It was assumed that a plant operated for 260 days per year and 18 h a day. The electricity cost was calculated to be 0.10 USD/kWh.

Under these conditions, the two-compartment system was estimated to treat 162.01 kg of Na_2SO_4 per year, consuming 304.58 kWh of electricity, which corresponded to an annual energy cost of 30.46 USD. Assuming that recovered NaOH and H_2SO_4 were sold as industrial-grade products (40 wt% and 95 wt%, respectively) and applying retail price standards, the value of the recovered products was found to offset 40.03 USD annually. Thus, the process net annual cost was found to be 4.43 USD.

The three-compartment system was estimated to treat 131.31 kg of Na_2SO_4 per year, consuming 260.00 kWh of electricity (annual energy cost = 26.00 USD). Assuming the same market value for recovered NaOH and H_2SO_4 as for the two-compartment system, the recovered product value was found to offset 30.92 USD, resulting in a net annual cost of 6.43 USD.

Although the net process cost of the two-compartment system was 2.00 USD less than that of the three-compartment system, the total capital investment for the two-compartment system is expected to be higher. This is due to the greater temperature increase observed during operation, which may require additional cooling equipment such as heat exchangers.

Moreover, H_2SO_4 recovered from the two-compartment system was mixed with Na_2SO_4, requiring additional post-treatment, such as methanol-induced precipitation, to obtain pure H_2SO_4 [36].

In conclusion, while both systems incur some processing costs, the BMED method has potential as an effective technology for treating Na_2SO_4-containing wastewater. In particular, under stricter future environmental regulations regarding Na_2SO_4 discharge, BMED could serve as a suitable process for NaOH and H_2SO_4 recovery from waste streams.

3.4. Comparison of Operational Stability

To evaluate the operational stability of the two systems for future recovery applications of NaOH and H_2SO_4 from Na_2SO_4-containing wastewater, a comparative analysis was conducted focusing on temperature variations, potential membrane degradation, and voltage stability (Table 7).

Table 7. Comparison between the operational stability of the two- and three-compartment systems.

Stability Factor	Two-Compartment System	Three-Compartment System
Membrane durability	The CEM is exposed to the acid and base, which adversely affect membrane durability	As the acid and base are separated, membrane durability is improved compared to the two-compartment system
Voltage stability	Owing to a reduced operating voltage, the system is more stable than the three-compartment system	The rapid increase in the voltage may affect membrane durability
Temperature stability	Higher than that of the three-compartment system	System stable at 25–28 °C
Operational stability	The two-compartment system exhibits superior voltage stability and electrical durability during long-term operation, whereas the three-compartment system offers better structural resistance to thermal and chemical stress	

First, thermal stabilities of the two systems were compared. The two-compartment system had fewer solution chambers and used a lower total solution volume compared to the three-compartment system. This resulted in higher generated heat per unit solution volume in the two-compartment system under the same current conditions, leading to a faster increase in the internal cell temperature. Thus, there is a higher risk of heat accumulation in the two-compartment system during extended operation, potentially compromising the overall system stability. In contrast, the separate acid and base chambers and high solution volumes used in the three-compartment system allow more even heat dissipation, resulting in a slower temperature increase and a better overall thermal stability.

Second, the two systems exhibited distinct membrane degradation risks. In the three-compartment system, the acid and base were confined to separate chambers, exposing each ion-exchange membrane to a single chemical environment. This separation minimized chemical deterioration. In contrast, the CEM in the two-compartment system was located between the feed and base compartments, where it was repeatedly exposed to H_2SO_4 and NaOH. This continuous pH cycling accelerated degradation and compromised membrane integrity over time. Furthermore, the anion exchange layer was prone to degradation under strongly alkaline conditions via Hofmann β-elimination and nucleophilic substitution reactions by OH^- ions [37,38]. Similarly, the cation exchange layer is vulnerable to acidic environments and radical-induced chain scission [39]. These intrinsic stability limitations under harsh pH conditions contribute to reduced membrane lifespan [40].

Lastly, the two-compartment system exhibited higher voltage stability than the three-compartment system. In the former, as the operation process progressed, H_2SO_4 was generated in the feed compartment and NaOH in the base compartment, both of which reduced the electrical resistance. This helped maintain moderately stable cell voltages during long-term operation, minimizing the risk of overvoltage across the membranes. In contrast, in the three-compartment system, continuous Na_2SO_4 desalination from the feed compartment led to a reduction in the ion concentration, which increased the electrical resistance and caused a sharp voltage increase. This voltage increase can result in an overvoltage stress on the membranes, negatively impacting their stability and potentially leading to electrical damage and reduced membrane lifespan [40].

In summary, the two-compartment system shows better electrical and voltage stabilities during prolonged operation, whereas the three-compartment system offers better resistance to thermal and chemical degradations owing to its more robust structural separation between the acid and base compartments.

4. Conclusions

Herein, the effects of the operating modes and key process parameters of BMED systems were investigated for the treatment of Na_2SO_4-containing wastewater generated from the recycling of spent LIBs. Economic feasibility and operational stability were also evaluated, leading to the following conclusions:

1. The two-compartment system initially showed higher Na^+ flux and NaOH recovery rate. However, after 240 min, the three-compartment system demonstrated superior performance. The three-compartment system enabled the recovery of a high-purity acid and base and showed moderately stable long-term performance in terms of CE and energy consumption.
2. With increasing current density, the Na^+ flux, NaOH recovery rate, and energy consumption increased. In most cases, the three-compartment system outperformed the two-compartment system. The CE continuously decreased in the two-compartment system, while it increased up to 360 A/m^2 and then declined in the three-compartment system. This was attributed to different heat accumulation and concentration polarization effects in each system. Although the two-compartment system exhibited lower energy consumption, additional thermal control may be required for extended operations.
3. With increasing Na_2SO_4 concentration, both systems exhibited higher concentrations of recovered NaOH and H_2SO_4 as well as improved CE. In the three-compartment system, the Na^+ flux decreased owing to reduced voltage at high Na_2SO_4 concentrations, whereas in the two-compartment system, the flux increased owing to the weakening of the concentration polarization effect. The two-compartment system was less energy efficient at low concentrations, but it showed improved energy efficiency at high concentrations.
4. With increasing initial concentrations of NaOH and H_2SO_4, the three-compartment system showed a decline in the Na^+ flux and recovery rate as well as increased energy consumption. In contrast, the two-compartment system exhibited an improved flux and recovery rate beyond a certain concentration threshold, with a corresponding decrease in the energy consumption. These results were attributed to the differences between the concentration gradients, back-diffusion effects, and water transport rates of the two systems, highlighting that recovery characteristics depend on the concentration conditions.
5. Increasing the initial volumes of NaOH and H_2SO_4 in the recovery compartments increased the Na^+ flux and NaOH recovery rates in both systems. At volumes of $\geq$1250 mL, the three-compartment system demonstrated superior recovery performance. Low initial volumes increased the concentrations of the recovered species; however, differences in the water transport and recovery rates caused variations in the final concentrations. Energy consumption substantially decreased in the three-

compartment system with the increasing volumes, suggesting that volume control is critical for process optimization.

6. Both configurations successfully recovered NaOH and H_2SO_4 from the Na_2SO_4-containing wastewater, with estimated annual process costs of USD 4.43 and USD 6.43 for the two- and three-compartment systems, respectively. Despite the lower processing cost of the two-compartment system, its capital investment is expected to be higher because it may require thermal management and post-treatment. Under future stricter environmental regulations, BMED can be an effective strategy for the recovery of resources and treatment of Na_2SO_4-containing wastewater.
7. The operational stability assessment confirmed the higher voltage stability and electrical reliability of the two-compartment system and its susceptibility to membrane degradation due to heat accumulation and repeated acid or base contact. In contrast, the three-compartment system exhibited better structural resistance to thermal and chemical stress but experienced voltage increases in long-term operations, which can compromise membrane integrity.

Author Contributions: Conceptualization, J.A. and Y.C.; methodology, M.S., Y.L. and J.K.; validation, J.A. and Y.C.; formal analysis, M.S., Y.L. and J.C.; investigation, M.S., Y.L. and J.K.; resources, M.S., J.K. and J.C.; data curation, M.S. and Y.L.; writing—original draft preparation, M.S.; writing—review and editing, J.A. and Y.C.; visualization, M.S., Y.L. and J.C.; supervision, J.A.; project administration, J.A.; funding acquisition, J.A. All authors have read and agreed to the published version of the manuscript.

Funding: This research was supported by the Technology development project to improve secondary battery circulation usability (Development of pollutants reduction technology generated in the lithium ion batteries recycling process) through the Korea Environmental Industry & Technology Institute funded by the Ministry of Environment (RS-2024-00345911).

Data Availability Statement: The original contributions presented in this study are included in the article. Further inquiries can be directed to the corresponding author.

Conflicts of Interest: Author Yeonchul Cho was employed by SungEel HiTech Co., Ltd. The remaining authors declare that the research was conducted in the absence of any commercial or financial relationships that could be construed as a potential conflict of interest.

Abbreviations

The following abbreviations are used in this manuscript:

AEM	Anion exchange membrane
C.E	Current efficiency
CEM	Cation exchange membrane
BPM	Bipolar membrane
BMED	Bipolar membrane electrodialysis
LIBs	Lithium-ion batteries

References

1. Giza, K.; Pospiech, B.; Gęga, J. Future technologies for recycling spent lithium-ion batteries (LIBs) from electric vehicles—Overview of latest trends and challenges. *Energies* **2023**, *16*, 5777. [CrossRef]
2. Yoo, K. Lithium ion battery recycling industry in South Korea. *Resour. Recycl.* **2023**, *32*, 13–20. [CrossRef]
3. Ahn, J.-W.; Cho, Y.-C. Current status and prospect of waste lithium ion battery (LIB) recycling technology by hydrometallurgical process. *Resour. Recycl.* **2023**, *32*, 3–17. [CrossRef]
4. Verbaan, N.; Naidoo, R. A Review of Hydrometallurgical Flowsheets Considered for the Treatment of Black Mass. Presented via iQHub. Available online: https://www.slideshare.net/slideshow/review-of-hydrometallurgical-flowsheets-for-treatment-of-black-mass/256586490 (accessed on 9 September 2025).

5. Ding, Z.; Li, J.; Huang, Y.; Sun, Y.; Lin, H.; Li, J.; Zhuge, X.; Ren, Y. The Green Recycling and Reuse of the Spent Lithium-Ion Battery with Nickel Cobalt Manganate as the Cathode. Preprint at SSRN. Available online: https://ssrn.com/abstract=4739651 (accessed on 18 August 2025).
6. Yu, K.S.; Yu, H.Y. Method for Preparing a Precursor for Cathode Active Material of Secondary Battery and Apparatus Thereof. KR Patent 10-2023063, 19 September 2019.
7. Ekholm, P.; Lehtoranta, J.; Taka, M.; Sallantaus, T.; Riihimäki, J. Diffuse sources dominate the sulfate load into Finnish surface waters. *Sci. Total Environ.* **2020**, *748*, 141297. [CrossRef]
8. Reinsel, M.A. A new process for sulfate removal from industrial waters. In Proceedings of the National Meeting of American Society for Surface Mining and Reclamation, Scottadale, AZ, USA, 13–19 August 1999; pp. 546–550. [CrossRef]
9. Kuldeep, B.; Badenhorst, W.D.; Kauranen, P.; Pajari, H.; Ruismäki, R.; Mannela, P.; Murtomäki, L. Bipolar membrane electrodialysis for sulfate recycling in the metallurgical industries. *Membranes* **2021**, *11*, 718. [CrossRef]
10. Shin, H.S.; Jung, J.Y.; Bae, B.U.; Paik, B.C. Phase-separated anaerobic toxicity assays for sulfate and sulfide. *Water Environ. Res.* **1995**, *67*, 802–807. [CrossRef]
11. Joo, J.; Moon, J.-K.; Jang, Y. A study on the desalination process for high-concentration Na_2SO_4 using pilot-scale bipolar membrane electrodialysis. *Desalination Water Treat.* **2025**, *322*, 101222. [CrossRef]
12. Atia, T.A.; Elia, G.; Hahn, R.; Altimari, P.; Pagnanelli, F. Closed-loop hydrometallurgical treatment of end-of-life lithium ion batteries: Towards zero-waste process and metal recycling in advanced batteries. *J. Energy Chem.* **2019**, *35*, 220–227. [CrossRef]
13. Parnamae, R.; Mareev, S.; Nikonenko, V.; Melnikov, S.; Sheldeshov, N.; Zabolotskii, V.; Hamelers, H.V.M.; Tedesco, M. Bipolar membranes: A review on principles, latest developments, and applications. *J. Membr. Sci.* **2021**, *617*, 118538. [CrossRef]
14. Kim, H.; Kim, S.; Lee, B.; Kim, M.; Kim, G.; Kim, C. Valorization of Na_2SO_4 in wastewater from spent lithium-ion battery recycling via BMED. *Chem. Eng. J.* **2025**, *504*, 158834. [CrossRef]
15. Strathmann, H. Electrodialysis and related processes. In *Membrane Separations Technology: Principles and Applications*; Noble, R.D., Stern, S.A., Eds.; Elsevier: Amsterdam, The Netherlands, 1995; Volume 2, pp. 213–278.
16. Strathmann, H. Electrodialysis, a mature technology with a multitude of new applications. *Desalination* **2010**, *264*, 268–288. [CrossRef]
17. Cho, Y.-C. A study on Na_2SO_4 Treatment Using Bipolar Electrodialysis System. Ph.D. Thesis, Daejin University, Pocheon-si, Republic of Korea, 2024.
18. Xue, S.; Wu, C.; Wu, Y.; Chen, J.; Li, Z. Bipolar membrane electrodialysis for treatment of sodium acetate waste residue. *Sep. Purif. Technol.* **2015**, *154*, 193–203. [CrossRef]
19. Lee, Y.; Seo, M.; Ahn, J. Regeneration of NaOH from the spent Na_2SO_4 solution by two-compartment bipolar membrane electrodialysis. *Korean J. Met. Mater.* **2025**, *63*, 281–290. [CrossRef]
20. Kroupa, J.; Kinčl, J.; Cakl, J. Recovery of H_2SO_4 and NaOH from Na_2SO_4 by electrodialysis with heterogeneous bipolar membrane. *Desalination Water Treat.* **2015**, *56*, 3238–3246. [CrossRef]
21. Wang, D.; Meng, W.; Lei, Y.; Li, C.; Cheng, J.; Qu, W.; Wang, G.; Zhang, M.; Li, S. The novel strategy for increasing the efficiency and yield of the bipolar membrane electrodialysis by the double conjugate salts stress. *Polymers* **2020**, *12*, 343. [CrossRef]
22. Giesbrecht, P.K.; Freund, M.S. Recent Advances in Bipolar Membrane Design and Applications. *Chem. Mater.* **2020**, *32*, 8060–8090. [CrossRef]
23. Luo, Y.; Liu, Y.; Shen, J.; Van der Bruggen, B. Application of Bipolar Membrane Electrodialysis in Environmental Protection: A Review. *Membranes* **2022**, *12*, 864. [CrossRef]
24. Tanaka, Y. Ion exchange membranes: Preparation, properties, and applications. In *Ion Exchange Membranes: Fundamentals and Applications*, 2nd ed.; Elsevier: Amsterdam, The Netherlands, 2015; Volume 12, pp. 217–268.
25. Tanaka, Y. Concentration polarization in ion-exchange membrane electrodialysis: The events arising in an unforced flowing solution in a desalting cell. *J. Membr. Sci.* **2003**, *217*, 65–80. [CrossRef]
26. Venugopal, K.; Dharmalingam, S. Desalination efficiency of a novel bipolar membrane based on functionalized polysulfone. *Desalination* **2012**, *296*, 37–45. [CrossRef]
27. Moon, S.H. *Electrochemical Processes of Ion Exchange Membranes*; GIST PRESS: Gwangju, Republic of Korea, 2021; pp. 1–296.
28. León, T.; Shah, S.A.; López, J.; Culcasi, A.; Jofre, L.; Cipollina, A.; Cortina, J.L.; Tamburini, A.; Micale, G. Electrodialysis with bipolar membranes for the generation of NaOH and HCl solutions from brines: An inter-laboratory evaluation of thin and ultrathin non-woven cloth-based ion-exchange membranes. *Membranes* **2022**, *12*, 1204. [CrossRef] [PubMed]
29. Simons, R.; Khanarian, G. Water dissociation in bipolar membranes: Experiments and theory. *J. Membr. Biol.* **1978**, *38*, 11–30. [CrossRef]
30. Strathmann, H. *Ion-Exchange Membrane Separation Processes*; Membrane Science and Technology Series; Elsevier: Amsterdam, The Netherlands, 2004; Volume 9, pp. 89–146.
31. Zhu, S.; Kingsbury, R.S.; Call, D.F.; Coronell, O. Impact of Solution Composition on the Resistance of Ion Exchange Membranes. *J. Membr. Sci.* **2018**, *554*, 39–47. [CrossRef]

32. Bazinet, L.; Geoffroy, T.R. Electrodialytic Processes: Market Overview, Membrane Phenomena, Recent Developments and Sustainable Strategies. *Membranes* **2020**, *10*, 221. [CrossRef] [PubMed]
33. Zhu, M.; Tian, B.; Luo, S.; Chi, Y.; Aishajiang, D.; Zhang, Y.; Yang, M. High-value conversion of waste Na_2SO_4 by a bipolar membrane electrodialysis metathesis system. *Resour. Conserv. Recycl.* **2022**, *186*, 106556. [CrossRef]
34. Brauns, E. Finite elements-based 2D theoretical analysis of the effect of IEX membrane thickness and salt solution residence time on the ion transport within a salinity gradient power reverse electrodialysis half cell pair. *Desalination Water Treat.* **2013**, *51*, 6429–6443. [CrossRef]
35. Gao, W.; Fang, Q.; Yan, H.; Wei, X.; Wu, K. Recovery of acid and base from sodium sulfate containing lithium carbonate using bipolar membrane electrodialysis. *Membranes* **2021**, *11*, 152. [CrossRef]
36. Huang, X.; Song, M.; Wang, H.; Lv, S.; Qi, S.; Wang, C.; Zhang, H.; Ruan, X. Waste salt separation by antisolvent crystallization process: Mechanism study of Na_2SO_4–NaCl–solvent ternary phase diagrams and its life cycle assessment. *Desalination* **2025**, *560*, 118985. [CrossRef]
37. Tufa, R.A.; Blommaert, M.A.; Chanda, D.; Li, Q.; Vermaas, D.A.; Aili, D. Bipolar Membrane and Interface Materials for Electrochemical Energy Systems. *ACS Appl. Energy Mater.* **2021**, *4*, 7419–7439. [CrossRef]
38. Blommaert, M.A.; Aili, D.; Tufa, R.A.; Li, Q.; Smith, W.A.; Vermaas, D.A. Insights and Challenges for Applying Bipolar Membranes in Advanced Electrochemical Energy Systems. *ACS Energy Lett.* **2021**, *6*, 2539–2548. [CrossRef]
39. Jeong, J.H.; Shin, E.K.; Jeong, J.J.; Na, I.C.; Chu, C.H.; Park, K.P. Degradation of Electrode and Membrane in Proton Exchange Membrane Fuel Cell After Water Electrolysis. *Korean Chem. Eng. Res.* **2014**, *52*, 695–700. [CrossRef]
40. Tanaka, Y. *Ion Exchange Membrane Electrodialysis: Fundamentals, Desalination, Separation*; Nova Science Publishers: New York, NY, USA, 2010; pp. 1–308.

metals

MDPI

Article

Recovery of Cu and Fe from a Sphalerite Concentrate by the MnO_2–KI Leaching Oxidation System

Aleksandar Jovanović [1,*], Dimitrije Anđić [1], Mladen Bugarčić [1,2], Ivana Jelić [1], Nela Vujović [1], Corby Anderson [3] and Miroslav Sokić [1]

1 Institute for Technology of Nuclear and Other Mineral Raw Materials, Boulevard Franše d'Eperea 86, 11000 Belgrade, Serbia; d.andjic@itnms.ac.rs (D.A.); m.bugarcic@itnms.ac.rs (M.B.); i.jelic@itnms.ac.rs (I.J.); n.petronijevic@itnms.ac.rs (N.V.); m.sokic@itnms.ac.rs (M.S.)
2 Milan Blagojević-Namenska AD, mr Radosa Milovanovica 2A, 32240 Lučani, Serbia
3 Kroll Institute for Extractive Metallurgy, Mining Engineering Department, Colorado School of Mines, Golden, CO 80401, USA; cganders@mines.edu
* Correspondence: a.jovanovic@itnms.ac.rs; Tel.: +381-11-3691-722

Abstract: This study examined the leaching behavior of copper and iron from a sphalerite concentrate in sulfuric acid utilizing an ensemble MnO_2–KI oxidizing system. The temperature was shown to significantly influence the leaching kinetics, with the efficiency notably improving between 40 °C and 80 °C. The introduction of KI affected the balance between sulfur passivation and oxidant availability, facilitating increased leaching efficiencies. At 3 wt% KI, maximum recoveries of 82.1% Cu and 85.3% Fe were achieved, which indicates a notable decrease in surface passivation. Kinetic study analysis revealed low activation energies of 28.90 kJ mol^{-1} for copper and 18.94 kJ mol^{-1} for iron, indicating that both processes proceed readily at moderate temperature regimes. Despite being diffusion-controlled, the mechanisms of dissolution are different: iron leaching is more complicated, involving pyrite oxidation, sulfur layer formation, transformation to marcasite, and ultimately iron (III) release, whereas copper leaching involves direct interaction of chalcopyrite with the oxidants, similar to the behavior of sphalerite.

Keywords: advanced oxidation leaching; Cu and Fe extraction; acid pH region; hydrometallurgy

1. Introduction

A crucial industrial metal, copper finds use extensively in electronics, plumbing, telecommunications, and electrical wiring [1]. It is critical for contemporary technology and infrastructure because of its superior thermal and electrical conductivity as well as its resistance to corrosion. To fulfill rising worldwide demand and promote advances in technology, copper must be extracted efficiently from chalcopyrite ($CuFeS_2$), the main copper-bearing material. Chalcopyrite often occurs in a dispersed form alongside other sulfide minerals (such as pyrite FeS_2, galena PbS, and sphalerite ZnS) in polymetallic ore deposits [2,3]. These minerals are typically concentrated by flotation and then processed. However, as richer ores become depleted, hydrometallurgical processes are attracting increased attention for treating lower-grade and complex sulfide concentrates [2,4]. Metals can be extracted from such challenging concentrates using acidic leaching techniques without emitting harmful smelting gases (like SO_2). It is important to note that acid leaching of chalcopyrite and its flotation concentrate cannot be fully understood without considering the broader context of sulfide mineral extraction in acidic media. In this regard, the leaching behavior of sphalerite and pyrite should also be taken into account, as their mechanisms share many similarities with those of chalcopyrite [3,5]. Chalcopyrite concentrate is leached in sulfuric acid with the use of oxidizing agents in an ordinary hydrometallurgical process to solubilize copper for recovery [5–7]. Compared to roasting, these techniques can reduce air pollution by producing byproducts like sulfuric acid or sulfate salts and achieving high metal recoveries. However, due to its refractory nature, the rate of chalcopyrite leaching is extremely slow under ambient conditions. Thus, it requires the employment of potent oxidizing agents to promote its oxidative leaching [8,9]. Despite chalcopyrite's abundancy among all other copper minerals, it is known to be the most stable copper mineral with its structurally face-centered tetragonal lattice [10]. Unlike many other copper sulfides, chalcopyrite tends to passivate during leaching by forming a solid product layer (e.g., sulfur or polysulfides) on its surface, which limits contact with the leaching solution and impedes further reaction [5,7,11]. This inherent inertness of chalcopyrite has made the development of enhanced leaching strategies a persistent challenge in hydrometallurgy.

Several oxidizing agents and techniques have been employed to improve chalcopyrite leaching in acidic media. Hydrogen peroxide (H_2O_2) is a strong oxidizer that can significantly accelerate the initial dissolution of sulfide minerals by oxidizing sulfide (S^{2-}) to both sulfates and elemental sulfur [12,13]. In practice, the use of H_2O_2 alone can be limited by the rapid decomposition of H_2O_2 and the formation of a passive sulfur layer on the mineral surface [14]. Recent studies suggest that the formation, structure, and persistence of this sulfur layer are not solely determined by the type or strength of the oxidant, but rather by the rate at which the oxidant is supplied, as well as by the formation of hydrogen sulfide as a key intermediate that subsequently transforms into elemental sulfur [15–17]. To address this, various studies demonstrated that leaching chalcopyrite in H_2SO_4–H_2O_2 systems is greatly enhanced by additives that prevent formation of compact sulfur layers on the surface [13,18], for example, the addition of certain compounds such as ethylene glycol [19,20], iodine/chloride and iodide/ferric sulfate media [21,22], amino acids [23], etc. Also, certain studies were performed in order to prove that kinetics may be augmented with mechanical [24] or microwave assistance [25], as it has been shown that those techniques inhibit or disrupt the sulfur film, thereby maintaining an active reaction interface. In addition to hydrogen peroxide, other oxygen-based oxidizing agents such as ozone and molecular oxygen can also be effectively utilized in chalcopyrite leaching. These oxidants can operate under atmospheric conditions, but their efficiency is often enhanced under elevated pressure in systems such as autoclaves [16,17,26–29].

Iron (III) salts (such as $FeCl_3$, $Fe_2(SO_4)_3$, and $Fe(NO_3)_3$) are widely used oxidizing agents for various sulfide concentrates. In acidic sulfate media, ferric ions can oxidize the sulfide in chalcopyrite, converting sulfide sulfur to elemental sulfur while Fe^{3+} is reduced to Fe^{2+}. This redox couple (Fe^{3+}/Fe^{2+}) is well-known, and it has been extensively studied in sulfide leaching. The main reason for such interest is the fact that ores rich in chalcopyrite are almost always followed by pyrite. Another reason comes from the biotechnological studies of sulfide leaching where it has been found that iron oxidizing bacteria indirectly leach sulfides (via the Fe^{3+}/Fe^{2+} couple) [30]. Fe^{2+} generated during leaching can be re-oxidized to Fe^{3+} by external oxidants (e.g., oxygen) or via bio-oxidation by *Acidithiobacillus* sp. bacteria. This creates a cyclic process where a small amount of Fe^{3+} can cycle repeatedly, thus oxidizing more sulfides from chalcopyrite [31,32]. Ferric salt leaching, for instance, is very effective at dissolving chalcopyrite; however, chloride-based leaching is highly corrosive and poses materials-handling challenges while iron (III) ions are prone to form jarosite even on pH values above 2.5 [33].

Persulfate compounds (e.g., sodium persulfate, $Na_2S_2O_8$) have been explored as alternative oxidants due to their ability to generate sulfate radical species. Upon activation (e.g., by heat, light or sonication), persulfate yields sulfate radicals ($\bullet SO_4^-$), which are highly reactive oxidizing agents capable of oxidizing disulfides and sulfur to higher oxidation states [34]. Free radicals such as hydroxyls and sulfates can oxidize chalcopyrite and other sulfides more aggressively than molecular oxidants, thereby enhancing the metal leaching rate. However, such strong oxidation agents can lead to a low leaching rate when used on poor raw materials such as ores, especially when there is a possibility to expend radicals on side reactions, needless to mention that they are also costly [34,35].

Industrially practiced nitrogen species catalyzed pressure leaching has also been researched and successfully applied for treatment of low-grade chalcopyrite concentrates. This highly oxidizing system also offers rapid kinetics [5].

While the above oxidizing agents and techniques can improve the leaching rate and/or leaching efficiency of chalcopyrite, they often come with significant drawbacks. Many require high reagent doses or stringent conditions, leading to increased operational costs and safety hazards. Some produce undesirable by-products (for example, chlorine gas in chloride systems or NO_x in nitrite and nitrate systems) that pose environmental challenges. Processing polymetallic concentrates containing multiple sulfide minerals and interactions during leaching can lead to severe passive action. For instance, the leaching of one sulfide (e.g., pyrite or sphalerite) may generate intermediate polysulfide or sulfate species that precipitate or form films on another mineral's surface. The formation of insoluble elemental sulfur layers or precipitates like jarosite on chalcopyrite surfaces is a well-documented impediment that hinders further leaching [36]. Such a passivation phenomenon helps explain why many conventional oxidizing systems fail to achieve high copper recoveries from chalcopyrite when multiple sulfides are present. Additionally, certain catalytic approaches (e.g., adding trace metals) that work in idealized tests are not always practical. For example, the presence of silver within a sulfide matrix (such as silver incorporated in pyrite) can greatly accelerate chalcopyrite oxidation by the formation of Ag_2S, which helps electron transfer (through the Ag^+/Ag_2S redox couple) [37]. Most of the concentrates found in genuine environments, however, lack enough silver to take advantage of this action; therefore, it is not viable to purposefully add silver salts to the leach. All these considerations underscore the need for alternative oxidation systems that can overcome chalcopyrite's refractory behavior, especially for complex feeds where multiple sulfides and gangue minerals may interfere with the leaching process.

A growing body of research has demonstrated that iodine (typically introduced as iodide, which is oxidized in situ to iodine) can dramatically enhance the leaching of chalcopyrite under mild conditions. In conventional ferric sulfate media, adding a small amount of KI has been shown to increase copper extraction rates significantly—for example, one study reported a 47% higher Cu recovery at 50 °C when 0.5 g/L KI was added, along with a sharp reduction in the apparent activation energy of chalcopyrite dissolution. Iodine is a powerful oxidant that attacks chalcopyrite to produce elemental sulfur, and it participates in a regenerative redox cycle: iodide is continuously oxidized to iodine (by agents like Fe^{3+}, MnO_2 or NO_2) [38], and iodine in turn oxidizes chalcopyrite (thus extracting Cu^{2+} and reverting to I^-). Several investigations confirmed that this I^-/I_2 shuttle can overcome or delay passivation layers on chalcopyrite, thereby sustaining faster leach kinetics even at ambient temperatures [38,39]. A notable application is the JX Iodine Process, in which ferric sulfate solutions with low-concentration iodide were irrigated over chalcopyrite ore in column tests—the presence of iodine markedly improved copper recoveries compared to ferric leaching alone, although the degree of improvement varied with ore mineralogy [38]. Likewise, patented iodine-leaching methods have demonstrated that chalcopyrite (and even refractory sulfidic minerals like enargite and pyrite) can be leached effectively at room temperature by employing an I^-/I_2 cycle with continuous regeneration of the oxidant [40,41]. Overall, these studies have established that iodine/iodide can serve as an efficient recyclable oxidant, yielding higher copper dissolution rates than conventional oxidants and enabling leaching of chalcopyrite under relatively mild conditions. In contrast to previous studies on pure chalcopyrite, the present work applied a MnO_2–KI oxidation system to a complex sphalerite concentrate (containing chalcopyrite and pyrite) and systematically examined how varying the iodine dosage and temperature affects the leaching efficiency and kinetics for both copper and iron. Detailed kinetic and thermodynamic calculations were performed in order to follow the oxidative tests of the initial concentrate.

2. Materials and Methods

The base material used for all advanced leaching processes was the sphalerite concentrate, received from a polymetallic ore deposit flotation plant ("Rudnik", Rudnik, Serbia). The collected sample was crushed and homogenized by a vibro mill (KHD Humboldt Wedag AG, Köln, Germany). During the leaching tests and the determination of the structural properties, the full range of granulometric classes was employed.

2.1. Structural Characterization

The initial concentrate and the obtained samples during oxidation leaching were collected and carefully characterized in order to determine the presence of the constituents and the surface morphology.

X-ray diffraction (XRD, Philips PW 1710/1820, Eindhoven, The Netherlands) was used to detect the phase arrangement by determining the mineral compositions of the concentrate and leaching residues. An automated diffractometer with a Cu tube running at 40 kV and 30 mA in the 2θ range of 4–65° was employed for gathering XRD patterns.

Scanning electron microscopy (SEM, JEOL JSM-7001F, Tokyo, Japan) equipped with an energy-dispersive spectrometer (EDS, Oxford Xplore 15, High Wycombe, UK) was used to depict surface functionalities and the mapping of sample surfaces.

In order to recognize metallic and non-metallic minerals, qualitative ore microscopic (OM) studies were conducted using a polarizing microscope (Carl Zeiss-Jena Axioscope 5 Pol, Jena, Germany) under reflected light. A color camera (Axiocam 105 color, White Plains, NY, USA) was connected to the microscope. With the "Multiphase" module of the Carl Zeiss AxioVision SE64 Rel. 4.9.1 software package, photomicrographs were gathered.

2.2. Leaching Experiments

Advanced leaching oxidation tests were performed according to our previous work (Figure 1) [42]. Leaching experiments were conducted in a closed glass reactor (Zhengzhou Great Wall, Xingyang, China). The reactor was adapted with a helix stirrer and temperature controller.

The reaction system consisted of base concentrate (10.0 g) and aqueous sulfuric acid solution (Zorka p. a, Šabac, Serbia), with a constant agitating rate of 300 rpm. Several operational variables were altered in the leaching tests, including temperature (40, 60, 70, and 80 °C), duration (0–180 min), and the presence of additional oxidizers–manganese dioxide (Alkaloid, purity > 95%, Skopje, North Macedonia) and potassium iodide 1–3 wt% of the initial concentrate (Merck, purity > 95%, Darmstadt, Germany). Manganese dioxide was dosed on a stoichiometric basis relative to the total sulfide-sulfur in the concentrate targeting complete oxidation of sulfide minerals to sulfate. Potassium iodide (KI), used as an auxiliary oxidant, was investigated up to a practical upper limit of 3 wt% of the initial concentrate mass to balance reactivity and cost. At determined time periods, acidic samples were collected and analyzed by atomic absorption spectrophotometry (AAS, Perkin Elmer 703, Shelton, CT, USA). To reduce the risk of measurement error, all leaching tests were conducted in triplicate, and the reported quantities of iron and copper are the average of the three measurements. The iron and copper extraction rates (%) were computed using AAS measurements.

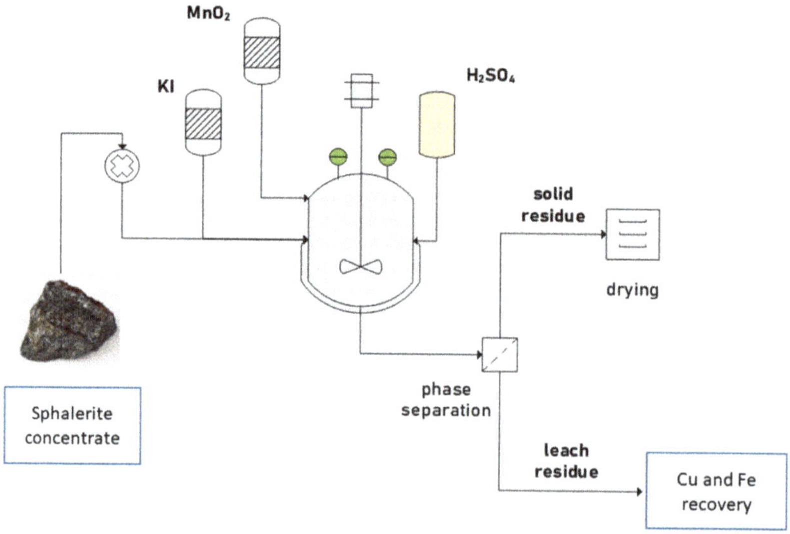

Figure 1. State of art—Technological scheme of sphalerite concentrate advanced leaching treatment (modified from [42]).

The ratio of the chemical elements in the initial sample was determined by the AAS technique. The presence of the detected elements in descending order (Zn—46.91, S—28.87, Fe—6.77, Cu—0.806, Cd—0.338, Pb—0.287, As—1.8×10^{-4}, all in wt%) was presented in our previous work [42]

The reaction solution was filtered to remove the leaching residues after each experiment was complete. After being gathered and thrice cleaned with distilled water, the solid residues were dried in an oven. Following preparation, the samples were carefully examined.

Kinetic models can be used to better describe various chemical reactions. Hence, we used the modified Sharps method of decreased half reaction time [43] to linearize the experimental results, utilizing various equations. Numerous articles provide a full description of the concept [43–45].

The Arrhenius equation, which is commonly used to calculate the activation energy (Ea), often provides important insights into the mechanisms driving the reactions:

$$k = A \times \exp(-Ea/RT) \quad (1)$$

where A is the pre-exponential constant, k is the rate constant (s^{-1}), Ea is the activation energy ($J \cdot mol^{-1}$), R is the gas constant ($8.314\ J \cdot mol^{-1} \cdot K^{-1}$), and T is the absolute temperature (K). The activation energy can be computed by plotting ln (k) versus 1/T, which provides the slope of the linear fit.

3. Results and Discussion

3.1. Material Characterization

To determine chemical, morphological and mineralogical properties, all collected specimens were thoroughly analyzed by XRD, AAS, SEM-EDS, and OM.

3.1.1. Optical Microscopy

From the OM results, nearly 80% of the sphalerite was found to exist as free grains (Figure 2a,b), with the remainder belonging to intergrowths with other minerals. In this concentrate, sphalerite is most commonly linked to chalcopyrite, which is followed by galena and chalcopyrite impregnations. The determined sphalerite's grain size is in the range 10–400 μm. Gold and silver particles were also found in the observed concentrate (Figure 2b).

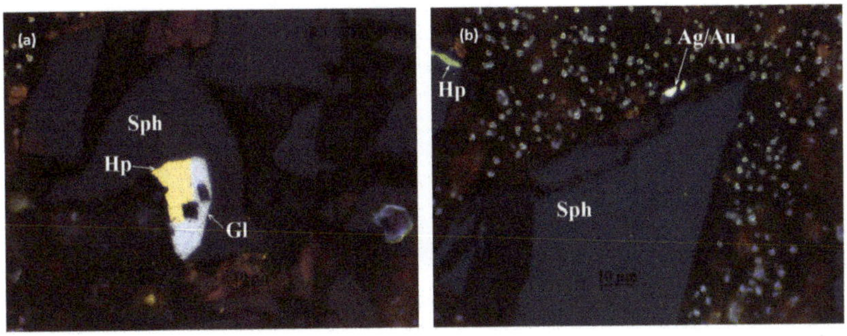

Figure 2. (**a**) Complex intergrowth of sphalerite (Sph), chalcopyrite (Hp), and galena (Gl) in the sample; (**b**) native silver/gold in addition to sphalerite.

Chalcopyrite can appear as oval, micronic impregnations within sphalerite grains or as complex assemblages with galena and sphalerite minerals. It can also occur as individual grains, which makes up about 65–70% of its frequency. There was no detectable free gold or silver linked to chalcopyrite. A material specimen undergoing optical assessment contained a trace amount of gold or silver (Figure 2b).

3.1.2. XRD Analysis

The X-ray diffraction technique was used to analyze the initial polycrystalline concentrate. The analyzed sample was found to contain the following minerals: chalcopyrite, pyrite, quartz, and sphalerite [42]. Sphalerite is with certainty the most frequent mineral, while quartz, pyrite, and chalcopyrite are far less abundant.

3.1.3. Field Emission SEM

Initial concentrate particles possess clean and sharp edges, as seen in the SEM image (Figure 3). The particles have a diameter between 50 and 100 μm.

Figure 3. SEM micrograph of initial specimen.

3.2. Leaching Assays

3.2.1. Advanced Oxidative Leaching of Initial Concentrate with MnO_2/KI

Impact of Temperature on Leaching Degree

The leaching efficiency of iron (Fe) and copper (Cu) from Zn concentrate increases with temperature, rising from 40 °C to 80 °C, in accordance with chemical kinetic principles. Elevated temperatures provide additional thermal energy, which enhances particle mobility and increases the frequency of the effective collisions between the sulfide minerals and the oxidizing agent (I_2). As a result, the reaction rate accelerates, leading to more efficient iron and copper extraction into the leaching solution.

The mineral phase distribution resulting from leaching at lower temperature is presented in Figure 4.

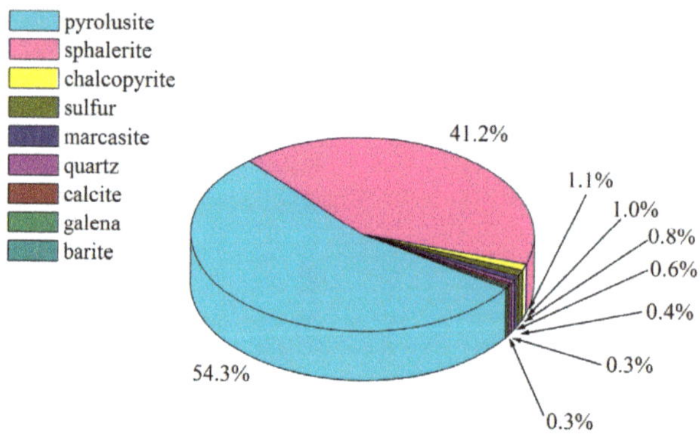

Figure 4. Pie chart of quantitative mineralogical analysis of the sample; experiment parameters (3 wt% KI + stoichiometric amount of MnO_2 at 40 °C, 180 min) [42].

The analyzed sample (Figure 4) contains the following mineral phases: pyrolusite, sphalerite, chalcopyrite, marcasite, sulfur and galena, as well as gangue minerals (quartz, calcite, and barite). Pyrolusite is the dominant mineral, followed by sphalerite present in a lower proportion, while quartz, chalcopyrite, and marcasite are present in minor quantities. All other minerals are present in traces.

The quantitative mineralogical composition is as follows [42]: pyrolusite—54.3 wt%, sphalerite—41.2 wt%, chalcopyrite—1.1 wt%, marcasite—1.0 wt%, quartz—0.8 wt%, sulfur—0.6 wt%, calcite—0.4 wt%, galena—0.3 wt%, and barite—0.3 wt%. These values were determined using software processing, aligned with the results of chemical analysis.

The structural and textural characteristics, along with the mineral occurrence forms, are presented in Figure 5. Reflected-light microscopy analysis (Figure 5) indicates that pyrolusite is the predominant mineral, followed by sphalerite. Chalcopyrite, marcasite, and quartz are notably less abundant, while galena and sulfur appear only in trace amounts.

From the reflected light microscopy images (Figure 5) it can be observed that pyrolusite is predominantly found in a lath-like habitus, reaching lengths of up to 200 μm, and less commonly appears as irregular grains measuring up to 100 μm. Sphalerite typically occurs as irregular, liberated grains—often significantly altered and eroded—up to 150 μm in size. It frequently exhibits signs of "chalcopyrite disease" and contains chalcopyrite inclusions reaching up to 50 μm. Minor intergrowths of sphalerite with chalcopyrite, or marcasite are also observed. Apart from these occurrences, chalcopyrite is occasionally present as liberated, irregular grains up to 150 μm in length. Marcasite is found as free grains up to 150 μm in length.

Figure 5. Reflected light microscopy image of the sample at 40 °C: (**a**) liberated grains of chalcopyrite (Hp) and pyrolusite (Pl), and sphalerite (Sph) with chalcopyrite inclusions; (**b**) intergrowth of sphalerite (Sph) and marcasite (Mc), liberated grains of sphalerite, chalcopyrite (Hp) and pyrolusite (Pl); (**c**) chalcopyrite (Hp) disease" and inclusion in sphalerite (Sph).

Apart from the above, irregular galena grains up to 60 μm in length were also identified in the sample. The occurrence of galena in the sphalerite concentrate is anticipated, given that the original ore contains both sphalerite and galena, which are often challenging to separate due to their intergrowth—particularly when grain sizes are below 80 μm. Consequently, the presence of galena in the leaching residue is to be expected.

The SEM image of the leaching residue obtained at 40 °C is presented in Figure 6. The chemical compositions of the minerals obtained by EDS analyses are shown in Table 1. As observed in Figure 6, the particles of the concentrate leached at a lower temperature exhibit signs of partial decomposition, evidenced by roughened and/or more rounded edges and surface alterations.

The mineral phase distribution resulting from leaching at higher temperature, 80 °C, is shown in Figure 7.

The analyzed sample (Figure 7) contains the following mineral phases: pyrolusite, sphalerite, chalcopyrite, marcasite, sulfur and galena, as well as gangue minerals (quartz, calcite and barite). Pyrolusite is the dominant mineral, followed by sphalerite in smaller amounts, while quartz, chalcopyrite, sulfur, and marcasite are present in minor quantities. All other minerals are present in traces.

The quantitative mineralogical composition [42] is as follows: pyrolusite—57 wt%, sphalerite—38 wt%, chalcopyrite—1.1 wt%, sulfur—1.0 wt%, marcasite—1.0 wt%, quartz—0.9 wt%, calcite—0.4 wt%, galena—0.3 wt%, and barite—0.3 wt%. These values were determined using software processing (Carl Zeiss AxioVision SE64 Rel. 4.9.1 with the "Multiphase" module), aligned with the results of chemical analysis.

The structural and textural characteristics, along with the mineral occurrence forms analysed by reflected-light microscopy, are presented in Figure 8.

Figure 6. SEM photomicrograph of the leaching residue sample at 40 °C.

Table 1. EDS chemical analyses of spectra obtained from the SEM image (Figure 6) (Spectra 50–52).

Element	Spectrum 50 (Sphalerite)		Spectrum 51 (Pyrolusite)		Spectrum 52 (Chalcopyrite)	
	Weight%	Atomic%	Weight%	Weight%	Weight%	Atomic%
O			30.83	60.19		
Al			0.22	0.25		
Si			0.60	0.67		
S	33.38	49.81	0.28	0.27	35.18	50.28
Mn	0.42	0.37	66.89	38.04	0.41	0.34
Fe	12.05	10.33			29.69	24.36
Cu	0.36	0.27			33.02	23.82
Ni			0.23	0.12		
Zn	53.30	39.02	0.95	0.45	1.71	1.20
Cd	0.48	0.20				
Total	100.00	100.00	100.00	100.00	100.00	100.00

Reflected-light microscopy observations (Figure 8) indicate that pyrolusite is the predominant mineral, followed by sphalerite. Chalcopyrite, marcasite, sulfur, and quartz are notably less abundant, and galena appears only in trace amounts. Pyrolusite primarily appears in a lath-like habitus, reaching lengths of up to 200 μm, with occasional irregular grains reaching 100 μm. Sphalerite typically occurs as irregular, eroded grains up to 250 μm and frequently exhibits "chalcopyrite disease" or contains chalcopyrite inclusions up to 50 μm in length. Less commonly, sphalerite and chalcopyrite are intergrown. Apart from these occurrences, chalcopyrite is occasionally present as liberated, irregular grains up to 100 μm. Marcasite is found as liberated grains up to 150 μm, while sulfur appears in the form of irregular grains. Gangue minerals mainly consist of quartz, with lesser amounts of feldspar, barite, and calcite.

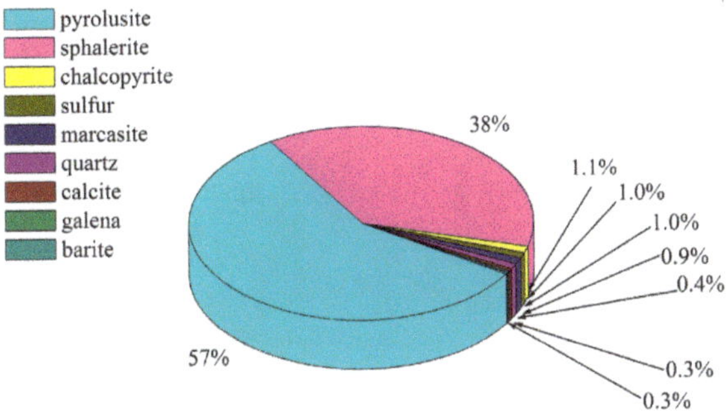

Figure 7. Pie chart of quantitative mineralogical analysis of the sample—experimental parameters (3 wt% KI + stoichiometric amount of MnO_2 at 80 °C, 180 min).

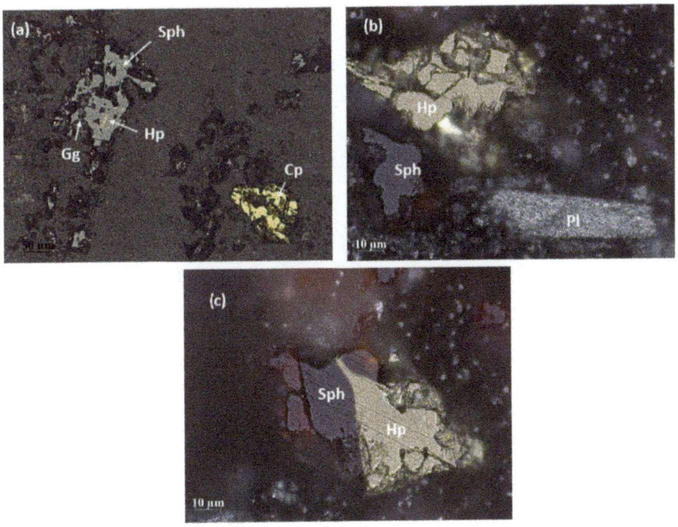

Figure 8. Reflected-light microscopy image of sample at 80 °C: (**a**) intergrowth of sphalerite (Sph) and gangue (Gg), "chalcopyrite disease" (Hp) in sphalerite, liberated chalcopyrite grain; (**b**) liberated grains of sphalerite (Sph), chalcopyrite (Hp), and pyrolusite (Pl); (**c**) intergrowth of sphalerite (Sph) and chalcopyrite (Hp); (**d**) liberated marcasite (Mc) grain.

The SEM image of the residue formed after leaching at 80 °C is shown in Figure 9, while the chemical composition of minerals obtained by EDS analyses is shown in Table 2.

As shown in the previously presented results, there are differences in the mineral content between the initial sample and those leached at higher and lower temperatures. In the initial sample, pyrite occurs as the primary iron-bearing mineral. The leaching treatments induced changes in the chemical environment, leading to oxidative conditions under which pyrite undergoes partial alteration. This process results in a decrease in pH and an increase in the concentration of the dissolved iron. Consequently, these conditions facilitate the transformation of pyrite into the less stable secondary phase, marcasite. Additionally, marcasite may be formed as a result of iron (Fe) leaching from sphalerite through a mechanism analogous to that observed in the case of pyrite. When conditions are favorable for the formation of iron sulfide with an orthorhombic FeS_2 structure, marcasite can precipitate from the solution. This process is typically associated with acidic environments and is often influenced by the presence of other reactive iron sulfide minerals, such as pyrite. Furthermore, during the leaching process, sphalerite undergoes decomposition, leading to the partial release of chalcopyrite, which occurs within the sphalerite matrix as inclusions or impregnations—a phenomenon known as "chalcopyrite disease".

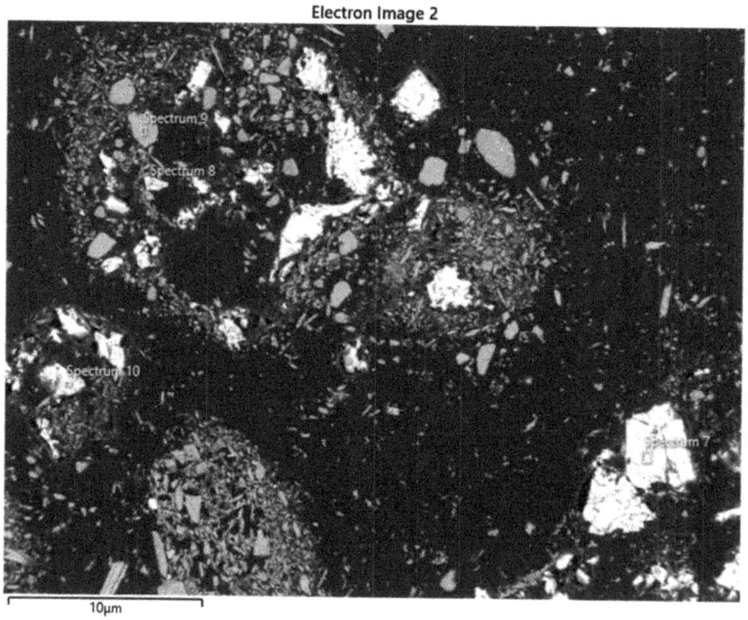

Figure 9. SEM photomicrograph of the leaching residue sample at 80 °C.

Table 2. EDS chemical analyses of spectra obtained from the SEM image (Figure 9) (Spectra 7–10).

Element	Spectrum 7 (Sphalerite)		Spectrum 8 (Chalcopyrite)		Spectrum 9 (Pyrolusite)		Spectrum 10 (Marcasite)	
	Weight %	Atomic %	Weight %	Weight %	Weight %	Atomic %	Weight %	Atomic %
O					30.39	59.90		
S	32.47	48.68	34.27	49.19	0.33	0.33	38.88	52.56
Mn	0.83	0.73	2.29	1.92	69.28	39.77	0.66	0.52
Fe	11.91	10.26	29.71	24.48			60.46	46.92
Cu	1.98	1.50	33.29	24.11				
Zn	52.81	38.84	0.43	0.31				
Total	100.00	100.00	100.00	100.00	100.00	100.00	100.00	100.00

The occurrence and behavior of sulfur are discussed in detail in the study ref. [42].

The amount sof copper (Cu) and iron (Fe) leaching from the concentrate over the time are shown in Figure 10, with determined α (%).

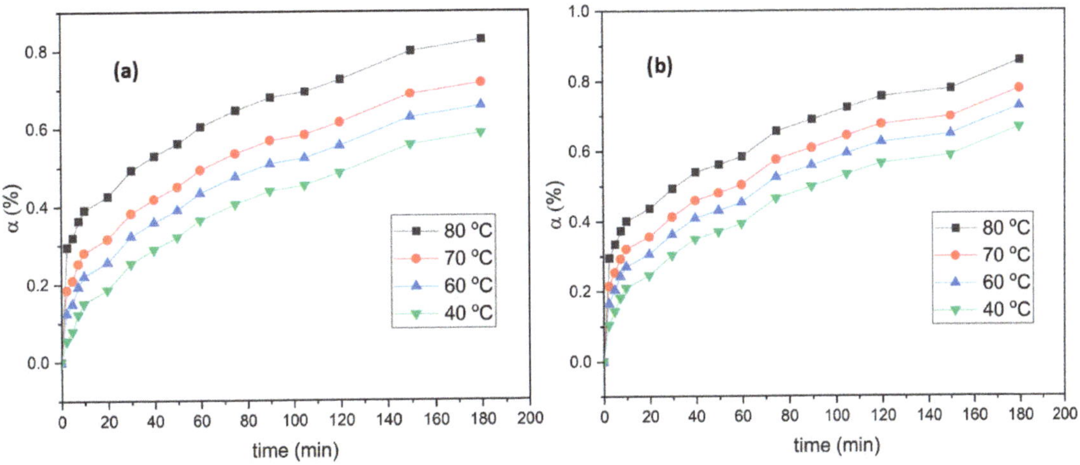

Figure 10. Temperature and reaction time effects on Cu (**a**) and Fe (**b**) leaching percentage (3 wt% KI + stoichiometric amount of MnO_2).

The reaction kinetics and extraction efficacy were significantly impacted by rise in the temperature (Figure 10). After 3 h, the maximum recoveries for copper and iron were 82.1% and 85.3%, respectively. The formation of the passivation layers on the surface of chalcopyrite and marcasite grains may be the cause of the lower leaching percentages of copper and iron at lower temperatures [46,47].

With a sufficient MnO_2 dosage, the most probable mechanisms for copper and iron leaching from chalcopyrite and pyrite, respectively, are as follows:

$$2CuFeS_2 + 17MnO_2 + 18H_2SO_4 = 2CuSO_4 + Fe_2(SO_4)_3 + 17MnSO_4 + 18H_2O \quad (2)$$

$$2FeS_2 + 15MnO_2 + 14H_2SO_4 = Fe_2(SO_4)_3 + 15MnSO_4 + 14H_2O \quad (3)$$

Both reactions (Equations (2) and (3)) take place in the presence of potassium iodide and without it; however, there are side reactions that may impede or even stop these two reactions. Passivation layers that could form include sulfur, anglesite, and barite. The presence of iodine that readily forms in the chemical reaction (Equation (4)) impacts the erosion of the sulfur passivation layer and thus enhance further leaching of chalcopyrite.

$$MnO_2 + 2KI + 2H_2SO_4 = MnSO_4 + K_2SO_4 + I_2 + 2H_2O \quad (4)$$

The product of reaction (4) is iodine (I_2) which reacts further by capturing an iodide ion (I^-) [48], and forms the triiodide anion as reported in the literature [49]. In the subsequent reaction (Equation (5)), iodine leaches the copper and is reduced back to iodide anion.

$$CuFeS_2 + 2I_2 = Cu^{2+} + Fe^{2+} + 2S + 4I^- \quad (5)$$

While hydrometallurgical studies typically consider only reaction (5) for chalcopyrite oxidation by iodine, fundamental inorganic chemistry indicates that Cu^{2+} can also oxidize iodide ions to iodine, forming copper (I) iodide [50]. Although CuI is insoluble in water, its formation under such oxidative conditions is unlikely, as Cu^+ is readily re-oxidized to the more stable Cu^{2+} state, rendering reaction (5) effectively the only viable pathway.

Besides the Cu^{2+} that is formed in reaction (5) another very important product is the Fe^{2+} ion. Ferrous ion is also unstable and in oxidative media it oxidizes to ferric ion, as reaction (6):

$$2Fe^{2+} + MnO_2 + 2H_2SO_4 = 2Fe^{3+} + SO_4{}^{2-} + MnSO_4 + 2H_2O \quad (6)$$

The ferric ion participates in several redox reactions and it can leach pyrite as well [51,52]. This is very important since the results of a recent study concluded that pyrite grains remain intact after 14 days of leaching chalcopyrite concentrate in the presence of iron (III) sulfate and potassium iodide [39]. The consecutive reactions proposed to account for iron (III) leaching from pyrite are presented in the subsequent reactions (Equations (7)–(9)):

$$FeS_2(\text{pyrite}) + xFe^{3+} = FeS^{+x}{}_{2-x}(\text{pyrite}) + xFe^{2+} + xS \quad (7)$$

$$FeS^{+x}{}_{2-x}(\text{pyrite}) + xH_3O^+ = Fe_{1-x}S_{2-2x}(\text{marcasite}) + xHS^- + xFe^{3+} + xH_2O \quad (8)$$

$$H^+ + HS^- + MnO_2 + H_2SO_4 = MnSO_4 + S + 2H_2O \quad (9)$$

Marcasite formed in reaction (8) continues to oxidize further in the same manner as pyrite in reaction (7) leading to progressive ferric leaching.

3.2.2. Influence of KI Amount on Leaching Effectiveness

The quantity of potassium iodide (KI) in the reaction solution at 80 °C shows a non-linear correlation with the leaching rate for copper (Cu) and iron (Fe) from sphalerite concentrates (Figure 11). This unusual trend points to an intricate interaction between the KI amount and the following chemical transformations that affects the oxidizing agents' accessibility and efficacy during the observed leaching.

At a low KI content (1 wt% relative to the mass of sphalerite), the system operates under stable equilibrium conditions, where the iodine (I_2) formed in reaction (4) promotes more efficient oxidation in reaction (5), without the occurrence of harmful side-effects. At the above KI concentration, the degree of copper and iron leaching is high and amounts to 71.4% and 75.3%, respectively. This indicates that elemental sulfur, which was formed during the leaching reaction at a KI content of 1 wt%, does not significantly interfere with the leaching process of the above metals.

With an increase in the concentration of KI to 2 wt%, there is a unexpected reduction in the degree of leaching of copper (59.9%) and iron (63.1%). This can be attributed to the formation of elemental sulfur (S), which forms a passive layer on the surface of the mineral, preventing the further penetration of oxidizing agents and thereby hindering the leaching process of these metals.

When the KI concentration increases to 3 wt%, there is a sudden increase in the degree of leaching of copper (82.1%) and iron (85.3%). At higher iodine concentrations, the oxidant becomes sufficient not only to react with chalcopyrite but also to react with the elemental sulfur passivation layer formed in reactions (5), (7), and (9), converting it into soluble sulfur species with higher oxidation states. These include the formation of aqueous sulfur dioxide ($SO_{2(aq)}$), followed by the oxidation of thiosulfate to tetrathionate, which is ultimately oxidized further to sulfate ions. The oxidation of sulfur containing ion species was described in detail in our previous study [42].

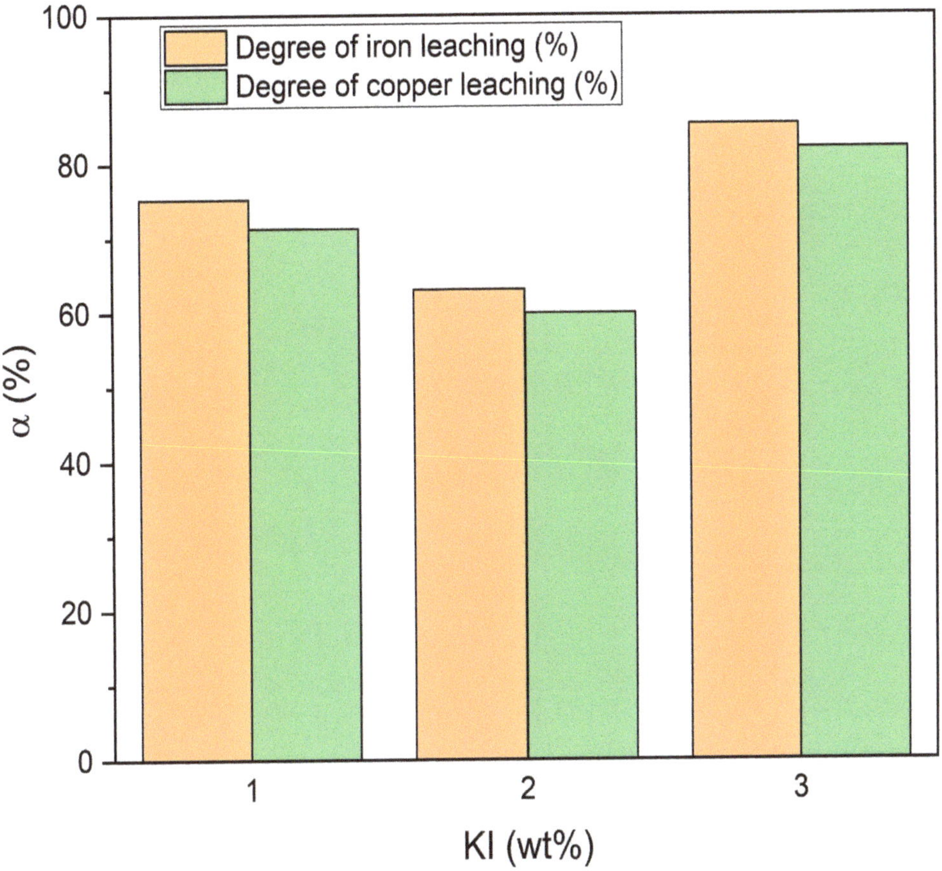

Figure 11. Impact of altering KI concentration on the leaching rate of Cu and Fe at 80 °C.

When the initial potassium iodide concentration is higher, more iodine is produced which can break down the passivation layer, thus re-exposing the sphalerite surface to the acid and making the leaching process more efficient. In addition, at this level the KI system exhibits a higher redox potential, and due to the excess iodide ions, the equilibrium shifts towards the formation of free iodine. This provides a sufficient amount of active oxidizing agent to sustain the process, which overcomes the limitations observed at 2 wt% KI.

3.3. Leaching Kinetics

To model the leaching kinetics, various kinetic models encompassed within the modified Sharp's method were employed to describe the system's behavior, as depicted in Figure 12. Among these, the Jander equation [43] for three-dimensional diffusion through a product layer (depicted by navy blue triangle symbols on Figure 12) provided the best fit to the experimental data at 80 °C, yielding a coefficient of determination (R^2) of 0.98. This high correlation indicates that the Jander model accurately describes the leaching process under the studied conditions. Consequently, this model was selected for the fitting of all leaching experimental data obtained at various temperatures.

The results suggest that the rate of the leaching reaction is limited by the diffusion of copper, iron, and sulfide species through a product layer—most likely composed of elemental sulfur formed on the surface of the sphalerite particles. This product layer acts as a barrier, impeding mass transfer and slowing down the leaching rate [53]. The formation of such a sulfur layer is consistent with the findings reported in the literature, where elemental sulfur deposition on sphalerite particles during leaching processes has been observed to significantly decrease the leaching yield of the target metals [54,55].

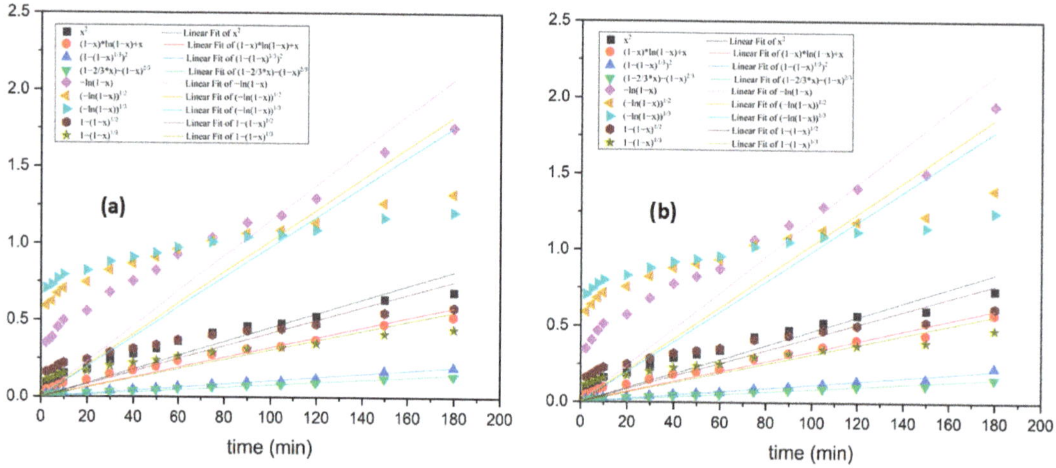

Figure 12. Modified Sharp's approach for Cu (**a**) and Fe (**b**) leaching related to experimental results.

Employing the Jander formula, the fitted data collected from the advanced leaching of sphalerite concentrate with MnO_2 and KI throughout the temperature range of 40 to 80 °C are shown in Figure 13. The proposed model shows a high coefficient of determination with the experimental data over the whole temperature range ($R^2 > 0.97$ for Fe and $R^2 > 0.98$ for Cu), suggesting a strong agreement.

The estimated leaching rate constants employing the modified Sharp method are 0.0012 min^{-1} and 0.0011 min^{-1} at 80 °C, and 0.0005 min^{-1} and 0.0003 min^{-1} at 40 °C for Fe and Cu, respectively. The obtained results from Figure 13 indicate that raising the temperature considerably accelerates the oxidation rate, increasing it by more than twice at 80 °C.

The obtained curves from Figure 13 present a good correlation of the laboratory results and the Jander equation. Therefore, the proposed model well depicts three-dimensional diffusion through a product layer, indicating that species migration via a product layer is the rate-limiting phase. The passivation layer of elemental sulfur developed on the sphalerite concentrate particle surface could be the slowest and the limitation step. Each temperature causes this layer to develop, albeit its quantity and effect on the rate of leaching may fluctuate. The existence of sulfur residues is verified even at the lowest leaching temperature (40 °C), while the crystallization of sulfur is only verified at the highest leaching temperature (80 °C). Increased leaching levels and elevated temperatures promote crystallization of sulfur onto the grain surface. Hence, this phenomenon can enhance the passivation effect by generating a more coherent and impermeable layer [42]. By adding enough potassium iodide (>3 wt%), this problem can be lessened as the concentration of iodine in the leaching solution will rise. Since iodine efficiently transforms elemental sulfur toward soluble sulfur compounds, the elevated iodine level permits the oxidation of the produced sulfur passivation layer. Increased leaching rates persist even at higher temperatures because of this oxidation, which lowers the mass transfer barrier and improves the leaching efficiency; this has already been noted in available studies [56,57].

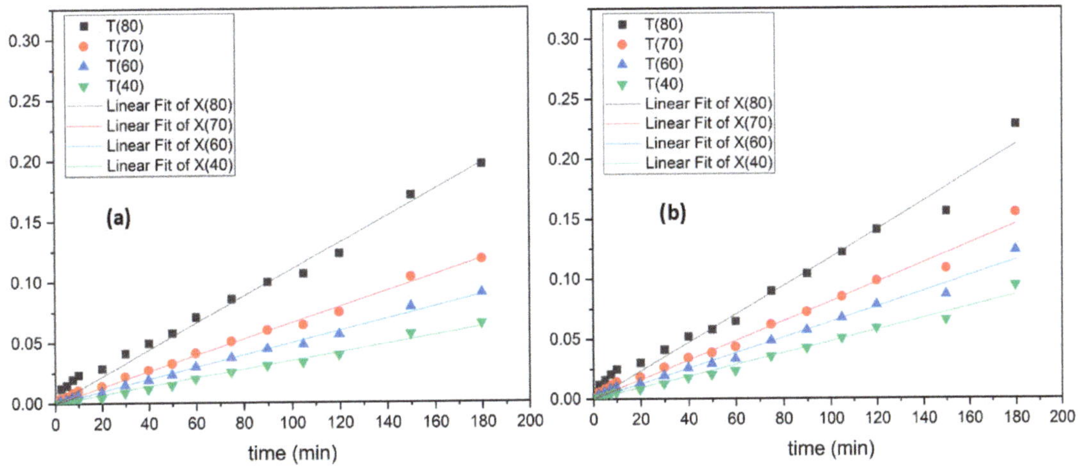

Figure 13. Jander's model for experimental results for Cu (**a**) and Fe (**b**) leaching at different temperatures.

3.4. Determination of Activation Energy for Concentrate Leaching

The activation energy (Ea), determined by utilizing the linearized version of the Arrhenius equation, is the last part in the kinetic analyses. The findings from Figure 14 display a high linear association ($R^2 > 0.95$ for Cu recovery and $R^2 > 0.98$ for Fe recovery). Equation (10) was used for estimating the activation energy based on the slope of this linear function:

$$Ea = -R \times Slope \tag{10}$$

As previously validated, the computed values of 28.90 and 18.96 kJ/mol for Cu and Fe extraction show that diffusion is the leaching reaction's limiting phase [54–56].

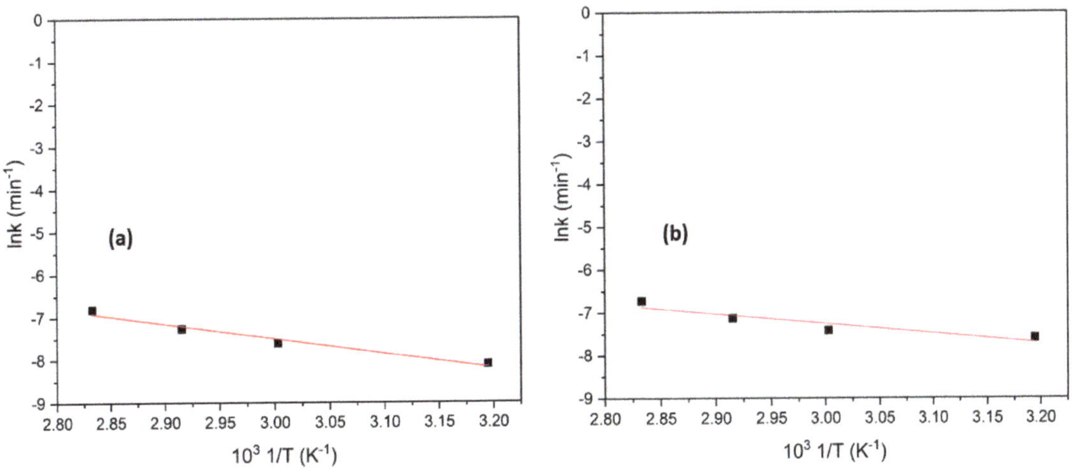

Figure 14. Determined activation energies for Cu (**a**) and Fe (**b**) recovery (T: 40–80 °C).

The calculated activation energy for the Cu and Fe leaching indicates that the oxidation process has a low energy barrier (activation energy of 28.90 kJ mol^{-1} for Cu). Given its low activation energy, copper leaching is a diffusion-controlled process, which is consistent with the value found by Kaplun et al. [58]. The activation energy of iron leaching is even lower (18.94 kJ mol^{-1}); such a low value is not obtained in any literature available on the study of pyrite leaching. However, such low values are obtained in the study of Asta and Acero for leaching marcasite in acidic media [59]. This observation is consistent with the absence of pyrite in the leaching residue, where only marcasite was detected with a sub-stoichiometric sulfur content (Table 2). Most likely Fe^{3+} ions interact with pyrite in the first step oxidizing the disulfides and the low pH value influences non-stoichiometric polymorph to be crystalized as marcasite. Once marcasite is formed, the strong oxidation agent (pyrolusite) assisted by iodine as a participant of the sulfur oxidation oxidize the marcasite, and iron (III) is leached further.

Such low activation energies suggest that leaching proceeds rapidly at moderate temperatures due to the low energy barrier of the oxidation process. This observation is consistent with the experimental results, where the Jander equation provided the best fit. The low activation-energy values further confirm that the leaching of both copper and iron is governed by a diffusion-controlled mechanism.

4. Conclusions

In our study we presented the Cu and Fe recovery from sphalerite concentrates in acidic solution, applying the auxiliary combination of oxidants—MnO_2 together with KI. The influence of temperature variation had a significant impact on Cu and Fe recovery; there were quicker reaction kinetics from 40 °C to 80 °C and significantly increased leaching efficiency.

The addition of KI resulted in a complex interaction between sulfur passivation and oxidant accessibility. Cu and Fe leaching achieved 82.1% and 85.3% at 3 wt% KI, thus a reduction of the passivation layer on the grain surface of the sphalerite concentrate.

Furthermore, low activation energies of 28.90 kJ mol^{-1} for Cu and 18.94 kJ mol^{-1} for Fe were found in the kinetic evaluation of the system $MnO_2/KI{+}H_2SO_4$. Although both copper and iron leaching kinetics are diffusion-controlled processes, the mechanism of their leaching differs significantly. Chalcopyrite interacts directly with both oxidation agents, thus copper leaching is similar to zinc leaching from sphalerite. Iron leaching is more complex and includes the interaction of pyrite with ferric ions, the formation of a sulfur passivation layer, and pyrite alteration to marcasite which eventually lead to the final decomposition of marcasite and iron (III) leaching.

The procedure is more energy-efficient and profitable because of the low activation energies, and this indicates that the leaching reactions continue readily at comparatively lower temperatures. Higher Cu and Fe recovery rates are made possible by the improved reaction kinetics, which also increase the efficacy in general.

In summary, the present investigation demonstrates that the temperature and the KI amount have a significant influence on Cu and Fe recovery from sphalerite, with the maximum Cu and Fe leaching rate achieved at 80 °C and 3 weight percent KI.

Author Contributions: Conceptualization, M.B., A.J., and M.S.; methodology, M.S., C.A., and A.J.; software, I.J., A.J., and D.A.; validation, M.B., M.S., A.J., and N.V.; formal analysis, M.B., A.J., and I.J.; investigation, D.A., M.B., and A.J.; resources, M.S.; data curation, A.J., I.J., C.A., and D.A.; writing—original draft preparation, M.B. and A.J.; writing—review and editing, I.J., M.S., and C.A.; visualization, A.J.; M.B., D.A., and N.V.; supervision, M.S.; project administration, M.S. and M.B. All authors have read and agreed to the published version of the manuscript.

Funding: This work was supported by the Ministry of Science and Technological Development and Innovation of the Republic of Serbia (Contract No. 451-03-136/2025-03/200023).

Data Availability Statement: The data presented in this study are available on request from the corresponding author due to privacy.

Conflicts of Interest: Author Mladen Bugarčić was employed by the company Milan Blagojević-Namenska AD. The remaining authors declare that the research was conducted in the absence of any commercial or financial relationships that could be construed as a potential conflict of interest.

References

1. Jaramillo, K.; Gómez, M.; Colet-Lagrille, M. From Copper Ores to E-Waste Processing: Can Urban Mining Practices Be Actually Integrated into Idle Capacity in Acid-Ferric Hydrometallurgical Facilities? *Miner. Eng.* **2025**, *229*, 109355. [CrossRef]
2. Bogdanović, G.D.; Petrović, S.; Sokić, M.; Antonijević, M.M. Chalcopyrite Leaching in Acid Media: A Review. *Metall. Mater. Eng.* **2020**, *26*, 177–198. [CrossRef]
3. Sokić, M.D.; Bugarčić, M.; Jovanović, A.A. Sphalerite Leaching in Acid Media: A Review. *Metall. Mater. Data* **2023**, *1*, 33–43. [CrossRef]
4. Moskalyk, R.R.; Alfantazi, A.M. Review of Copper Pyrometallurgical Practice: Today and Tomorrow. *Miner. Eng.* **2003**, *16*, 893–919. [CrossRef]
5. Gok, O.; Anderson, C.G. Dissolution of Low-Grade Chalcopyrite Concentrate in Acidified Nitrite Electrolyte. *Hydrometallurgy* **2013**, *134–135*, 40–46. [CrossRef]
6. Shiri, H.R.; Mokmeli, M.; Ghadamgahi, S.M.; Babakhani, A. Deep Eutectic Solvents (DESs) for Chalcopyrite Concentrate Extraction: Leaching, Optimization and Kinetics Mechanism. *J. Environ. Chem. Eng.* **2025**, *13*, 117779. [CrossRef]
7. Ali, Z.; Wilkes, N.; Raza, N.; Omar, M. Modified Hydrometallurgical Approach for the Beneficiation of Copper from Its Low-Grade Ore. *ACS Omega* **2025**, *10*, 14826–14834. [CrossRef]
8. Flores, D.J.; Graber, T.A.; Angel-Castillo, A.H.; Hernández, P.C.; Taboada, M.E. Use of Hydrogen Peroxide as Oxidizing Agent in Chalcopyrite Leaching: A Review. *Metals* **2025**, *15*, 531. [CrossRef]
9. Aghazadeh-Ghomi, M.; Adli-Mehr, A.; Mozammel, M. A Kinetic Study on the Atmospheric Leaching of Chalcopyrite Concentrate in Sulfuric Acid-Hydrogen Peroxide-Isopropanol Medium. *Metall. Mater. Trans. B Process Metall. Mater. Process. Sci.* **2025**, *56*, 300–306. [CrossRef]
10. Panda, S.; Akcil, A.; Pradhan, N.; Deveci, H. Current Scenario of Chalcopyrite Bioleaching: A Review on the Recent Advances to Its Heap-Leach Technology. *Bioresour. Technol.* **2015**, *196*, 694–706. [CrossRef]
11. Behnajady, B.; Najafi, M.; Karimi, S. A New Approach to Direct Chemical Leaching of Sungun Chalcopyrite Concentrate via Green Deep Eutectic Solvent Choline Chloride-ρ-Toluenesulfonic Acid and MD Simulation. *J. Taiwan Inst. Chem. Eng.* **2025**, *172*, 106118. [CrossRef]
12. Adebayo, A.O.; Ipinmoroti, K.O.; Ajayi, O.O. Dissolution Kinetics of Chalcopyrite with Hydrogen Peroxide in Sulphuric Acid Medium. *Chem. Biochem. Eng. Q.* **2003**, *17*, 213–218.
13. Agacayak, T.; Aras, A.; Aydogan, S.; Erdemoglu, M. Leaching of Chalcopyrite Concentrate in Hydrogen Peroxide Solution. *Physicochem. Probl. Miner. Process.* **2014**, *50*, 657–666.
14. Laubertová, M.; Velgosová, O.; Pirošková, J.; Briančin, J. Effects of Microwave Energy and MnO_2 from Deep-Sea Polymetallic Nodules as an Oxidizing Agent on the Leaching of Chalcopyrite Concentrate. *Minerals* **2025**, *15*, 914. [CrossRef]
15. Du, Y.; Wang, L.; Zhang, W.; Chen, Y.; Ma, B.; Wang, C. Disparity in Leaching Behavior of Representative Copper Sulfide Minerals with Dilute H_2SO_4 and the Underlying Microscopic Mechanism. *Appl. Surf. Sci.* **2025**, *706*, 163413. [CrossRef]
16. Cháidez, J.; Parga, J.; Valenzuela, J.; Carrillo, R.; Almaguer, I. Leaching Chalcopyrite Concentrate with Oxygen and Sulfuric Acid Using a Low-Pressure Reactor. *Metals* **2019**, *9*, 189. [CrossRef]
17. Ren, M.; Zhang, J.; Xu, Z.; Wang, J.; Qiu, J.; Xie, Y. Accelerated Metal Separation from Chalcopyrite Assisted with Ozone. *Chem. Eng. Process. Process Intensif.* **2025**, *212*, 110263. [CrossRef]
18. Petrović, S.J.; Bogdanović, G.D.; Antonijević, M.M.; Vukčević, M.; Kovačević, R. The Extraction of Copper from Chalcopyrite Concentrate with Hydrogen Peroxide in Sulfuric Acid Solution. *Metals* **2023**, *13*, 1818. [CrossRef]
19. Ruiz-Sánchez, Á.; Lapidus, G.T. Study of Chalcopyrite Leaching from a Copper Concentrate with Hydrogen Peroxide in Aqueous Ethylene Glycol Media. *Hydrometallurgy* **2017**, *169*, 192–200. [CrossRef]
20. Ruiz-Sánchez, A.; Lázaro, I.; Lapidus, G.T. Improvement Effect of Organic Ligands on Chalcopyrite Leaching in the Aqueous Medium of Sulfuric Acid-hydrogen Peroxide-Ethylene Glycol. *Hydrometallurgy* **2020**, *193*, 105293. [CrossRef]
21. Taboada, M.E.; Jamett, N.E.; Moraga, G.A.; Hernández, P.C.; Graber, T.A. Obtention of Suitable Pregnant Leach Solution (PLS) for Copper Solvent Extraction Plants from Copper Concentrate Using Hydrogen Peroxide and Iodine in a Sulfuric Acid–Chloride Medium. *Metals* **2024**, *14*, 817. [CrossRef]
22. Granata, G.; Miura, A.; Liu, W.; Pagnanelli, F.; Tokoro, C. Iodide-Assisted Leaching of Chalcopyrite in Acidic Ferric Sulfate Media. *Hydrometallurgy* **2019**, *186*, 244–251. [CrossRef]
23. Nurtazina, N.; Uvarov, N.; Azhigulova, R.; Tyapkin, P. Chalcopyrite Leaching by Amino Acid Solutions in the Presence of Hydrogen Peroxide. *Physicochem. Probl. Miner. Process.* **2022**, *58*, 157067. [CrossRef]

24. Laskar, C.; Dakkoune, A.; Julcour, C.; Bourgeois, F.; Biscans, B.; Cassayre, L. Case-Based Analysis of Mechanically-Assisted Leaching for Hydrometallurgical Extraction of Critical Metals from Ores and Wastes: Application in Chalcopyrite, Ferronickel Slag, and Ni-MH Black Mass. *Comptes Rendus. Chimie* **2024**, *27*, 37–52. [CrossRef]
25. Zandevakili, S.; Akhondi, M.R. Microwave-Assisted Leaching for Copper Recovery from the Chalcopyrite Concentrate of Sarcheshmeh Copper Complex. *Int. J. Min. Geo-Eng.* **2022**, *56*, 277–284.
26. Karimov, K.; Tretiak, M.; Rogozhnikov, D. The Dissolution Behavior of Pyrite and Chalcopyrite in Their Mixture During Low-Temperature Pressure Oxidation: A Kinetic Analysis. *Materials* **2025**, *18*, 551. [CrossRef]
27. Aydoğan, S.; Aras, A.; Canbazoğlu, M. Oxidative Ammonia Leaching of Sphalerite Concentrate. *J. Fac. Eng. Arch. Selcuk Univ.* **2005**, *20*, 55–62.
28. Mubarok, M.Z.; Sukamto, K.; Ichlas, Z.T.; Sugiarto, A.T. Direct Sulfuric Acid Leaching of Zinc Sulfide Concentrate Using Ozone as Oxidant under Atmospheric Pressure. *Miner. Metall. Process.* **2018**, *35*, 133–140. [CrossRef]
29. Wang, J.; Faraji, F.; Ghahreman, A. Evaluation of Ozone as an Efficient and Sustainable Reagent for Chalcopyrite Leaching: Process Optimization and Oxidative Mechanism. *J. Ind. Eng. Chem.* **2021**, *104*, 333–344. [CrossRef]
30. Conić, V.; Stanković, S.; Marković, B.; Božić, D.; Stojanović, J.; Sokić, M. Investigation of the Optimal Technology for Copper Leaching from Old Flotation Tailings of the Copper Mine Bor (Serbia). *Metall. Mater. Eng.* **2020**, *26*, 209–222. [CrossRef]
31. Stanković, S.; Goldmann, S.; Kraemer, D.; Ufer, K.; Schippers, A. Bioleaching of a Lateritic Ore (Piauí, Brazil) in Percolators. *Hydrometallurgy* **2024**, *224*, 106262. [CrossRef]
32. Bakhti, A.; Moghimi, H.; Bozorg, A.; Stankovic, S.; Manafi, Z.; Schippers, A. Comparison of Bioleaching of a Sulfidic Copper Ore (Chalcopyrite) in Column Percolators and in Stirred-Tank Bioreactors Including Microbial Community Analysis. *Chemosphere* **2024**, *349*, 140945. [CrossRef]
33. Cruells, M.; Roca, A. Jarosites: Formation, Structure, Reactivity and Environmental. *Metals* **2022**, *12*, 802. [CrossRef]
34. Dakubo, F.; Baygents, J.C.; Farrell, J. Peroxodisulfate Assisted Leaching of Chalcopyrite. *Hydrometallurgy* **2012**, *121–124*, 68–73. [CrossRef]
35. Turan, M.D.; Arslanoğlu, H.; Altundoğan, H.S. Optimization of the Leaching Conditions of Chalcopyrite Concentrate Using Ammonium Persulfate in an Autoclave System. *J. Taiwan Inst. Chem. Eng.* **2015**, *50*, 49–55. [CrossRef]
36. Jorjani, E.; Ghahreman, A. Challenges with Elemental Sulfur Removal during the Leaching of Copper and Zinc Sulfides, and from the Residues; a Review. *Hydrometallurgy* **2017**, *171*, 333–343. [CrossRef]
37. Li, L.; Soleymani, M.; Ghahreman, A. New Insights on the Role of Lattice-Substituted Silver in Catalytic Oxidation of Chalcopyrite. *Electrochim. Acta* **2021**, *369*, 137652. [CrossRef]
38. Fukano, Y.; Miura, A. Chalcopyrite Leaching with Iodine (JX Iodine Process) for Various Ore Types. *Hydrometallurgy* **2021**, *206*, 105752. [CrossRef]
39. Winarko, R.; Dreisinger, D.B.; Miura, A.; Fukano, Y.; Liu, W. Characterization of the Solid Leach Residues from the Iodine-Assisted Chalcopyrite Leaching in Ferric Sulfate Media. *Hydrometallurgy* **2024**, *226*, 106302. [CrossRef]
40. Manabe, M. Method of Leaching Copper Sulfide Ore with the Use of Iodine. U.S. Patent No. 8,163,063, 24 April 2012.
41. Kuwano, K.; Abe, A.; Manabe, M.; Miura, A. Method of Leaching Copper Sulfide Ore. U.S. Patent No. 20110229385, 4 September 2012.
42. Bugarčić, M.; Jovanović, A.; Anđić, D.; Jelić, I.; Miletić, M.; Marković, B.; Sokić, M. Kinetics of Sphalerite Leaching by MnO_2-KI Oxidation System in Sulfuric Acid. *Metals* **2025**, *15*, 50. [CrossRef]
43. Sharp, J.H.; Brindley, G.W.; Achar, B.N.N. Numerical Data for Some Commonly Used Solid State Reaction Equations. *J. Am. Ceram. Soc.* **1966**, *49*, 379–382. [CrossRef]
44. Ojeda, C.; Jaques, A.; Aracena, A. Optimal Sampling Times for Leaching Experiments. *Metall. Mater. Trans. B Process Metall. Mater. Process. Sci.* **2022**, *53*, 1082–1088. [CrossRef]
45. Fu, W.; Meng, L.; Qu, J. Sintering Mechanism and Leaching Kinetics of Low-Grade Mixed Lithium Ore and Limestone. *Metals* **2024**, *14*, 1075. [CrossRef]
46. Liu, X.; Du, Z.; Sun, C.; Zhang, N. A Review on the Electrochemical Analysis of Sulfide Minerals—Pyrite, Chalcopyrite, and Galena. *Green Smart Min. Eng.* **2025**, *2*, 18–31. [CrossRef]
47. Liu, X.J.; Liao, Y.; Ma, H.; Liu, Q. Electrochemical Characterizations and Galvanic Effect of Chalcopyrite Leaching and Passivation-A Review. *Miner. Eng.* **2024**, *210*, 108673. [CrossRef]
48. Ramette, R.W.; Sandford, R.W. Thermodynamics of Iodine Solubility and Triiodide Ion Formation in Water and in Deuterium Oxide. *J. Am. Chem. Soc.* **1955**, *77*, 1068. [CrossRef]
49. Winarko, R.; Dreisinger, D.B.; Miura, A.; Fukano, Y.; Liu, W. Iodine-Assisted Chalcopyrite Leaching in Ferric Sulfate Media: Kinetic Study under Fully Controlled Redox Potential and PH. *Hydrometallurgy* **2022**, *208*, 105797. [CrossRef]
50. Kauffman, G.B.; Fang, L.Y.; Viswanathan, N.; Townsend, G. Purification of Copper (I) Iodide. *Inorg. Synth.* **1984**, *22*, 101–103.
51. Breed, A.W.; Harrison, S.T.L.; Hansford, G.S. A Preliminary Investigation of the Ferric Leaching of a Pyrite/Arsenopyrite Flotation Concentrate. *Miner. Eng.* **1997**, *10*, 1023–1030. [CrossRef]

52. Li, L.; Ghahreman, A. The Synergistic Effect of Cu^{2+}-Fe^{2+}-Fe^{3+} Acidic System on the Oxidation Kinetics of Ag-Doped Pyrite. *J. Phys. Chem. C* **2018**, *122*, 26897–26909. [CrossRef]
53. Picazo-Rodríguez, N.G.; Soria-Aguilar, M.D.J.; Martínez-Luévanos, A.; Almaguer-Guzmán, I.; Chaidez-Félix, J.; Carrillo-Pedroza, F.R. Direct Acid Leaching of Sphalerite: An Approach Comparative and Kinetics Analysis. *Minerals* **2020**, *10*, 359. [CrossRef]
54. Weisener, C.G.; Smart, R.S.C.; Gerson, A.R. A Comparison of the Kinetics and Mechanism of Acid Leaching of Sphalerite Containing Low and High Concentrations of Iron. *Int. J. Miner. Process* **2004**, *74*, 239–249. [CrossRef]
55. Weisener, C.G.; Smart, R.S.C.; Gerson, A.R. Kinetics and Mechanisms of the Leaching of Low Fe Sphalerite. *Geochim. Cosmochim. Acta* **2003**, *67*, 823–830. [CrossRef]
56. Debernardi, G.; Carlesi, C. Chemical-Electrochemical Approaches to the Study Passivation of Chalcopyrite. *Miner. Process. Extr. Metall. Rev.* **2013**, *34*, 10–41. [CrossRef]
57. Hackl, R.P.; Dreisinger, D.B.; Peters, E.; King, J.A. Passivation of Chalcopyrite during Oxidative Leaching in Sulfate Media. *Hydrometallurgy* **1995**, *39*, 25–48. [CrossRef]
58. Kaplun, K.; Li, J.; Kawashima, N.; Gerson, A.R. Cu and Fe Chalcopyrite Leach Activation Energies and the Effect of Added Fe^{3+}. *Geochim. Cosmochim. Acta* **2011**, *75*, 5865–5878. [CrossRef]
59. Asta, M.P.C.; Acero, P. Dissolution Kinetics of Marcasite at Acidic PH. *Eur. J. Mineral.* **2010**, *22*, 49–61. [CrossRef]

metals

MDPI

Article

Uranium(VI), Thorium(IV), and Lanthanides(III) Extraction from the Eudialyte Concentrate Using the N,O-Hybrid Heterocyclic Reagents

Alfiya M. Safiulina [1,2,*], Alexey V. Lizunov [1], Alexey V. Ivanov [3], Nataliya E. Borisova [3,*], Petr I. Matveev [3], Sergey M. Aksenov [4] and Dmitry V. Ivanets [2,†]

[1] JSC "A.A. Bochvar Advanced Research Institute of Inorganic Materials", Moscow 123098, Russia; alex_avl@rambler.ru
[2] Institute of Materials of Modern Energy and Nanotechnology, Mendeleev University of Chemical Technology of Russia, Moscow 125047, Russia
[3] Faculty of Chemistry, Lomonosov Moscow State University, Moscow 119991, Russia; phthal_iv@org.chem.msu.ru (A.V.I.)
[4] FRC Kola Science Centre RAS, Apatity 184209, Russia; s.aksenov@ksc.ru
* Correspondence: amsafiulina@bochvar.ru (A.M.S.); borisova.nataliya@gmail.com (N.E.B.)
† Current address: State Atomic Energy Corporation "Rosatom", Moscow 119017, Russia.

Abstract: N,O-donor hybrid heterocyclic extractants have great potential for separation of actinides from lanthanides in spent nuclear fuel reprocessing processes. We demonstrate that this type of reagents can be used for primary concentration of actinides contained in eudialyte, a promising mineral containing a heavy group of lanthanides. With respect to lanthanide ions, the efficiency of their extraction decreases in the series L3 >> L1 > L2, and the extraction of actinides decreases in the series L1 ≈ L3 >> L2. For the extractant L2 based on 2,2′-bipyridine-6,6′-dicarboxylic acid diamide, the efficiency of lanthanide purification from U, Th exceeds 50. The structure and stereochemical features of the ligands do not have a significant effect on the composition of the formed complexes. The solvation numbers are close to 1 for all range f-elements studied, except for thorium, which indicates the predominant formation of complexes with the composition ratio of 1:1. The solvation numbers 1.4–1.5 are observed for thorium(IV), and the established values indicate the formation of a mixture of complexes with the composition ratios of 1:1 and 2:1.

Keywords: lanthanides; uranium; thorium; eudialyte ore; extraction

1. Introduction

Interest in the elements of the lanthanide group and other rare metals is rapidly growing in global science and industry. As a result, new technologies are being developed and high-tech products, based on rare and rare-earth metals (*RM*s and *REE*s), are being created. The dynamics of RM consumption growth are driven by their application in both traditional sectors (metallurgy, composite materials, electrical engineering and electronics, and nuclear energy) and in new high-tech industries [1–5]. The *REE* market represents an economic and geopolitical system based on the interaction of various entities (companies, corporations, countries, and governments) at different stages of the technological chain: exploration and extraction—processing—high-tech production—management of the global supply chain [6–8].

The scarcity of rare metals, along with increasing prices, has made it essential to develop new methods for producing rare metal concentrates, including the processing of low-grade ores and man-made waste [9–11]. Traditionally, the processes for producing metal concentrates involve extraction stages that separate and concentrate *REE*s from aqueous solutions [12,13], typically using phosphorus-containing extractants [14–16].

The most productive natural source of strategic metals (REE, Zr, Hf, Nb, Ta) are the minerals of alkaline rocks and carbonatites [17–19]. Of particular interest are the large deposits of eudialyte-group minerals (EGMs) in Lovozero alkaline massif at the Kola Peninsula (Russia) or other countries (South Africa, Brazil, Canada, Denmark) [20,21]. Among them, the Lovozero alkaline massif contains eudialyte ores, which attract interest as a valuable source of medium-heavy and heavy rare earth metals, as well as zirconium and hafnium. Moreover, because of the complex chemical composition of EGMs, the eudialyte deposits could be considered as complex ores, also containing manganese, niobium, scandium, as well as radioactive metals.

The off-balance reserves of eudialyte ores at the Alluaiv site of the Lovozero deposit are estimated to be 7.3 million tons, which is approximately equivalent to 1 million tons of ZrO_2. The average eudialyte content in these ores is around 12%, with coarse and medium-grained varieties containing up to 20–25%. Despite the wide distribution of eudialyte ores, there are no industrial-scale operations exploiting deposits of this type. However, technologies for eudialyte mineral processing are actively developed. In Australia, for instance, the technological processing for eudialyte ores from the Toongi deposit (which contains tantalum, niobium, REE, and zirconium) has been developed over the past two decades. Moreover, the preparations are also underway for the development of eudialyte ore deposits of Norra Karr (Sweden) and Tanbreez (Greenland).

EGMs are complex zirconium and calcium silicates, which are characterized by the following general formula: $[N(1)_3N(2)_3N(3)_3N(4)_3N(5)_3]\{M(1)_6M(2)_3M(3)M(4)Z_3(Si_9O_{27-3x}(OH)_{3x})_2(Si_3O_9)_2Ø_{0–6}\}X(1)X(2)$, where $M(1)$ = ${}^{VI}Ca$, ${}^{VI}Mn^{2+}$, ${}^{VI}REE$, ${}^{VI}Na$, ${}^{VI}Fe^{2+}$; $M(2)$ = ${}^{IV,V}Fe^{2+}$, ${}^{V,VI}Fe^{3+}$, ${}^{V,VI}Mn^{2+}$, ${}^{V,VI}Na$, ${}^{IV,V}Zr$; $M(3)$ and $M(4)$ = ${}^{IV}Si$, ${}^{VI}Nb$, ${}^{VI}Ti$, ${}^{VI}W^{6+}$; Z = ${}^{VI}Zr$, ${}^{VI}Ti$; Ø = O, OH; $N(1)$–$N(5)$ are extra-framework cations (Na, H_3O^+, K, Sr, *Ln*, Y Ba, Mn^{2+}, Ca) or H_2O; $X(1)$ and $X(2)$ are extra-framework water molecules, halide (Cl^-, F^-) and chalcogenide (S^{2-}) anions, and anionic groups ($CO_3{}^{2-}$, $SO_4{}^{2-}$); x = 0–1. The coordination numbers of cations are indicated with Roman numerals [22–24]. The crystal structure of EGMs is represented by a dense microporous framework built up by zirconium(IV) in octahedral environment, calcium octahedral rings and two types of silicate tetrahedral rings ($[Si_3O_9]$ and $[Si_9O_{27}]$) (Figure 1), while the pores of the framework are filled by alkaline, alkaline- and rare-earth elements, as well as additional halide and chalcogenide anions and water molecules. The high chemical rank of R = 11–13 of EGMs makes influences the searching of the efficient schemes of the decomposing of EGMs and the extraction of RM and REM.

Many technologies have now been developed for processing eudialyte ores, containing 25–27% of eudialyte, 52% of feldspar, loparite and nepheline, and 20% of aegirine, which can be processed using combined flotation–gravity–magnetoelectric enrichment schemes to produce eudialyte and loparite concentrates [25–27]. The resulting eudialyte concentrates, with minimal radioactivity, contain 54 wt.% SiO_2, 11–13 wt.% ZrO_2, 1.8–2.5 wt.% REE_2O_3, 0.21–0.27 wt.% HfO_2, 0.06–0.1 wt.% Ta_2O_5, 0.6–1% wt.% Nb_2O_5 and TiO_2, as well 0.95–1.48 wt.% SrO [25–28]. Existing technologies for extracting Zr and *REE* from eudialyte concentrates are based on their multi-stage decomposition using acids and alkalis, followed by extraction separation of valuable components from the digestion solutions using extractants of various classes [25,29–33].

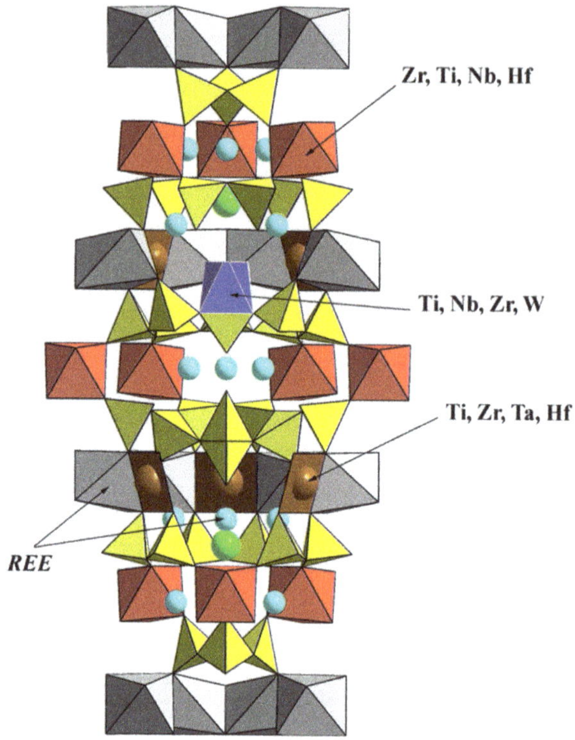

Figure 1. The general view of the crystal structure of EGMs and the distribution of RM and REM over the different crystallographic sites.

During the processing of REE-containing raw materials of both natural and technogenic origin, low-level radioactive waste is typically generated, primarily due to the presence of uranium and thorium at concentrations ranging from 0.5 to 2 mg/L [17,20]. However, current environmental legislation in the Russian Federation mandates the transfer of these *REE*s for storage with subsequent burial, which significantly reduces the economic viability of producing concentrates and other products based on *RM* and *REE*. Consequently, it appears advisable to pre-treat these solutions to allow for the selective separation of radioactive uranium and thorium into a separate fraction, enabling the resulting REE concentrate to be sent for further processing. To address this issue, a technological method based on the extraction–sorption separation of radionuclides from solutions is often employed. This operation allows for the highly selective separation of uranium(IV) and thorium(IV), resulting in solutions with significantly lower concentrations down to 0.1 μg/L [32,33].

As a practical solution to the problem of selective extraction of radionuclides, it is necessary to use specialized extractants. In particular, the derivatives of bipyridine dicarboxylic acid amides (L1 and analogs) have previously been successfully used for joint extraction of both (IV) and (VI) valence actinide [34].

In this work, we conducted studies of three types of tetradentate N-heterocyclic extractants (Scheme 1). Ligands **L2** and **L3** are phenatroline derivatives that reproduce tetradentate coordination core of **L1**. The extraction properties of **L2** have not been described previously. **L3** is a 2,9-alkyl-substituted diphosphonate, the extraction properties of which were previously characterized with respect to some actinides and lanthanides in model experiments [35]. The extraction capacity of polydentate extractants, with respect to actinides and lanthanides, was investigated in relation to the development of a method for the separation of radioactive elements during the production of *REE*s from eudialyte ore concentrates.

Name		Structure
N,N′-diethyl-N,N′-di(4-ethylphenyl)-diamide of 2,2′-bipyridyl-6,6′-dicarboxylic acid	**L1**	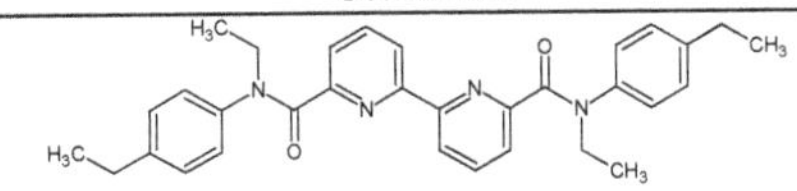
Tetraoctyl (1,10-phenanthroline-2,9-diyl)-bis phosphine oxide	**L2**	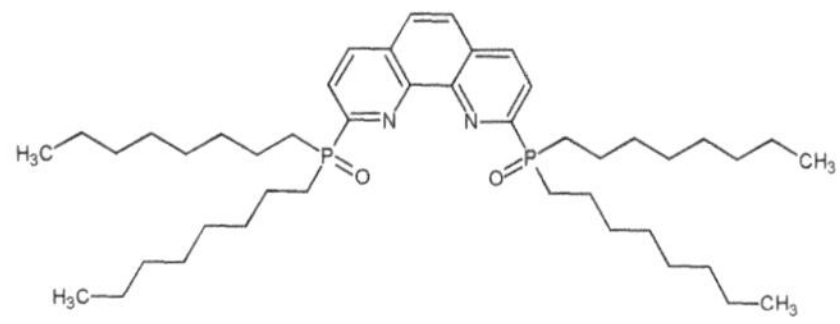
Tetrapentyl (1,10-phenanthroline-2,9-diyl)-bisphospho-nate	**L3**	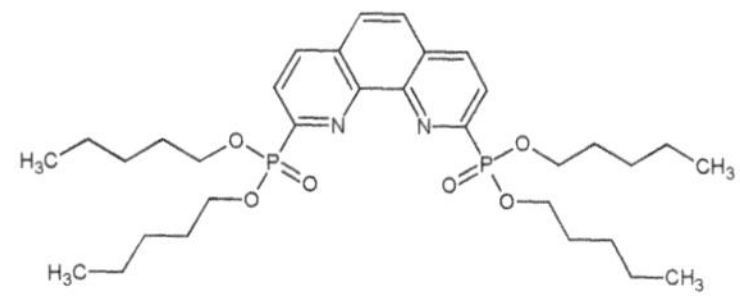

Scheme 1. Extractants used in this work.

2. Materials and Methods

2.1. General Remarks

Heterocyclic derivatives have been synthesized according to the previously published procedures (see Supplementary Materials). Meta-nitrobenzotrifluoride (F-3) (PIM-Invest) was used as the extractant, whereas HNO_3 (special purity) was used as the solvent. GSO 8363-2003 [36] uranium oxide, certified for uranium content of 84.784 $\pm$ 0.016%, as well as chemically pure grade nitrates $Ln(NO_3)_3 \times 6H_2O$ (Ln = La $-$ Nd, Sm $-$ Lu) and $Th(NO_3)_4 \times 5H_2O$ were used. The solutions were prepared by the volume–weight method: aqueous solutions were prepared in bidistilled water, while the solutions of nitrates of the studied elements were prepared by dissolving a weighed portion of the corresponding nitrate in a 0.01 mol/L HNO_3 solution. The concentration of the solution of lanthanide and actinide nitrates (0.1 mmol/L) was refined by ICP MS (PlasmaQuant MS Elite, Analytik Jena AG, Jena, Germany). The concentration of HNO_3 solutions was determined by potentiometric titration of 0.1 mol/L NaOH using an S470 SevenExcellence™ pH meter/conductometer (MettlerToledo, Columbus, OH, USA) with an accuracy of ±0.01 units. pH, while the concentration of the NaOH solution, was specified with a certified solution (FIXANAL, Sigma-Aldrich, St. Louis, MO, USA).

The eudialyte ore concentrate (enriched up to 40% of EGMs) was used. Before the breakdown, the granulometric and elemental composition of the concentrate had been determined. The granulometric composition of the concentrate was studied by the method of low-angle laser light scattering (LALLS, laser diffraction) using an Analyzette 22 Compact diffractometer (Fritsch, Pittsboro, NC, USA). The calculation of the particle size distribution was calculated using the Fraunhofer theory. The obtained average particle size was 120 μm, the sizes of ca. 90% of the particles were in the range between 30 and 200 μm.

Elemental analysis was performed using X-ray fluorescence spectrometry with an ARL Advant'X wavelength dispersive X-ray fluorescence spectrometer. The UniQuant software (version AA 83706) was used without any corrections. The obtained results (wt.%), excluding oxygen, are presented in Table 1. The weight percentages of the individual components relative to the sum of *REE*s are given in Table 2.

Table 1. Elemental composition of the eudialyte concentrate (enriched up to 40% of EGMs).

Element	Si	Al	Fe	Zr	Mn	Ca	Na	K	Ti	U
wt.%	49.97	11.65	11.55	8.67	3.81	3.13	2.63	2.01	1.96	0.012
Element	**Sr**	**Cl**	**Nb**	**Mg**	**P(V)**	**Hf**	**Ba**	**Mo**	**Ta**	**Th**
wt.%	1.25	0.675	0.503	0.309	0.292	0.122	0.064	0.039	0.031	0.023
Element	**S**	**Ni**	**La − Lu + Y**		**La − Nd**		**Sm − Lu +Y**		**Y**	
wt.%	0.023	0.017	1.22		0.77		0.45		0.403	

Table 2. Weight fractions of REEs' groups as well as some individual REEs in the eudialyte concentrate per total REEs weight.

Element	La − Nd	Ce	Sm − Lu + Y	Y
wt.%	63.2	0.40	36.8	0.40

Before conducting the extraction studies, the metals containing in the ore eudialyte concentrate were transferred into solution by leaching. Sulfuric acid leaching of eudialyte concentrate occurs with significant losses of zirconium and other rare elements due to the entry of these elements into silica gels [37,38]. The formation of poorly soluble silica gel also complicates the filtration process. The formation of poorly filterable silica gel sediment during the decomposition of EGMs is the main problem in the leaching of eudialyte concentrate. In order to intensify the extraction of REEs, as well as uranium(IV), thorium(IV), and other rare metals from the host EGMs into solution, it was decided to carry out leaching in nitric acid with the maximal possible concentration during heating. Leaching of the eudialyte ore concentrate (10 g) was carried out in 13.8 mol/L HNO_3 (S:L = 1:10), with stirring and a temperature of 80 °C for 72 h.

During nitric acid decomposition of eudialyte under the described conditions, *REEs*, zirconium(IV), as well as emanating uranium(VI) and thorium(IV) passed into solution. A released amorphous well-coagulated silica ensures good filterability of the pulp. It should also be noted that during leaching, zirconium passes into solution as Zr^{4+} ions. The content of oxo-cations ZrO^{2+} and $Zr_2O_3{}^{2+}$ in a nitric acid solution of high concentration of HNO_3 is minimal. Figure 2 shows the content of elements in the solution after nitric acid leaching of eudialyte ore concentrate; the content of nitric acid in the resulting solution is 12.2 mol/L.

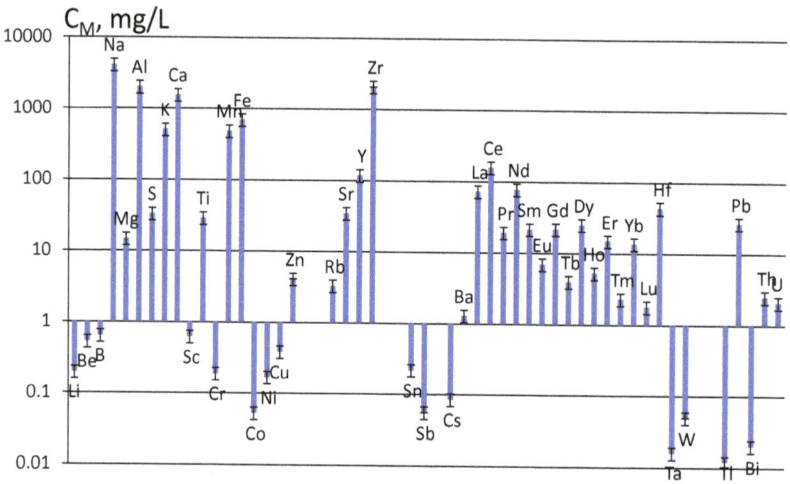

Figure 2. The element content in a nitric acid (12.2 mol/L) leaching solution of eudialyte ore concentrate (leaching conditions: S:L = 1:10, temperature 80 °C, time 72 h).

2.2. Liquid Extraction Experiments

The metal cation extraction was carried out in glass tubes with ground stoppers in a phase volume ratio of 1:1 at a temperature of 25 ± 1 °C. The composition of organic phase was similar for all three studied extractants: 0.05 mol/L in meta-nitrobenzotrifluoride.

The phases were mixed for 20 min in a rotator. The time of the establishment of the extraction equilibrium was checked by increasing the phase contact time to 120 min (the distribution coefficients had not been changed). Phase separation was performed by centrifugation. After phase separation, the metal cation concentration in the aqueous phase was determined by ICP-MS (PlasmaQuant MS Elite, Analytik Jena AG, Jena, Germany). For all cases, at least five independent experiments were performed. The total error of the obtained results was ~20%, taking into account the non-excluded and random components. The confidence interval of the determined metal concentrations in the experiment was 0.002 mmol/L. The distribution coefficients during extraction ($D = [M]_{org}/[M]_w$) were determined at constant concentrations of the extractant (0.05 mol/L in meta-nitrobenzotrifluoride) and initial concentrations of the metal in the experiment (0.1 mmol/L in the aqueous phase).

3. Results and Discussion

3.1. Phase Stability of the Extraction Systems

Since the eudialyte solution contains 12.2 mol/L nitric acid, experiments were previously conducted to assess the phase stability of such systems (Table 3). The selected solvent has a high density, as does a sufficiently (12.2 mol/L) concentrated solution of nitric acid, which leads to an increase in the phase separation time, or even makes it impossible to separate them, even when using centrifugation. Moreover, in the case of ligand **L3**, a precipitate was formed. Based on these data, further experiments were conducted with concentrations of 4–5 mol/L of nitric acid.

Table 3. Data on the phase stability of the systems "0.05 mol/L ligand in meta-nitrobenzotrifluoride—nitric acid".

Type of Ligand	HNO$_3$ (mol/L) 4	5	6	8	12
L1	Stable	Emulsion	Emulsion	Emulsion	Emulsion
L2	Stable	Stable	Emulsion	Emulsion	Emulsion
L3	Stable	Emulsion	Sediment and emulsion	Sediment and emulsion	Sediment and emulsion

"Stable" means that phase separation occurs in 15 s without the use of centrifugation; "emulsion" means that emulsion is stable for 3 min.

3.2. Extraction of Actinides and Lanthanides from Model Solutions

To assess the extraction capacity of the ligands under study, a model 4 mol/L HNO_3 nitrate solution was used. The concentrations of emanating uranium(VI) and thorium(IV) were 0.1 mmol/L, while the concentration of lanthanides in total was 0.1 mmol/L. The values of the established distribution coefficients for the extraction systems based on **L1**, **L2** and **L3** are presented in Figure 3.

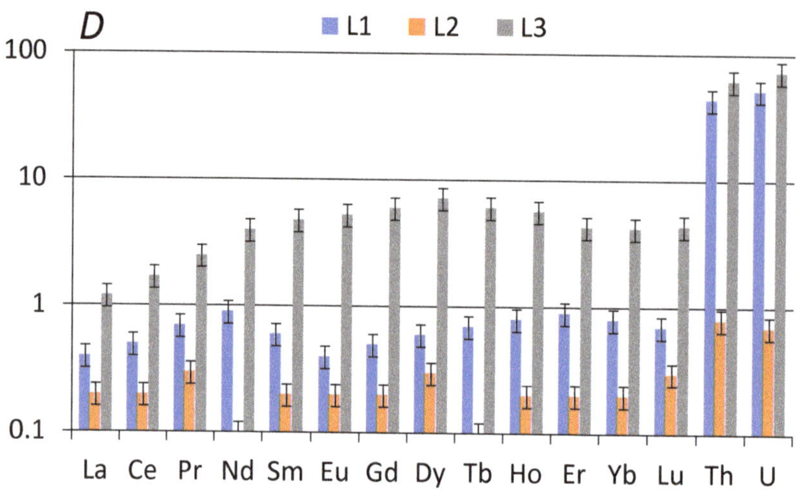

Figure 3. Distribution coefficients of lanthanides, uranium(VI) and thorium(IV) during extraction from 4 mol/L HNO_3 using heterocyclic derivatives **L1**, **L2** and **L3** (0.05 mol/L in meta-nitrobenzotrifluoride). Extraction conditions: O:W = 1:1, temperature 25 ± 1 °C, contact time 20 min.

The nature of the extractants significantly affects the extraction behavior of U(VI), Th(IV), and *REE*s. The extractant **L3** has the highest extraction efficiency both with respect to lanthanides and emanating U(VI) and Th(IV). **L3** organic ligand is certainly effective for separating all *f*-elements into the organic phase to obtain a bulk concentrate or purify raffinates in the case of additional waste conditioning. However, in the case of extractive separation of lanthanides and emanating U(VI) and Th(IV), the use of the extractant **L3** will lead to the accumulation of lanthanides in the organic phase. Nevertheless, the minimal separation factor SF (U, Th/REE) is not less than 8 (see Table 4), which potentially allows using L3 as an extractant for separating actinides in several extraction stages.

Table 4. Values of selectivity factors SF(U, Th/REE), calculated based on the values of the obtained distribution coefficients.

Type of Ligand	L1	L2	L3
SF(U, Th/REE)	50–100	4	8

Ligand **L2** has a low extraction capacity with respect to both actinide and lanthanide cations, and a significantly lower selectivity factor value, which does not allow us to speak about the prospects of this system.

The system based on ligand **L1** is characterized by the best characteristics. Under the experimental conditions, the condition $D(Ln) < 1$ and $D(\text{U, Th}) > 1$ is observed for the separated components. The selectivity factor reaches high values in the range of 50–110 (Table 4). The minimal factor is observed for the Th/Er pair which is equal to 50. These data point out the possibility of selective separation of actinide cations.

3.3. Determination of Solvation Numbers Using L

To assess the stoichiometry of the complex compounds *Ln*(III), U(VI) and Th(IV) passing into the organic phase, the solvation numbers were determined. They were obtained by analyzing the slope angle in the logD(Me)–logC(Me) dependencies. The obtained values are presented in Table 5, corresponding to the n value in the following equation:

$$Me^{x+}_{(aq.)} + x(NO_3)^-_{(aq.)} + nL_{(org.)} = [L_nMe(NO_3)_x]_{(org.)}$$

Table 5. Values of solvation numbers for the extraction of metal cations from 3 mol/L nitric acid. Organic phase: 0.01–0.1 mol/L in meta-nitrobenzotrifluoride. Temperature 25 ± 1 °C. Error of values ±0.1.

Extractant	*Ln*(III)	Th(IV)	U(VI)
L1	1.0–1.1	1.5	1.2
L2	1.1–1.2	1.5	1.1
L3	1.3–1.4	1.4	1.3

The structure and stereochemical features of the ligands do not have a significant effect on the composition of the formed complexes. For *Ln*(III), the range of values from the minimum to the maximum in the lanthanide series is represented. The highest solvation numbers are observed for thorium(IV), and the established values indicate the formation of a mixture of complexes with the composition ratios of 1:1 and 2:1. In other cases, the solvation numbers are close to 1, which indicates the predominant formation of complexes with a composition ratio of 1:1.

In order to find the best conditions for the efficient extraction of uranium(IV) and thorium(IV) into the organic phase, the effect of nitric acid on extraction using **L1** type ligand was studied (Figure 4). It should be noted that the extraction capacity of **L1** with respect to uranium(VI) and thorium(IV) is high throughout the entire studied range of nitric acid. The degree of the extraction at a concentration of 0.5 mol/L nitric acid is more than 90% for both uranium(VI) and thorium(IV). With an increase in the concentration of nitric acid in the solution, the distribution coefficients increase and the degree of extraction of emanating actinides increases accordingly. At a concentration of 4 mol/L HNO_3, uranium(VI) and thorium(IV) are extracted into the organic phase almost completely.

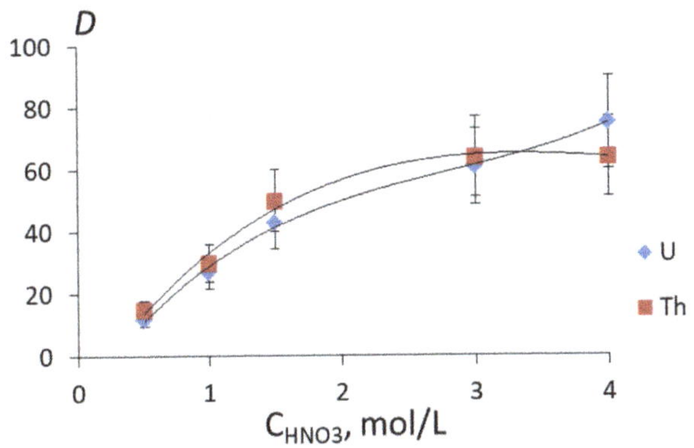

Figure 4. Dependence of the distribution coefficients of uranium(VI) and thorium(IV) on the concentration of nitric acid in the aqueous phase during extraction with a 0.05 mol/L solution of **L1** in meta-nitrobenzotrifluoride. Extraction conditions: O:W = 1:1, temperature 25 ± 1 °C, contact time 20 min.

In order to find conditions for efficient re-extraction of emanating uranium(IV) and thorium(IV) from the organic phase, the effect of sodium carbonate concentration in the aqueous phase on the extraction capacity of ligand **L1** with respect to uranium(VI) and thorium(IV) was studied (Figure 5). Carbonate anion traditionally forms stable complex compounds with actinides (IV, VI), and solutions containing carbonate anions can be effectively used to extract these cations from the organic phase [39]. In the case of loaded organic phase stripping with a sodium carbonate solution, almost all of the extracted metal cations are re-extracted into the aqueous phase. However, it is possible to achieve group separation of the elements by stripping the loaded organic phase with nitric acid solutions with concentrations varying from 0.1 to 0.5 mol/L [40]. In the case (Figure 6) of extraction from eudialyte ore concentrate leaching solutions, when Zr(IV), Hf(IV) and Ti(IV) are successfully extracted into the organic phase in addition to uranium(VI) and thorium(IV), at the reextraction stage, it is possible to separate these extracted elements by washing the loaded organic phase with a 0.1 mol/L nitric acid solution with a separation factor of SF[Zr(VI)/U(VI)]~100 for zirconium and uranium cations as an example. This effect can be explained by the better affinity of actinides to the ligand during complexation in comparison with Zr(IV), Hf(IV) and Ti(IV).

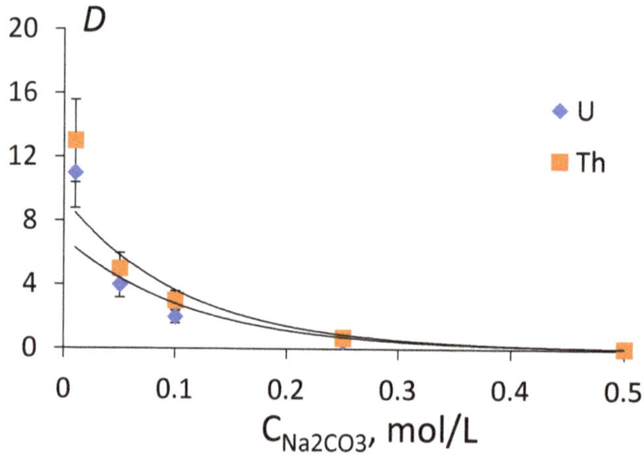

Figure 5. Dependence of the distribution coefficients of uranium(VI) and thorium(IV) on the concentration of sodium carbonate in the aqueous phase during extraction with a 0.05 mol/L solution of **L1** in meta-nitrobenzotrifluoride. Extraction conditions: O:W = 1:1, temperature 25 ± 1 °C, contact time 20 min.

As can be seen from Figure 5, with increasing sodium carbonate concentration in the aqueous solution, the extraction capacity of ligand **L1** (with respect to uranium and thorium) decreases. Thus, for efficient stripping of uranium and thorium together with Zr(IV), Hf(IV) and Ti(IV) from the organic phase, it is quite sufficient to use a solution containing 0.5 mol/L Na_2CO_3.

Thus, the following conditions were selected to conduct an experiment to study the extraction of *f*-elements from the solution of the eudialyte ore concentrate recovery:

- Organic phase: 0.05 mol/L solution of ligand **L1** in meta-nitrobenzotrifluoride;
- Aqueous phase: nitric acid content of 4 mol/L; it is necessary to dilute the concentrated solution of the eudialyte ore concentrate recovery by three times, due to the phase instability of the extraction systems at high nitric acid contents;
- Stripping of the organic phase with 0.1 mol/L nitric acid solution for selective re-extraction of Zr(IV), Hf(IV) and Ti(IV).
- Re-extraction of uranium and thorium—0.5 mol/L sodium carbonate solution.

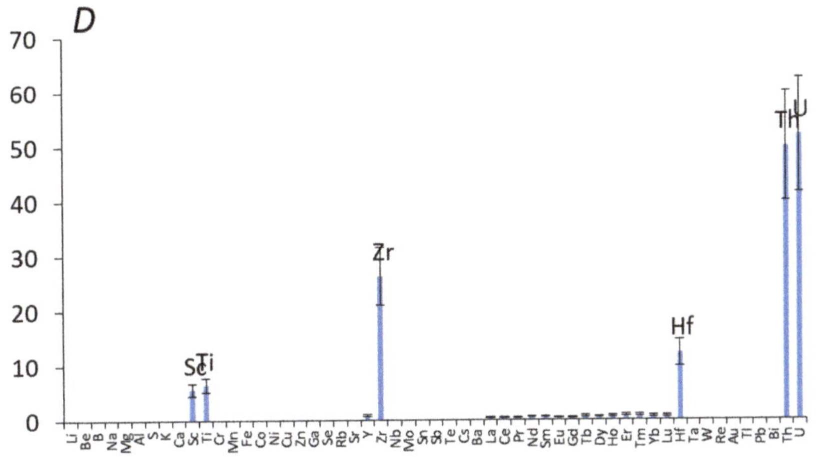

Figure 6. Distribution coefficients of elements during extraction from the solution of ore eudialyte concentrate recovery during extraction with a 0.05 mol/L solution of L1 in meta-nitrobenzotrifluoride. Extraction conditions: O:W = 1:1, temperature 25 ± 1 °C, contact time 20 min.

3.4. Extraction of Metals from Eudialyte Nitric Acid Solution

Under the above conditions (Section 3.3), experiments were conducted with the solution of eudialyte ore concentrate recovery. The established distribution coefficients of elements are presented in Figure 6. The general nature of the dependence reproduces the results of model experiments. The results also indicate the possibility of separating actinides from the nitric acid solution of ore eudialyte concentrate leaching.

When using amide bipyridine dicarboxylic acid derivatives in meta-nitrobenzotrifluoride, uranium(VI), thorium(IV), zirconium(IV), hafnium(IV), scandium(III), and titanium(III) are extracted from the solution of eudialyte concentrate recovery quantitatively, while rare earth elements are concentrated in the raffinate. The obtained results can become the basis for a new industrial technology for conditioning waste in the production of rare and rare earth metals from natural and man-made raw materials.

4. Conclusions

N,O-donor hybrid heterocyclic extractants have great potential for separation of actinides from lanthanides in spent nuclear fuel reprocessing processes; however, their ability has not been sufficiently studied in the field of processing natural raw materials containing emanating elements. In this work, it is shown that reagents capable of extracting a wide range of actinide ions in different valence states [35] can be used for primary concentration of actinides contained in eudialyte, a promising mineral containing a heavy group of lanthanides. Among heterocyclic N,O-donor hybrid extractants, the most effective are 2,2′-bipyridin-6,6′-dicarboxylic acid amide L1 and phenanthroline2,9-diphosphonic acid ester L3, while phenanthroline-2,9-bis(phosphine oxide) L2 exhibits moderate selectivity and efficiency of f-element extraction. With respect to lanthanide ions, the efficiency of their extraction decreases in the series L3 >> L1 > L2, and the extraction of actinides decreases in the series L1 ≈ L3 >> L2. For the extractant L2 based on 2,2′-bipyridine-6,6′-dicarboxylic acid diamide, the efficiency of lanthanide purification from U, Th exceeds 50. The structure and stereochemical features of the ligands do not have a significant effect on the composition of the formed complexes. For *Ln*(III), the range of values is from the minimum to the maximum in the lanthanide series is represented. The highest solvation numbers are observed for thorium(IV), and the established values indicate the formation of a mixture of complexes with the composition ratios of 1:1 and 2:1. In other cases, the solvation numbers are close to 1, which indicates the predominant formation of complexes with the composition ratio of 1:1. The most probable mechanism for the extraction of metal ions is the formation of mononuclear coordination compounds with actinide and lanthanide nitrates. When using amide bipyridine dicarboxylic acid derivatives in meta-nitrobenzotrifluoride, uranium(VI), thorium(IV), zirconium(IV), hafnium(IV), scandium(III), and titanium(III) are extracted from the solution of eudialyte concentrate recovery quantitatively.

Supplementary Materials: The following supporting information can be downloaded at https://www.mdpi.com/article/10.3390/met15050494/s1. References [41–43] are cited in the Supplementary Materials.

Author Contributions: Conceptualization, N.E.B. and A.M.S.; methodology, N.E.B., P.I.M. and A.V.I.; formal analysis, A.M.S. and A.V.L.; investigation, A.V.I., P.I.M., S.M.A. and A.V.L.; writing—original draft preparation, N.E.B., P.I.M., A.V.I. and D.V.I.; writing—review and editing, A.M.S., D.V.I. and S.M.A.; visualization, P.I.M. and S.M.A.; supervision, N.E.B. All authors have read and agreed to the published version of the manuscript.

Funding: The work was supported by the Ministry of Science and Higher Education of Russia (Grant No. 075-15-2024-534).

Data Availability Statement: The original contributions presented in this study are included in the article/Supplementary Material. Further inquiries can be directed to the corresponding authors.

Conflicts of Interest: Authors Alfiya M. Safiulina and Alexey V. Lizunov was employed by JSC "A.A. Bochvar Advanced Research Institute of Inorganic Materials". Author Dmitry V. Ivanets was employed by State Atomic Energy Corporation "Rosatom". The remaining authors declare that the research was conducted in the absence of any commercial or financial relationships that could be construed as a potential conflict of interest.

References

1. Zhiltsov, S.S. Policy of Western States in Central Asia: The Role of Rare-Earth Metals. *Post-Sov. Issues* **2024**, *11*, 8–19. [CrossRef]
2. Kryukov, V.A.; Yatsenko, V.A.; Kryukov, Y.V. Rare Earth Industry—How to take advantage of opportunities. *Gorn. Promyshlennost = Russ. Min. Ind.* **2020**, *5*, 68–84. (In Russian) [CrossRef]
3. Zepf, V. *Rare Earth Elements*; Springer Theses; Springer: Berlin/Heidelberg, Germany, 2013. [CrossRef]
4. Romanova, O.A.; Sirotin, D.V. Metal Industry Development in the Conditions of Formation of New Technological and Institutional Trends. *KnE Mater. Sci.* **2019**, *5*, 15–28. [CrossRef]
5. Schulz, K.J.; DeYoung, J.H., Jr.; Seal, R.R., II; Bradley, D.C. *Critical Mineral Resources of the United States-Economic and Environmental Geology and Prospects for Future Supply*; U.S. Geological Survey Professional Paper 1802; U.S. Geological Survey: Reston, VA, USA, 2017; 797p. [CrossRef]
6. Paulick, H.; Machacek, E. The global rare earth element exploration boom: An analysis of resources outside of China and discussion of development perspectives. *Resour. Policy* **2017**, *52*, 134–153. [CrossRef]
7. Machacek, E.; Richter, J.L.; Lane, R. Governance and Risk–Value Constructions in Closing Loops of Rare Earth Elements in Global Value Chains. *Resources* **2017**, *6*, 59. [CrossRef]
8. Samsonov, N.Y. Global Chains of Supply of Rare-Earth and Rare Metals as High-Tech Raw Materials Within the Framework of International Industrial Cooperation. *Prostranstvennaya Ekon. = Spat. Econ.* **2018**, *3*, 43–66. [CrossRef]
9. Toishybek, A.M.; Baigenzhenov, O.S.; Turan, M.D.; Kurbanova, B.; Merkibayev, Y.S. A review of recovery technologies of rare and rare earth metals from wastes generated in titanium and magnesium production. *Complex Use Miner. Resour.* **2023**, *327*, 64–73. [CrossRef]
10. Li, Y.; Zhang, T.; Dou, Z.; Xie, W.; Lan, C.; Li, G. Summary of the Research Progress on Advanced Engineering, Processes, and Process Parameters of Rare Earth Green Metallurgy. *Materials* **2024**, *17*, 3686. [CrossRef]
11. Jarosinski, A.; Kowalczyk, J.; Mazanek, C. Development of the Polish wasteless technology of apatite phosphogypsum utilization with recovery of rare earths. *J. Alloys Compd.* **1993**, *200*, 147–150. [CrossRef]
12. Soukeur, A.; Szymczyk, A.; Berbar, Y.; Amara, M. Extraction of rare earth elements from waste products of phosphate industry. *Sep. Purif. Technol.* **2021**, *256*, 117857. [CrossRef]
13. Hammache, Z.; Berbar, Y.; Bensaadi, S.; Trari, M.; Amara, M. Recovery of light rare earth elements by leaching and extraction from phosphate mining waste (Fluorapatite and Carbonate-Fluorapatite). *J. Afr. Earth Sci.* **2020**, *171*, 103937. [CrossRef]

14. Afonin, M.A.; Nechaev, A.V.; Yakimenko, I.A.; Belova, V.V. Extraction of Rare Earth Elements from Chloride Solutions Using Mixtures of P507 and Cyanex 272. *Compounds* **2024**, *4*, 172–181. [CrossRef]
15. Liu, Y.; Jeon, H.S.; Lee, M.S. Solvent extraction of Pr and Nd from chloride solution by the mixtures of Cyanex 272 and amine extractants. *Hydrometallurgy* **2014**, *150*, 61–67. [CrossRef]
16. Kuang, S.; Zhang, Z.; Li, Y.; Wei, H.; Liao, W. Synergistic extraction and separation of rare earths from chloride medium by the mixture of HEHAPP and D2EHPA. *Hydrometallurgy* **2017**, *174*, 78–83. [CrossRef]
17. Kogarko, L.N. Features of eudialyte ore formation in high-alkaline magmas of the Lovozero deposit (Kola peninsula). *Dokl. Earth Sci.* **2021**, *496*, 112–114. [CrossRef]
18. Wu, F.-Y.; Yang, Y.-H.; Marks, M.A.; Liu, Z.-C.; Zhou, Q.; Ge, W.-C.; Yang, J.-S.; Zhao, Z.-F.; Mitchell, R.H.; Markl, G. In situ U–Pb, Sr, Nd and Hf isotopic analysis of eudialyte by LA-(MC)-ICP-MS. *Chem. Geol.* **2010**, *273*, 8–34. [CrossRef]
19. Aksenov, S.M.; Rastsvetaeva, R.K.; Mitchell, R.H.; Chakrabarty, A. Crystal structure of manganese-rich variety of eudialyte from Suchina Hill, India, and manganese ordering in eudialyte-group minerals. *Crystallogr. Rep.* **2014**, *59*, 146–154. [CrossRef]
20. Schilling, J.; Wu, F.-Y.; McCammon, C.; Wenzel, T.; Marks, M.A.W.; Pfaff, K.; Jacob, D.E.; Markl, G. The compositional variability of eudialyte-group minerals. *Miner. Mag.* **2011**, *75*, 87–115. [CrossRef]
21. Kogarko, L.N. Geochemistry of Rare Earth Metals in the Ore Eudialyte Complex of the Lovozero Rare Earth Deposit. *Dokl. Earth Sci.* **2020**, *491*, 231–234. [CrossRef]
22. Rastsvetaeva, R.K.; Chukanov, N.V. Classification of eudialyte-group minerals. *Geol. Ore Depos.* **2012**, *54*, 487–497. [CrossRef]
23. Johnsen, O.; Ferraris, G.; Gault, R.A.; Grice, J.D.; Kampf, A.R.; Pekov, I.V. The nomenclature of eudialyte-group minerals. *Can. Miner.* **2003**, *41*, 785–794. [CrossRef]
24. Sjöqvist, A.S. The Tale of Greenlandite: Commemorating the Two-Hundredth Anniversary of Eudialyte (1819–2019). *Minerals* **2019**, *9*, 497. [CrossRef]
25. Chanturiya, V.A. Scientific substantiation and development of innovative processes for the extraction of zirconium and rare earth elements in the deep and comprehensive treatment of eudialyte concentrate. *J. Min. Inst.* **2022**, *256*, 505–516. [CrossRef]
26. Balinski, A.; Wiche, O.; Kelly, N.; Reuter, M.A.; Scharf, C. Separation of rare earth elements from contaminants and valuable components by in-situ precipitation during the hydrometallurgical processing of eudialyte concentrate. *Hydrometallurgy* **2020**, *194*, 105345. [CrossRef]
27. Balinski, A.; Atanasova, P.; Wiche, O.; Kelly, N.; Reuter, M.A.; Scharf, C. Recovery of REEs, Zr(+Hf), Mn and Nb by H2SO4 leaching of eudialyte concentrate. *Hydrometallurgy* **2019**, *186*, 176–186. [CrossRef]
28. Ma, Y.; Stopic, S.; Gronen, L.; Friedrich, B. Recovery of Zr, Hf, Nb from eudialyte residue by sulfuric acid dry digestion and water leaching with H2O2 as a promoter. *Hydrometallurgy* **2018**, *181*, 206–214. [CrossRef]
29. Voßenkaul, D.; Birich, A.; Müller, N.; Stoltz, N.; Friedrich, B. Hydrometallurgical Processing of Eudialyte Bearing Concentrates to Recover Rare Earth Elements Via Low-Temperature Dry Digestion to Prevent the Silica Gel Formation. *J. Sustain. Met.* **2016**, *3*, 79–89. [CrossRef]
30. Davris, P.; Stopic, S.; Balomenos, E.; Panias, D.; Paspaliaris, I.; Friedrich, B. Leaching of rare earth elements from eudialyte concentrate by suppressing silica gel formation. *Miner. Eng.* **2017**, *108*, 115–122. [CrossRef]
31. Ma, Y.; Stopic, S.; Gronen, L.; Milivojevic, M.; Obradovic, S.; Friedrich, B. Neural Network Modeling for the Extraction of Rare Earth Elements from Eudialyte Concentrate by Dry Digestion and Leaching. *Metals* **2018**, *8*, 267. [CrossRef]
32. Silin, I.; Dertmann, C.; Cvetković, V.S.; Stopic, S.; Friedrich, B. Prevention of Silica Gel Formation for Eudialyte Study Using New Digestion Reactor. *Minerals* **2024**, *14*, 124. [CrossRef]
33. Safiulina, A.M.; Semenov, A.A.; Lizunov, A.V.; Lesina, I.G.; Goryunov, E.I.; Goryunova, I.B.; Bodrin, G.V.; Brel, V.K.; Tananaev, I.G. Recovery and Separation of Rare Metals during the Processing of Eudialyte Concentrate with New Reagents of a Series of Phosphoryl Ketones. *Radiochemistry* **2022**, *64*, 713–720. [CrossRef]
34. Borisova, N.E.; Fedoseev, A.M.; Kostikova, G.V.; Matveev, P.I.; Starostin, L.Y.; Sokolova, M.N.; Evsiunina, M.V. Solvent Extraction and Conformation Rigidity: Actinide(IV) and Actinide(VI) Come Together. *Inorg. Chem.* **2022**, *61*, 20774–20784. [CrossRef] [PubMed]
35. Xu, L.; Yang, X.; Zhang, A.; Xu, C.; Xiao, C. Separation and complexation of f-block elements using hard-soft donors combined phenanthroline extractants. *Coord. Chem. Rev.* **2023**, *496*, 215404. [CrossRef]
36. *GSO 8363-2003*; APPROVED TYPE STANDARD SAMPLE OF URANIUM OXIDE COMPOSITION (COMMON URANIUM). JSC High-Tech Research Institute of Inorganic Materials named after academician A.A. Bochvar (JSC "VNIINM"): Moscow, Russia, 2014.
37. Chizhevskaya, S.V.; Chekmarev, A.M. Non-traditional methods of treating high-silicon ores containing rare elements. In Proceedings of the Hydrometallurgy '94, Cambridge, UK, 11–15 July 1994; pp. 219–228.
38. Altshuler, G.N.; Malyshenko, N.V.; Popova, A.N. The Ion-Exchange Properties of Polymeric Zirconium Phosphate and Zirconium Dioxide. *Inorg. Mater. Appl. Res.* **2018**, *9*, 746–750. [CrossRef]

39. Katz, J.J.; Seaborg, G.T.; Morss, L.R. *The Chemistry of the Actinide Elements*; Chapman and Hall Ltd.: Boca Raton, FL, USA, 1986; 990p, ISBN 10: 0412105500.
40. Goryunov, E.I.; Safiulina, A.M.; Morgalyuk, V.P.; Goryunova, I.B.; Molchanova, G.N.; Baulina, T.V.; Matrosov, E.I.; Petrovskii, P.V.; Nifant'ev, E.E.; Tananaev, I.G.; et al. N-(n-alkyl)-N′-diphenylphosphorylureas as a new type of extractants for preconcentration of actinides and lanthanides. *Russ. Chem. Bull.* **2008**, *57*, 383–388. [CrossRef]
41. Borisova, N.E.; Ivanov, A.V.; Matveev, P.I.; Smirnova, A.A.; Belova, E.V.; Kalmykov, S.N.; Myasoedov, B.F. Screening of the Structure of Americium Extractants Based on a 2,2′-Bipyridyl Scaffold: A Simple Way to a N2,O2-Tetradentate Ligands Library for Rational Design of An/Ln Extractants. *ChemistrySelect* **2018**, *3*, 1983–1989. [CrossRef]
42. Zakirova, G.G.; Mladentsev, D.Y.; Borisova, N.E. Palladium-Catalyzed C–P Cross-Coupling between (Het)aryl Halides and Secondary Phosphine Oxides. *Synthesis* **2019**, *51*, 2379–2386. [CrossRef]
43. Konopkina, E.A.; Pozdeev, A.S.; Kalle, P.; Kirsanov, D.; Smol'yanov, N.A.; Kirsanova, A.A.; Kalmykov, S.N.; Petrov, V.G.; Borisova, N.E.; Matveev, P.I. Sensing and extraction of hazardous metals by di-phosphonates of heterocycles: A combined experimental and theoretical study. *Dalton Trans.* **2023**, *52*, 12934–12947. [CrossRef]

metals

Article

Effect of Lead in Antimony and Tin Dissolution from Recycled Lead–Acid Battery Dross in Hydrobromic Acid Solution

Arturo Hirata-Miyasaki * and Corby G. Anderson

Kroll Institute for Extractive Metallurgy, Colorado School of Mines, Golden, CO 80401, USA; cganders@mines.edu
* Correspondence: ahiratamiyasaki@mines.edu

Abstract: Demand and prices for antimony have increased over the last few years. Recycling supplied 15% of domestic consumption in the US, while the remaining 85% was imported. Hydrometallurgical processes have long used alkaline sulfide systems and hydrochloric acid, closing doors on new approaches. Bromine compounds have been recently used to recover PGMs and REEs successfully; thus, antimony leaching with bromine compounds is theoretically feasible. This research was conducted to develop a viable technology for hydrobromic acid between 50 °C and 70 °C as a leaching reagent on dross through single- and two-stage leaching using design of experiment (DoE) and adding sustainability to current industrial processes while minimizing waste products in recycling processes. The preliminary results showed that bromine, specifically hydrobromic acid, can be used as a leaching reagent for antimony dissolution. By decreasing the lead content in the solids and increasing the concentration, temperature, and reaction time, antimony leaching from the dross was increased from 20% to 50%. The findings, coupled with acid regeneration, can be implemented as an alternative to other reagents in industrial plants.

Keywords: hydrometallurgy; hydrobromic acid; antimony

1. Introduction

Current antimony global reserves can supply short-term demand for the United States, but the development of antimony recycling is necessary to reduce dependency on China's antimony imports [1]. In the United States of America, many products are recycled, with car batteries being the most recycled. However, the focus of these plants is to recover lead, acid, and plastics, whereas strengthening and corrosion-resistant alloying metals, such as antimony, do not receive further treatment. While traditional pyrometallurgical and hydrometallurgical approaches have been widely used, the introduction of bromine-based lixiviants, particularly HBr, remains largely unexplored. This study focuses on the recovery of antimony from the byproducts of the recycling process, specifically dross from lead acid recycling plants using bromine compounds.

Lead dross, a byproduct of lead smelting and refining, contains valuable metals such as lead, tin, and antimony. Traditional processing methods rely on pyrometallurgy for high-temperature oxidation and reduction. Glencore's Nordenham plant has upgraded its process by using a Top-Submerged Lance (TSL) smelting furnace coupled with a Side-Blown Furnace (SBF) to treat lead-rich slag and dross. These pyrometallurgical stages have allowed for an increase in metal recovery [2]. However, hydrometallurgical techniques have gathered interest as a better alternative due to their capabilities for selective leaching and refining, and are viewed as a greener alternative. Recent advancements have significantly improved the efficiency, selectivity, and environmental sustainability of

https://doi.org/10.3390/met15040356

these methods. Drzazga and Ciszewski discovered that hydrometallurgical methods have enabled the recovery of valuable elements, such as germanium, that in some cases are lost in pyrometallurgical routes [3].

Antimony has been traditionally recovered through pyrometallurgical processes. Volatilization is used for low-grade sulfide ores (5–25% antimony) to produce antimony trioxide, which is later reduced in reverberatory furnaces to obtain antimony metal; intermediate-grade (25–40% antimony) ores go through blast furnace smelting to produce antimony metal; and rich-grade (45–60% antimony) ores are treated through iron precipitation to form iron sulfide and antimony metal [4].

Hydrometallurgy is advantageous because it can treat simple antimony feeds as well as complex ones. There are two main processes, followed by electrodeposition or precipitation. In summary, the first one uses an alkaline sulfide system (sodium sulfide and sodium hydroxide). Sodium sulfide (Na_2S) forms sodium thioantimoniate in solution and is later treated with a sodium polysulfide (Na_2S_2) formed when combining sodium hydroxide and elemental sulfur (Equations (1) and (2)). The second method uses hydrochloric acid (HCl) or ferric chloride to form antimony chloride. The solution containing the chloride complex is further refined through electrowinning to produce the antimony metal (Equations (3) and (4)).

$$3Na_2S + Sb_2S_3 \text{ (stibnite)} = 2Na_3SbS_3 \quad (1)$$

$$Na_3SbS_3 + Na_2S_2 = Na_3SbS_4 + Na_2S \quad (2)$$

$$Sb_2S_3 + 6HCl = 2SbCl_3 + 3H_2S \quad (3)$$

$$Sb_2O_3 \text{ (valentinite)} + 6HCl = 2SbCl_3 + 3H_2O \quad (4)$$

In 2021, Palden and Machiels used HCl and ethanol to demonstrate that antimony can be selectively leached from lead-rich dross with minimal lead dissolution. In the processes, the antimony-bearing feed dissolved into the solution while the lead remained as insoluble $PbCl_2$, achieving 90% antimony dissolution with only 0.4% of lead co-dissolved [5]. At the same time, Kim proposed a pyro-hydrometallurgical flowsheet to selectively recover Pb, Sn, and Sb from the lead-refining dross [6]. The pyrometallurgical process consisted of air oxidation followed by sulfuric acid sulfate roasting to convert lead into its oxidized form and isolate the antimony. Subsequently, leaching and selective precipitation were employed to recover each metal, achieving 95.4% and 86.3% of tin and antimony recovery from the dross, respectively.

Although significant progress has been made in lead dross processing over the last couple of years, pyrometallurgical and hydrometallurgical advancements have made use of the same reagents to increase the recovery of metals like antimony and tin [7,8]. However, some reagents have yet to be studied in depth to fully understand their capabilities and potential to selectively and effectively recover antimony from lead dross, as is the case for bromine compounds.

2. Hydrobromic Acid

Bromine is a great oxidizer that has been studied for the recovery of rare earth elements through bromination, in addition to the search for new alternatives for the recovery of precious metals [9–17]. In 1964, Sullivan and Cattoir prepared a high-purity vanadium metal by electrorefining commercial vanadium in a molten sodium bromide, potassium bromide, and vanadium dibromide electrolyte. Bromide electrolytes were used because of their low operating temperatures and the possibility of electrorefining vanadium to a lower iron and oxygen content [18]. In 1982, Groves and Beasley used bromine chloride for cyanide destruction. Although chlorine oxidation was effective in destroying the CN ion, excess chlorine was needed to drive the reaction to completion [19]. In 2020, Varvara, Dornanu, and Hao's leaching on waste-printed circuit board material on a bromine-based (potassium bromide, sodium bromide, liquid bromide, and hydrobromic acid) system provided an alternative for lixiviants in the hydrometallurgical route of metal recovery, for which we believe that an alternative lixiviant for antimony can be found in hydrobromic acid [20,21].

Hydrobromic acid (HBr) is one of the strongest mineral acids known, but there are few to no records showing the potential of HBr as a leaching reagent, especially for antimony [22,23]. Hydrochloric acid has been widely used as a reagent for antimony and tin, and limited work has been carried out on HBr as a leaching reagent for antimony and tin. Thus, developing a technology that utilizes HBr for antimony and tin is important.

Since bromine and chlorine are both halogens, we can assume that HBr and HCl behave similarly, as they can form a compound with antimony (Figures 1 and 2). We can replace HCl with HBr and obtain Equations (5)–(8).

$$Sb_2S_3 + 6HBr\ (g) = 2SbBr_3 + 3H_2S\ (g) \quad (5)$$

$$Sb_2O_3 + 6HBr\ (g) = 2SbBr_3 + 3H_2O \quad (6)$$

$$Sn + 2HBr\ (g) = SnBr_2 + H_2\ (g) \quad (7)$$

$$SnO_2 + 4HBr\ (g) = SnBr_2 + H_2O + 2Br \quad (8)$$

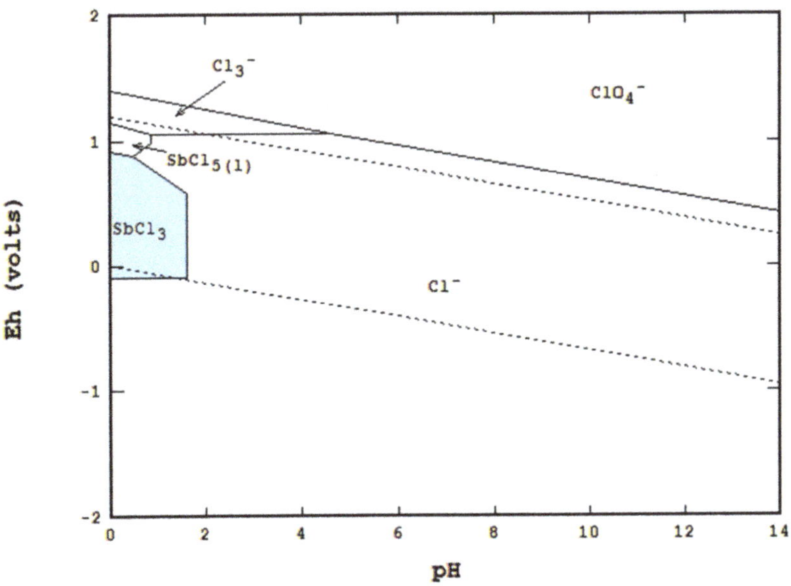

(a)

Figure 1. *Cont.*

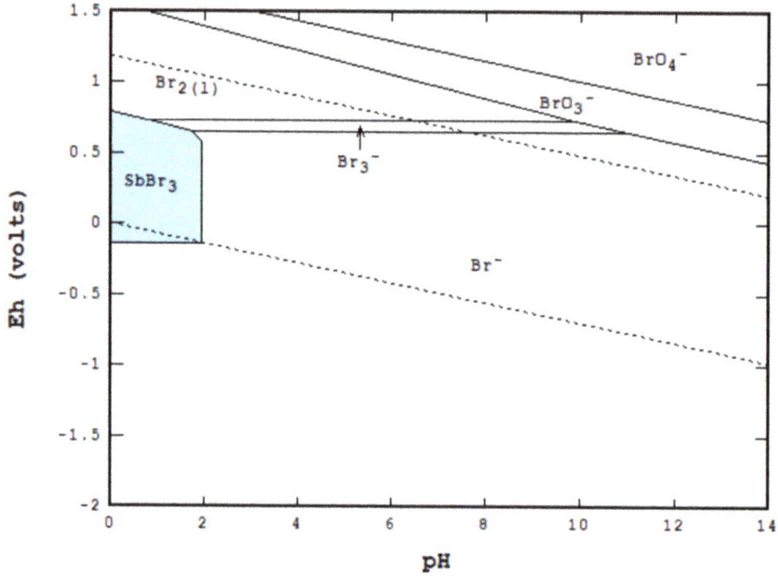

(b)

Figure 1. Eh-pH diagram for (**a**) Sb-Cl-H_2O and (**b**) Sb-Br-H_2O systems at 298.15 K.

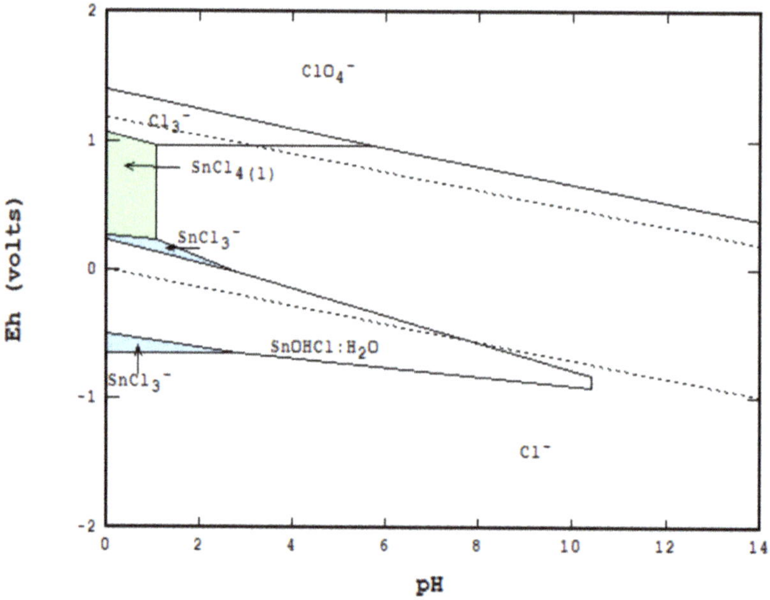

(a)

Figure 2. *Cont.*

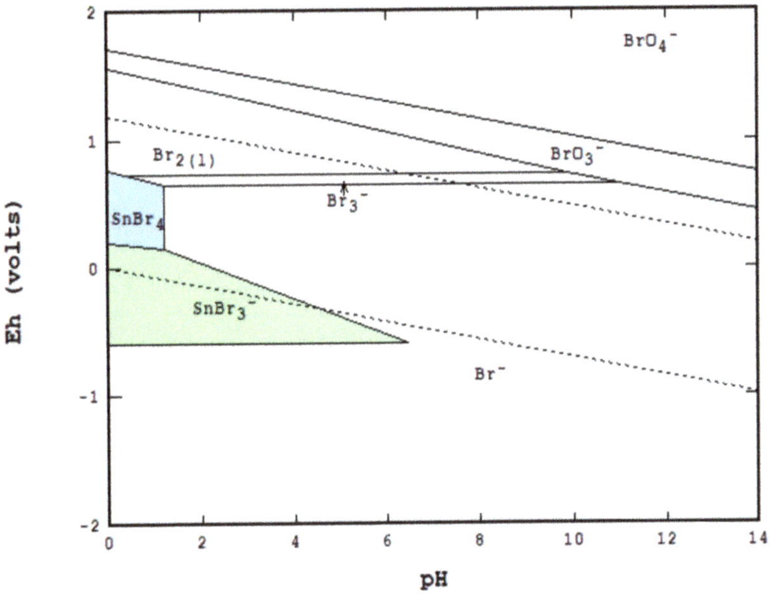

(**b**)

Figure 2. Eh-pH diagram for (**a**) Sn-Cl-H_2O and (**b**) Sn-Br-H_2O systems at 298.15 K.

From the figures, the formation of antimony tribromide ($SbBr_3$) seems to be stable in solution at pH ranges of 0 to 2 and Eh (V) ranges of 0 to 0.7 V. For tin, the formation of stannic bromide or tin (IV) bromide ($SnBr_4$) seems stable in solution at pH ranges of 0 to 1.75 and Eh (V) ranges of 0.25 to 0.75 V. These ranges are key for the formation of the complexes and should be kept in mind. Although the formation of antimony–bromide complexes can form, lead is still one of the main elements that reduces the amount of available bromide ions and bromine to obtain a higher recovery of antimony. Figure 3 demonstrates the theoretical effect of lead in the dissolution of antimony.

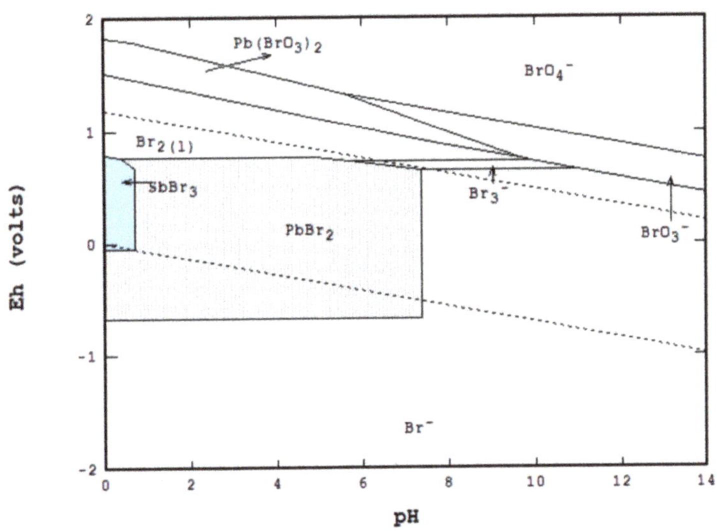

Figure 3. Eh-pH diagram for Sb-Pb-Br-H_2O system at 298.15 K.

3. Materials and Methods

While developing the procedure, the focus was on testing scenarios that would represent the general behavior of antimony in hydrobromic acid (HBr). The chosen variables were temperature, residence time, acid concentration, and liquid-to-solid ratio (Table 1). A second set of experiments was designed to determine the effect of hydrobromic acid (HBr) (MilliporeSigma, Burlington, MA, USA) on the sample after initially treating it with nitric acid (HNO_3) (MilliporeSigma, Burlington, MA, USA) to remove part of the lead content (Table 2). The quantitative analysis of this material prior to treatment is presented in Table 3.

Table 1. Experimental conditions for single-stage leaching.

Temperature (°C)	Time (Hours)	HBr (M)	Solids (Grams)
70	3	1	10
60	2	0.75	7.5
50	1	0.5	5

Table 2. Experimental conditions for nitric and hydrobromic acid leach.

HNO_3 Temp. (°C)	HNO_3 (M)	HBr Temp. (°C)	HBr (M)
30	1.5	70	2
25	0.5	60	1

Table 3. Sample concentrate analysis.

Element	Fe, %	Ca, %	Cu, %	Si, %	Al, %	Mg, %
	0.48	0.06	0.51	0.22	0.08	0.12
Element	**Na, %**	**Sn, %**	**Pb, %**	**Sb, %**	**Ni, %**	**O, %**
	0.36	0.24	83.21	3.4	0.37	10.94

Additionally, the sample was sent for mineral characterization. Figure 4 shows the backscatter images received from the SEM and AMICS (Automated Material Identification and Classification System) for the sample. The SEM and AMICS reports received labeled two multi-element phases as slag 4 and 5 due to how complex the samples received were. When comparing the SEM, AMICS, and safety data sheet provided by the supplier, we were able to determine that the phases for lead, tin, and antimony are lead monoxide (PbO), tin oxide (SnO), and antimony trioxide (Sb_2O_3).

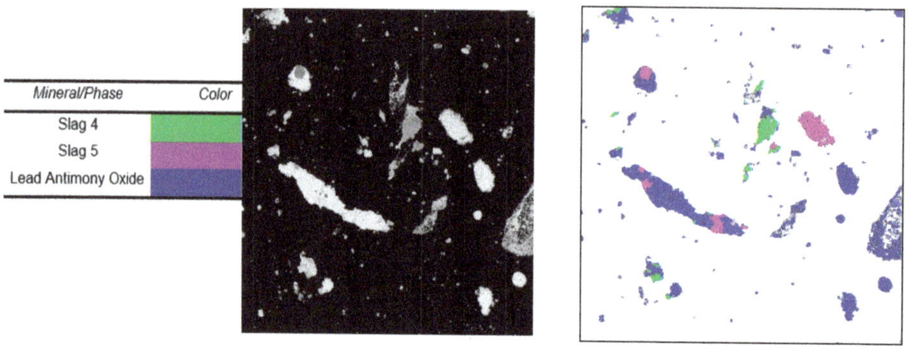

(a) (b)

Figure 4. BSE images from (**a**) SEM and (**b**) AMICS for dross sample.

Detailed information on the slag phase composition can be found in Table 4, which allows us to see that most elements are in an oxide phase.

Table 4. Slag and phase chemistry of dross samples.

Element	wt. (%)	
	Slag 4	Slag 5
Aluminum	4.0	-
Antimony	26.0	-
Calcium	3.0	-
Copper	-	21.0
Iron	-	20.0
Lead	4.0	-
Magnesium	6.0	-
Nickel	18.0	-
Oxygen	16.0	44.0
Silicon	11.0	-
Sodium	-	15.0
Sulfur	-	-
Tin	12.0	-

The bulk samples from the lead–acid battery recycling plant were crushed in a Massco single-toggle jaw crusher (Mine & Smelter Supply Co., Denver, CO, USA) for each experiment to a particle size of <2 cm, ground with a Bico VP-1989 ring (BICO Inc., Burbank, CA, USA), puck pulverizer to a particle size < 150 μm, and separated by particle size utilizing a Tyler RX-29 mechanical sieve shaker (W.S. Tyler, Mentor, OH, USA). Then, the samples were carefully split into smaller sample sizes adequate for one leach test. Solutions were prepared for the leaching test. The first leach test solutions were prepared from ACS 48% HBr diluted with di-ionized water. The second leach test solutions were prepared from ACS 68% HNO_3 diluted with di-ionized water. The solution was then added to the beaker and heated to the appropriate temperature.

For the first set of experiments, a total volume of 250 milliliters of hydrobromic acid was added to a 500 mL beaker and heated on a hot plate to the appropriate temperature (Table 1). Additionally, a magnetic stirring bar was used to agitate the solution to 400 RPM, which was monitored throughout the test. Once the temperature had been reached, the appropriate solids were added to the hot solution. The beaker was then sealed from the top and the leaching timer was started. Upon completion, the beaker was removed from the hot plate and the leach slurry was filtered and measured volumetrically while the solids were rinsed with di-ionized water, dried to remove excess moisture, and carefully weighed.

For the second set of experiments, a total volume of 250 milliliters of nitric acid was added to a 500 mL beaker and heated in a hot plate to the appropriate temperature with a magnetic stirring bar to agitate the solution to 400 RPM (Table 2). Once the temperature had been reached, 10 g of solid feed was added to the hot solution. The beaker was then sealed from the top and the leaching timer for 2 h was started. Once the two hours had passed, the beaker was removed from the hot plate and the leach slurry was filtered and measured volumetrically while the solids were rinsed with di-ionized water, dried to remove excess moisture, and carefully weighed. Once the solids were completely dried, we split the solids to have a representative sample of 0.5 g for elemental analysis while the rest went on to the next acid stage.

To continue the second set of experiments, the same procedure presented in the nitric acid stage was used; however, we changed the acid from nitric acid to hydrobromic acid and the concentrations used.

One of the best ways to represent the behavior of antimony in hydrobromic acid was to perform a design of experiment (DoE). The first set of experiments consisted of a total of 40 laboratory experiments (Table 5). For the DoE, temperatures of 50 °C, 60 °C, and 70 °C were selected to avoid higher temperatures; reaction times of 1, 2, and 3 h were selected for shorter times and test reaction rates; HBr concentrations of 0.5 M, 0.75 M, and 1 M were selected; however, there is a lack of studies with hydrobromic acid as a leaching reagent, for which we decided to use a slightly wider range of concentrations.

Table 5. Experimental design of tests conducted in single-stage leaching.

Trial	Temperature (°C)	Time (Hours)	HBr (M)	Solids (Grams)
1	70	1	1	10
2	70	2	1	7.5
3	50	3	0.75	10
4	60	1	0.5	10
5	70	2	0.75	7.5
6	70	2	0.75	10
7	70	2	0.5	5
8	50	1	0.5	10
9	70	1	0.75	5
10	50	2	0.5	10
11	70	2	1	5
12	70	1	0.5	7.5
13	50	1	1	5
14	70	1	0.5	7.5
15	50	2	1	10
16	50	3	0.5	5
17	70	3	0.75	7.5
18	70	3	1	5
19	60	2	1	5
20	60	3	0.5	10
21	60	2	0.5	7.5
22	50	3	1	7.5
23	70	3	1	10
24	60	1	0.5	5
25	50	1	0.75	7.5
26	60	1	1	10
27	70	2	0.5	10
28	70	3	0.5	10
29	60	3	1	7.5
30	50	3	0.75	10
31	70	1	0.75	10
32	50	2	0.75	7.5
33	50	2	0.5	7.5
34	50	2	0.75	5
35	60	3	0.75	5
36	50	1	1	7.5
37	70	3	0.5	7.5
38	60	2	1	7.5
39	60	2	0.75	10
40	60	1	0.75	7.5

The second set of experiments consisted of a total of 16 laboratory experiments (Table 6). For the DoE, the level in each variable was reduced from 3 to 2. The nitric acid values for both temperature and concentration were selected from Ichlas' paper on selective nitric acid leaching [24]. The hydrobromic acid values for temperature were selected based on the first set of experiments since they yielded better results, while the concentration selected was higher, since not much antimony was dissolved into the solution.

Table 6. Experimental design of tests conducted in two-stage leaching for 3-h reaction time.

Trial	HNO_3 Temp. (°C)	HNO_3 (M)	HBr Temp. (°C)	HBr (M)
1	30	0.5	60	2
2	25	0.5	60	2
3	30	0.5	70	1
4	30	1.5	60	2
5	30	1.5	60	1
6	25	1.5	70	2
7	25	0.5	70	1
8	25	1.5	60	2
9	30	0.5	70	2
10	25	1.5	70	1
11	30	0.5	60	1
12	30	1.5	70	2
13	25	0.5	60	1
14	30	1.5	70	1
15	25	0.5	70	2
16	25	1.5	60	1

3.1. Digestion Method for Elemental Analysis

The samples were dissolved using a two-acid method or aqua regia (3-part HCl and 1-part HNO_3). The process involved 0.5 g of the desired dry sample to be analyzed in a Teflon beaker. To ensure the best dissolution of the samples, two digestion cycles were used, with less than 0.05 g left behind as solids. For the first cycle, 24 mL of HCl (37% w/v) and 8 mL of HNO_3 (69% w/v) were added to the beaker with the sample and heated at 100 °C for 30 min on a hot plate. The open surface of the beaker was covered during the heating process with a glass watch to minimize the evaporation and release of toxic gases. Once 45 min had passed, the second cycle was started by adding 12 mL of HCl (37% w/v), 4 mL of HNO_3 (69% w/v), and 5 mL of deionized water to the beaker. The solution was covered again with the glass watch and heated to 100 °C for 45 min. To finish, the solution was cooled to room temperature and filtered to remove any undissolved particles in case there were some left. The filtered solution was collected in a 50 mL vial after recording the mass to be used in the dilution section.

3.2. Elemental Analysis

Atomization and absorbance measurements of the diluted solutions were performed with a PerkinElmer PinAAcle 900F Spectrometer (PerkinElmer Inc., Waltham, MA, USA) according to the user manual provided by the company. Each element has a characteristic concentration and linear range. For antimony, the range was established at 1–45 ppm. For lead, the range was established at 1–40 ppm. Finally, the tin was established at 1–400 ppm. To ensure the reproducibility and reliability of the results, the equipment was calibrated before each measurement cycle. The diluted solutions were run five times with a 2 s delay between each measurement, rejecting any value with a relative standard deviation (%RSD) greater than 2%, which was given by the instrument's software.

4. Results

Antimony bromide and tin bromide stable complex formation in solution required narrow Eh and pH values for them to happen. Tables 7 and 8 demonstrate such values through HBr leaching during their trial from the DoE. In single-stage HBr leaching, the final Eh decreases slightly while the pH increases compared to the initial values. In the two-stage HBr leaching, the same behavior can be observed. The final Eh decreases slightly while the pH increases compared to the initial values. The values do not change dramatically and are still within the ranges for which the stable complexes can form in solution. We are aware that human and equipment errors influence the data recorded, which should be kept in mind, but the difference between readings is not significant enough to be a concern.

Table 7. Eh and pH before and after single-stage treatment from design of experiment.

Trial	Eh (mV) Before	pH Before	Eh (mV) After	pH After
1	390	0.018	367.8	0.54
2	392	0.15	362.8	0.63
3	376.8	0.11	357.3	0.72
4	388.4	0.45	332.7	1.13
5	390	0.16	375.9	0.71
6	388	0.19	356.1	0.79
7	397.4	0.29	358.3	0.94
8	386.9	0.46	348.2	1.11
9	389.4	0.18	359.6	0.68
10	373.9	0.60	350.1	1.08
11	388	0.2	362.8	0.63
12	395.4	0.33	336.8	1.07
13	379	0.3	367.8	0.54
14	389.3	0.32	337.9	1.05
15	380	0.29	366.2	0.57

Table 7. *Cont.*

Trial	Eh (mV) Before	pH Before	Eh (mV) After	pH After
16	383.2	0.54	348.4	1.12
17	390.4	0.16	351.09	0.82
18	395.2	0.009	363.1	0.62
19	387	0.19	364.8	0.59
20	388.6	0.44	326.1	1.24
21	387	0.48	333.3	1.13
22	378	0.31	365.2	0.59
23	394.3	0.1	361.3	0.65
24	384.9	0.42	335.8	1.09
25	380.2	0.34	357.5	0.72
26	383	0.25	365.7	0.58
27	394.3	0.34	333.6	1.12
28	396.3	0.3	330.2	1.18
29	380	0.28	365.7	0.58
30	388.9	0.19	357.6	0.71
31	392	0.15	357.7	0.71
32	375.9	0.41	356	0.74
33	383.7	0.55	331.6	1.16
34	376.4	0.4	357.8	0.71
35	394.7	0.1	361.9	0.64
36	382.1	0.28	366.2	0.57
37	397.1	0.07	356	0.74
38	383	0.22	364.3	0.6
39	386	0.23	357.7	0.71
40	396.3	0.26	385.6	0.7

Table 8. Eh and pH before and after second-stage treatment from design of experiment.

Trial	Eh (mV) Before	pH Before	Eh (mV) After	pH After
1	490	0.08	467.8	0.44
2	492	0.05	462.8	0.53
3	476.8	0.11	457.3	0.62
4	488.4	0.05	432.7	0.93
5	490	0.12	475.9	0.61
6	488	0.03	456.1	0.69
7	497.4	0.19	458.3	0.74
8	486.9	0.03	448.2	0.91
9	489.4	0.08	459.6	0.48
10	473.9	0.13	450.1	0.98
11	488	0.15	462.8	0.53
12	495.4	0.03	436.8	0.97
13	479	0.13	467.8	0.44
14	489.3	0.12	437.9	0.95
15	480	0.09	466.2	0.57
16	483.2	0.14	448.4	0.92

4.1. Single-Stage Acid Leach

4.1.1. Effect of Hydrobromic Acid

The dissolution of antimony and tin was significantly determined by the hydrobromic acid concentration. As we increased the acid concentration, the antimony dissolution increased as the temperature increased at intervals of 50 °C, 60 °C, and 70 °C. The antimony dissolution greatly increases at 1 M compared to 0.5 M and 0.75 M because there are more bromine ions and bromine available to form the $SbBr_3$ complex [25]. The highest dissolution achieved was 21.48% under the conditions of 1 M and 3 h. The initial antimony grade was 3.2%, which is low, and achieving 21.48% recovery makes this process unfeasible due to the low recovery. Additionally, longer leaching times were not evaluated in this project, but there is a high possibility that the recovery of antimony would not surpass 30% recovery due to the amount of lead present in the system. Although the formation of antimony–bromide complexes had only a small area in the Pourbaix diagrams above, lead was still one of the main elements that reduced the amount of available bromide ions and bromine to obtain a higher recovery of antimony.

In terms of tin, the increase in bromine increased the dissolution of tin. The highest recovery of tin was 17.56% at 1 M of HBr at 3 h. This is lower than antimony's highest recovery, but this can be explained by the solubility of tin. Due to the high content of bromine, tin is assumed to be dissolved into tin–bromide complexes. Consequently, it is likely that a tin–bromide complex or antimony–bromide complex precipitated out of the solution and was lost during the filtration process, resulting in lower dissolution rates.

4.1.2. Effect of Temperature

Given that the boiling point of hydrobromic acid is 122 °C, the temperatures chosen in this study were lower than 122 °C in order to prevent the excessive loss of bromine, as the reaction is exothermic; thus, 50 °C, 60 °C, and 70 °C were chosen to perform the leaching experiments. Generally, the behavior was largely the same as the effect of acid concentration. In antimony, the increase in temperature gradually increases the dissolution rate with longer times. However, at a temperature of 70 °C, the dissolution of antimony at 1 h is lower at 12.48% compared to the 1 M of HBr at 1 h at 14.20%, but the dissolution is higher at a temperature of 50 °C at 8.64% compared to the 0.5 M of HBr at 1 h at 7.69%. This could be because when the solids come into contact with the acid, the high acid concentration has a faster reaction rate than high temperatures, but the lower high temperature has a faster reaction rate than the lower acid concentration. Additionally, at a temperature of 50 °C, the dissolution of antimony after 3 h starts to slow down, but this is not as visible at a temperature of 70 °C.

The excessive amount of bromine has been an ongoing challenge for bromine to be used as a leaching reagent [3–7]; thus, the effect of the solid–liquid ratio was studied. The quantities of solids chosen in this study were 5 g, 7.5 g, and 10 g per 250 mL of leaching solution. The effect of the solid–liquid ratio in antimony recovery follows the same trend seen with the other conditions studied. However, the highest dissolution of antimony was 18.79%, which is significantly lower compared to the experiments performed before. The increase in solids in the solution decreased the dissolution. In terms of tin, the increase in solids in the solution increased the dissolution of it. Consequently, the dissolution of tin did not increase between 1 and 2 h, like when testing the effect of temperature. The dissolution increase in tin between these two leaching times was >1%, but at 3 h of leaching time, the increase was <5%. Figures 5 and 6 demonstrate the highest recoveries of antimony and tin per variable tested.

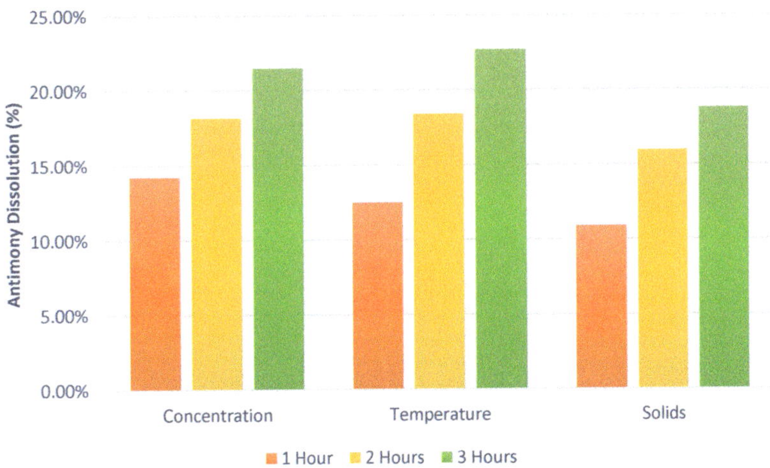

Figure 5. Antimony dissolution from single-stage hydrobromic acid leach.

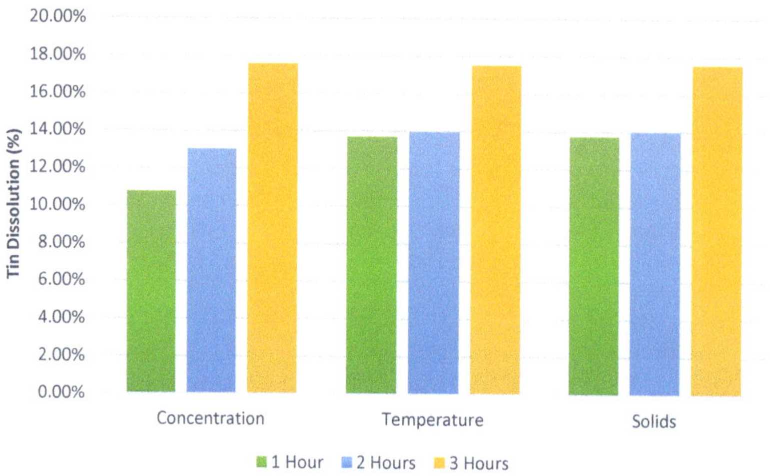

Figure 6. Tin dissolution from single-stage hydrobromic acid leach.

However, the recoveries are low, which can be attributed to lead content. Lead is one of the main components in the sample, and it reduced the amount of available bromide ions and bromine to obtain a higher recovery of antimony and tin.

4.2. Nitric and Hydrobromic Acid Leach

Although antimony and tin can form a bromine complex, lead has an easier path to form a bromine complex than the elements previously mentioned; thus, reducing the lead concentration should potentially increase the recovery of both. We tested this theory with the parameters in Table 2. After treating the sample with nitric acid leach, a spectrum of lead dissolution was achieved while retaining most of the antimony and tin in the solid residue (Table 9).

Table 9. Elemental recoveries (%) for nitric acid leaching tests after 3 h from DoE of Table 6.

Trial	Tin	Antimony	Lead
1	0.0220%	0.0317%	22.66
2	0.0000%	0.0961%	55.37
3	0.0000%	0.0283%	23.10
4	0.5106%	1.2207%	53.63
5	0.0000%	0.2129%	26.21
6	0.7602%	0.7628%	71.79
7	0.0000%	0.1105%	57.15
8	0.0000%	0.2122%	35.54
9	0.0000%	0.1213%	45.61
10	0.0000%	0.1576%	33.88
11	0.0000%	0.0716%	43.37
12	0.0000%	0.2416%	27.30
13	0.0000%	0.0413%	27.00
14	0.2988%	1.7016%	53.96
15	0.0000%	0.0494%	30.34
16	0.0506%	1.0982%	59.88

For the recovery data, it can be seen that lead is being removed from the feed at the nitric acid leach, but there is still a considerable amount that moves on to the hydrobromic acid leach. Since we achieved the goal of reducing the lead concentration without affecting antimony, in theory, treating with HBr should produce a higher dissolution of antimony and tin. However, the dissolution of antimony and tin does not look great (Table 10). There are some cases where the recovery greatly increased compared to the results from the first set of experiments, but in general, they are not great. The primary element of interest, antimony, is still not successfully transferred to the aqueous phase. Although there is room for improvement, reducing the amount of lead does increase the antimony dissolution. In Figure 7, the surface response for antimony from the DoE is shown. It can be seen that there is a considerable difference depending on the values associated with each variable.

Table 10. Elemental recoveries (%) for two-stage leaching tests after 3 h from DoE of Table 6.

Trial	Antimony	Tin	Lead *
1	19.44	11.98	22.66
2	52.52	20.32	55.37
3	18.86	10.37	23.10
4	8.26	15.57	53.63
5	3.14	8.88	26.21
6	2.10	16.01	71.79
7	5.23	13.07	57.15
8	21.78	16.24	35.54
9	6.24	21.89	45.61
10	2.65	7.75	33.88
11	7.58	13.89	43.37
12	3.74	18.56	27.30
13	21.6	2.07	27.00
14	9.81	7.56	53.96
15	23.08	16.64	30.34
16	40.16	6.17	59.88

* Values displayed are from nitric acid leach only.

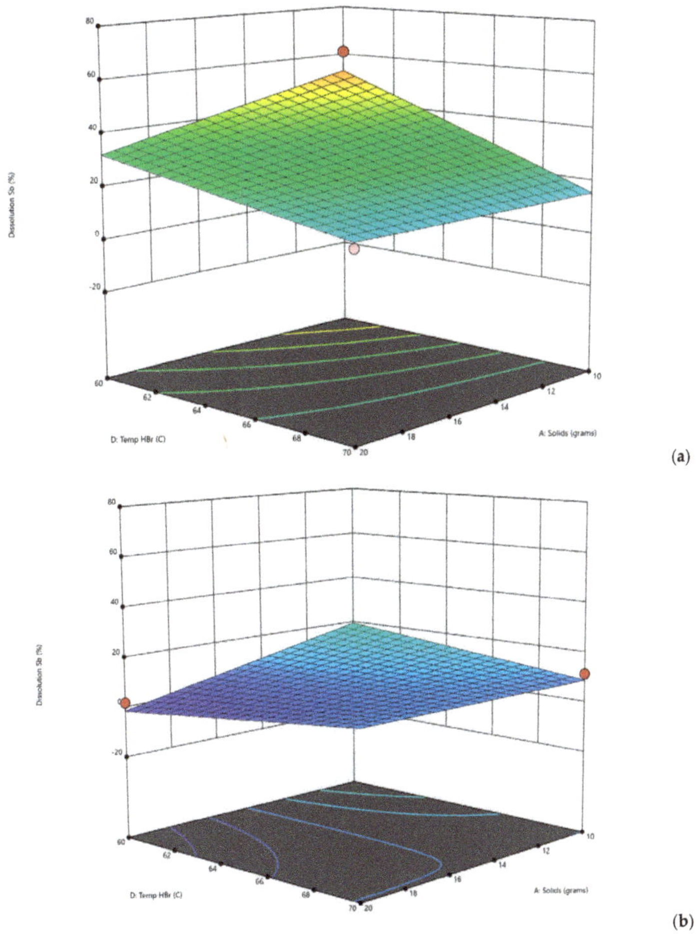

Figure 7. Surface response for antimony at (**a**) minimum HNO_3 conditions and maximum HBr concentrations, and (**b**) maximum HNO_3 conditions and minimum HBr concentrations.

For tin, there was no considerable increase in dissolution after decreasing the lead concentration (Figure 8). The surface response follows the same behavior as antimony's surface response. The higher concentrations of HBr are necessary for a good dissolution of antimony and tin; however, this would not be economically feasible, as initially discovered by Shaffer on precious metals in 1882. The values obtained cannot be considered as successful due to the low dissolution rates, but if we compared the results of the single HBr leach stage to the two-stage acid leach of HNO_3 and HBr, we can see, at least for antimony, that there is room to gather better results if further experimentation and optimization is performed.

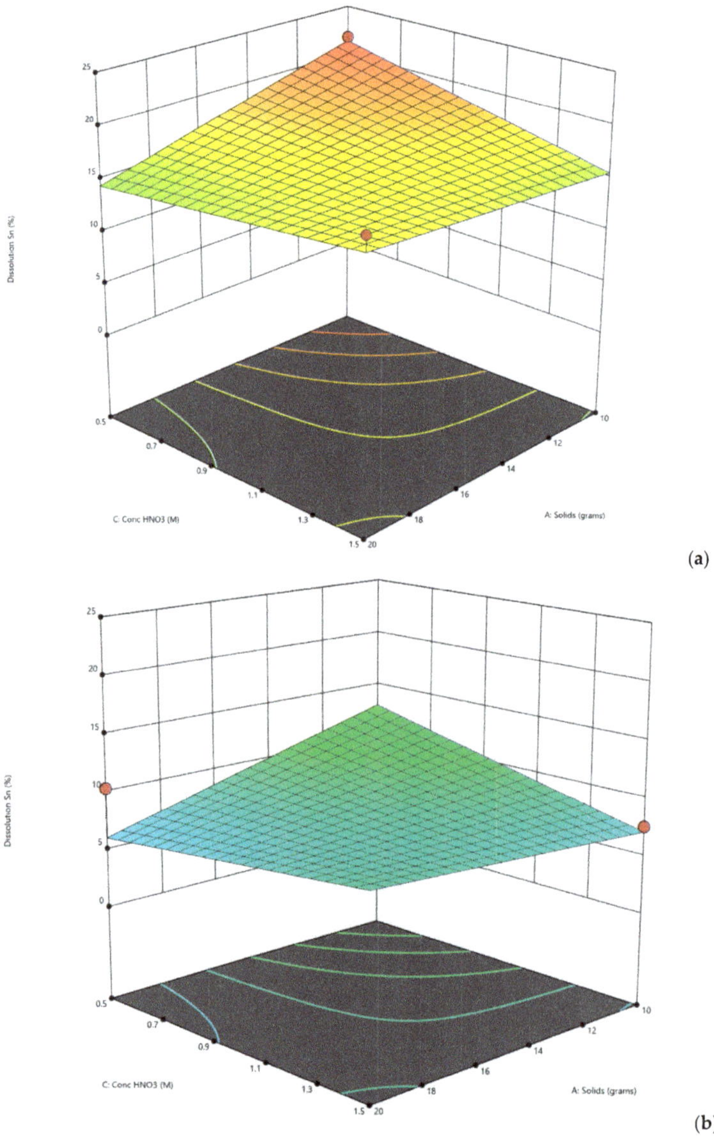

Figure 8. Surface response for tin at (**a**) maximum HBr concentration and (**b**) minimum HBr concentration.

5. Conclusions

This study demonstrated the effectiveness of hydrobromic acid (HBr) as a leaching reagent for recovering antimony from recycled lead–acid battery dross. Key parameters such as acid concentration, solid-to-liquid ratio, reaction time, and temperature were investigated, revealing that lead dissolution occurs prior to any significant reaction between antimony and bromine.

In single-stage leaching, antimony dissolution reached a maximum of 21.48% under optimal conditions (1 M HBr, 3 h, 70 °C), while tin dissolution reached 17.56%. These dissolution rates were limited by lead interference, which consumed bromine and reduced the efficiency. The introduction of a two-stage acid leach was proposed to increase dissolution rates for antimony. In the initial leaching stage, 1.5 M of HNO_3 at room temperature was found to be the optimal condition and necessary to reduce lead concentration. In the second stage, the optimal conditions for HBr leaching were determined to be 2 M HBr at 70 °C, which enhanced the antimony dissolution by 2.5 times compared to the untreated dross. Despite this improvement, the high residual lead content (over 80% in the original sample) remained a significant barrier to further increasing antimony recovery. Additionally, tin recovery did not see significant enhancement, indicating the need for further process.

Overall, the findings suggest that while HBr can selectively dissolve antimony and tin, its effectiveness is hindered by the high lead content and the associated loss of bromide ions. To enhance antimony and tin recoveries, future studies should explore additional process optimizations, such as alternative oxidants, extended leaching durations, or selective precipitation methods, to minimize lead interference and improve extraction efficiency. Additionally, given that high HBr concentrations are required to achieve dissolution rates above 90%, the economic feasibility of HBr regeneration should be examined to reduce operational costs and improve process sustainability. By refining these parameters, HBr-based leaching could become a more viable and sustainable alternative for recovering critical metals from lead–acid battery recycling waste streams.

Author Contributions: Conceptualization, C.G.A. and A.H.-M.; methodology, C.G.A. and A.H.-M.; formal analysis, A.H.-M.; investigation, A.H.-M.; data curation, A.H.-M.; writing—original draft preparation, A.H.-M.; writing—review and editing, C.G.A.; supervision, C.G.A. All authors have read and agreed to the published version of the manuscript.

Funding: This research was funded by the Center for Resource, Recovery, and Recycling (CR3).

Data Availability Statement: The original contributions presented in this study are included in the article. Further inquiries can be directed to the corresponding author.

Conflicts of Interest: The authors declare no conflict of interest.

References

1. Carlin, J.F., Jr. *Antimony: U.S. Geological Survey Mineral Commodity Summaries*; USGS: Reston, VA, USA, 2020; pp. 18–19.
2. Llamas, A.; Kandalam, A.; Zschiesche, C. Understanding the side-blown furnace slag system at Glencore Nordenham's lead plant—From theory to industrial application. In Proceedings of the 12th International Conference of Molten Slags, Fluxes and Salts MOLTEN 2024 Proceedings, Brisbane, Australia, 17–19 June 2024.
3. Drazga, M.; Cizcewski, M. Leaching of liquation-feeding furnace dross as a first step for germanium recovery. *BMC Res. Notes* **2024**, *17*, 180.
4. Anderson, C.G. The metallurgy of antimony. *Geochemistry* **2012**, *72* (Suppl. S4), 3–8.
5. Palden, T.; Machiels, L. Antimony Recovery from Lead-Rich Dross of Lead Smelter and Conversion into Antimony Oxide Chloride. *ACS Sustain. Chem. Eng.* **2021**, *9*, 5074–5084.
6. Woo Kim, K.; Sangyun, S. A study on pyro-hydrometallurgical process for selective recovery of Pb, Sn, and Sb from lead dross. *J. Hazard. Mater.* **2021**, *417*, 126071.

7. Ibrahim, A.; Aboelgamel, M. Production of high-grade antimony oxide from smelter slag via leaching and hydrolysis process. *J. Sep. Purif. Technol.* **2025**, *354*, 129355.
8. Liang, K.; Ma, B.; An, Y.; Zhang, Y.; Chen, Y.; Wang, C. A Summary of Smelting and Secondary Recovery Process of Antimony. *JOM* **2025**, *77*, 2666–2683. [CrossRef]
9. Shaeffer, C.A. Process of Extracting Gold and Silver from Their Ores. U.S. Patent 267723, 21 November 1882.
10. Hinman, B.C. Process of Extracting Gold. U.S. Patent 545890, 10 September 1895.
11. Dadgar, A. Inorganic Perbromide Compositions and Methods of Use Thereof. U.S. Patent 5607619A, 4 March 1997.
12. Hess, P.L. Gold and Silver Recovery Processes by Electrolytic Generation of Active Bromine. U.S. Patent 4904358A, 27 February 1990.
13. Cui, H. Hydrometallurgical Treatment of E-Scrap. Ph.D. Dissertation, Colorado School of Mines, Golden, CO, USA, 2018.
14. Wang, C. *Antimony: Its History, Chemistry, Mineralogy, Geology, Metallurgy, Uses, Preparations, Analysis, Production, and Valuation; with Complete Bibliographies*, 2nd ed.; Charles Griffin & Company, Ltd.: London, UK, 1919; 217p.
15. Greenwood, N.N.; Earnshaw, A. *Chemistry of the Elements*, 2nd ed.; Butterworth-Heinemann: Oxford, UK, 1997.
16. Iizuka, A. *Separation of Arsenic and Recovery of Antimony and Tin from Silicate Slag in a Nonferrous Smelter*; Canadian Institute of Mining, Metallurgy and Petroleum: Montreal, QC, Canada, 2019.
17. Mineralogical Society of America. *Handbook of Mineralogy*; Mineralogical Society of America: Chantilly, VA, USA, 2023; Volume III.
18. Sullivan, T.A.; Cattoir, F.R. *Electrorefining Vanadium in Molten Bromide Electrolyte*; Report of Investigations 6631; US Department of the Interior, Bureau of Mines: Washington, DC, USA, 1965.
19. Groves, K.O.; Beasly, D.K. The Use of Bromine Chloride for Cyanaide Destruction. In Proceedings of the Society of Mining Engineers of AIME, Dallas, TX, USA, 13–15 September 1982.
20. Scott, A. Preparation of Pure Hydrobromic Acid. *J. Chem. Soc. Trans.* **1900**, *77*, 648–651.
21. Goldschmidt, F. A Simple Method for the Preparation of Hydrobromic Acid. *J. Soc. Chem. Ind.* **1947**, *66*, 444–445.
22. McCall, A.S.; Cummings, C.F.; Bhave, G.; Vanacore, R.; Page-McCaw, A. Bromine Is an Essential Trace Element for Assembly of Collagen IV Scaffolds in Tissue Development and Architecture. *Cell* **2014**, *157*, 1380–1392. [CrossRef] [PubMed]
23. Brauer, G. *Handbook of Preparative Inorganic Chemistry*, 2nd ed.; Academic Press: New York, NY, USA, 1963; Volume 1, p. 285.
24. Ichlas, Z.T.; Rustandi, R.A.; Mubarok, M.Z. Selective nitric acid leaching for recycling of lead-bearing solder dross. *J. Clean. Prod.* **2020**, *264*, 121675.
25. US National Library of Medicine. Antimony Tribromide. National Center for Biotechnology Information. PubChem Compound Database. Available online: https://pubchem.ncbi.nlm.nih.gov/compound/Antimony-tribromide (accessed on 27 February 2025).

metals

MDPI

Article

Innovative Methods for Intensifying the Processing of Zinc Clinker: Synergy of Microwave Treatment and Ultrasonic Leaching

Bagdaulet Kenzhaliyev, Tatiana Surkova, Ainur Berkinbayeva, Zhazira Baltabekova, Kenzhegali Smailov *, Yerkezhan Abikak, Shynar Saulebekkyzy, Nazerke Tolegenova, Tursynkul Omirbek * and Zamzagul Dosymbaeva

The Institute of Metallurgy and Ore Beneficiation, Satbayev University, Almaty 050013, Kazakhstan; bagdaulet_k@satbayev.university (B.K.); t.surkova@satbayev.university (T.S.); a.n.berkinbayeva@satbayev.university (A.B.); zh.baltabekova@satbayev.university (Z.B.); a.abikak@satbayev.university (Y.A.); sh.saulebekkyzy@satbayev.university (S.S.); 000219600614-m@stud.satbayev.university (N.T.); z.dossymbayeva@satbayev.university (Z.D.)

* Correspondence: k.smailov@satbayev.university (K.S.); 971202400632-m@stud.satbayev.university (T.O.); Tel.: +7-747-6536293 (K.S.); +7-747-6287095 (T.O.)

Abstract: This study presents an innovative approach to processing refractory zinc-bearing clinker through the synergistic application of microwave thermal treatment and ultrasonic-assisted leaching. Microwave irradiation induces phase transformations in the clinker, improving its reactivity and facilitating subsequent zinc dissolution, while ultrasonic cavitation enhances mass transfer by disrupting passivation layers. Key process parameters, including acid concentration, temperature, pulp density, and leaching time, were systematically investigated using response surface methodology (RSM) and central composite design (CCD). The results demonstrate that the optimized process conditions led to a significant increase in zinc recovery from refractory materials.

Keywords: zinc; clinker; microwave roasting; phase transformation; ultrasonic; leaching

1. Introduction

In recent decades, the significant depletion of high-grade ores has inevitably driven the search for and development of innovative methods to extract metals from unconventional raw material sources [1–5]. These sources include technogenic waste from polymetallic ores, dumps of substandard and off-grade ores, as well as resources from marine and oceanic deposits [6–10].

In industrially developed countries, the utilization rate of industrial waste reaches 70–80% [11,12]. For Kazakhstan, a major producer of mineral products with significant mining potential, the issue of industrial waste recycling holds particular importance [13–15]. The low utilization rate of technogenic raw materials in the country is primarily due to a lack of technologies and equipment for processing various types of waste [16]. Notably, the production cost of marketable products derived from industrial waste is generally lower than that of products obtained from traditionally mined ores. One of the key advantages of technogenic dumps is their readiness for processing, as the ores have already been extracted, washed, and disintegrated [17,18].

In the CIS countries, the annual volume of mined solid minerals is approximately 3.5 billion cubic meters. Including preparatory and beneficiation operations, this volume increases to around 5 billion cubic meters. Alongside this, approximately 1.5 billion cubic meters of rock are mined annually, most of which, after beneficiation, is stockpiled in dumps and tailings storage facilities to enable the extraction of primary minerals from the Earth's crust [19].

Amid declining metal content in ores [20] and increasing volumes of refractory raw materials [21,22], transitioning to the comprehensive processing of low-grade, off-grade, and hard-to-reach mineral resources becomes both economically viable and scientifically justified [23]. Additionally, the use of technogenic raw materials as a secondary source of valuable elements is gaining relevance. Such materials are found in dumps and tailings storage facilities, which is particularly characteristic of countries like Kazakhstan, the USA, Finland, Poland, Canada, China, and South Africa. This approach significantly enhances the efficiency of mineral resource utilization while minimizing environmental impact and optimizing the economic performance of the mining and processing industries [24].

One of the most significant types of waste deserving particular attention is clinker, a technogenic byproduct of zinc production [25]. Clinker contains a variety of valuable compounds, including gold, silver, and copper. Of particular importance is the presence of zinc compounds, a highly valued non-ferrous metal widely used in the metallurgical industry [26]. The economic and technological feasibility of processing complex and refractory zinc-containing raw materials is driven by the possibility of extracting these valuable elements, which not only promotes the rational utilization of resources but also reduces the environmental burden through industrial waste recycling [27].

Currently, both pyrometallurgical [28] and hydrometallurgical [29,30] methods are employed to process zinc-containing technogenic raw materials. The most widely used pyrometallurgical technique is roasting, or reduction-distillation roasting, conducted at high temperatures of 1100–1300 °C with the addition of coke, constituting 35–45% of the processed material's mass [31]. This high-temperature process facilitates zinc evaporation and the formation of clinker enriched with valuable components, which are subsequently subjected to sulfuric acid leaching for further extraction. However, in Kazakhstan, an estimated 4.5 to 5.7 million tons of zinc industry waste has accumulated, underscoring the critical need for effective waste management and resource recovery strategies [32].

Traditional smelting processes are often associated with significant metal losses due to the volatility of zinc compounds and increased slag formation. The zinc recovery rate typically ranges from 70% to 85%, depending on furnace conditions and flux composition [33]. Despite the widespread application of pyrometallurgical processes, this method has notable disadvantages, including significant CO_2 emissions [34], high energy consumption, and the complexity of technological processes [35]. These shortcomings make it less suitable for processing high-silica zinc-containing technogenic raw materials due to the complex composition and structure of these materials' matrices. As a result, such raw materials often remain unprocessed and stored in sludge fields and tailings facilities, leading to substantial environmental and economic challenges [36].

Given the pressing issues surrounding the processing of refractory zinc-containing raw materials, the development of innovative and highly efficient technologies has become imperative. Currently, promising research directions include the application of microwave irradiation and ultrasonic treatment as preliminary processing stages.

Microwave irradiation, due to its mechanism of selective dielectric heating [26,37,38], induces localized thermal gradients and microstructural disruptions. This significantly enhances the reactivity and extractability of target elements [39]. Empirical studies have demonstrated that microwave treatment substantially improves the kinetics of leaching and extraction processes. Microcracks and mineral phase transformations in the matrix improve the penetration of leaching reagents, leading to more efficient metal recovery.

This approach also conserves energy, as targeted heating reduces the overall energy input required for subsequent processing stages [40,41].

Ultrasonic methods, driven by cavitation effects [42,43], play a pivotal role in enhancing the efficiency of metal leaching from ores and industrial waste [44]. Cavitation induced by ultrasound improves reagent transport, increases contact surface area, and promotes the penetration of leaching agents into the material, significantly intensifying metal extraction processes and reducing waste generation [45]. The interaction of ultrasonic waves with the treated medium leads to the dispersion of solid particles, the formation of microbubbles, and the generation of microjets, which reduce the thickness of the diffusion layer and accelerate dissolution reactions. The combination of ultrasonic treatment with oxidative leaching substantially enhances process kinetics while lowering activation energy [46–48].

The integration of ultrasonic technology with microwave irradiation represents an innovative approach that significantly improves the efficiency and transformation degree of materials during processing. These technologies profoundly influence the physicochemical properties of materials, creating favorable conditions for the intensification of chemical reactions and the optimization of process parameters. This integrated method facilitates the development of highly efficient and resource-saving technologies that align with modern requirements for environmental sustainability and the optimization of mineral and technogenic material processing. Such an approach offers promising prospects for the more complete extraction of metals from complex multicomponent systems, which is of critical importance for hydrometallurgy and related industries.

The aim of this study is to conduct a theoretical and experimental analysis of the synergistic effects of microwave treatment and ultrasonic cavitation on the kinetics of zinc clinker leaching in a sulfuric acid medium. Special attention is given to the detailed characterization of mass transfer mechanisms and phase transformations within the clinker matrix induced by the high-energy effects of localized cavitation collapse zones and selective dielectric heating. This comprehensive approach enables the clarification of the roles of passivation layer activation, agglomerate destruction, and the increase in reactive surface area in the kinetic patterns of leaching. The obtained results expand the fundamental understanding of the interaction between ultrasound and microwaves with multicomponent systems and provide a basis for the development of energy-efficient hydrometallurgical technologies for processing complex technogenic materials.

2. Materials and Methods

2.1. Materials

The subject of this study is refractory technogenic waste from zinc production in the form of clinker. Results of X-ray fluorescence analysis reveal that the clinker contains significant amounts of iron (37.53%), calcium (3.81%), silicon (4.58%), oxygen (41.64%), copper (1.04%), zinc (over 1.2%), and other elements, as detailed in Table 1.

Table 1. X-ray fluorescence analysis of the clinker (reprinted from Ref. [49]).

Elemental Content, %											
O	Na	Mg	Al	Si	P	S	Cl	K	Ca	Ti	Cr
41.644	0.173	1.030	0.912	4.581	0.055	0.807	0.011	0.109	3.807	0.101	0.020
Mn	Fe	Ni	Cu	Zn	As	Sr	Zr	Mo	Sb	Ba	Pb
0.110	37.532	0.033	1.037	1.217	0.138	0.043	0.012	0.026	0.034	0.825	0.154

The X-ray fluorescence (XRF) analysis of the clinker provides a quantitative assessment of its elemental composition, offering crucial insights into the chemical nature of the material. This compositional data is fundamental for selecting appropriate processing methods and optimizing the extraction of valuable components. The results from the XRF analysis formed the basis for subsequent experimental investigations, focusing on the influence of microwave irradiation and ultrasonic treatment on structural and phase transformations. Particular emphasis was placed on evaluating reagent penetration within the clinker matrix and its impact on the leaching kinetics and efficiency of valuable metal recovery.

2.2. Analytical Techniques

This study utilizes advanced analytical techniques to investigate the phase composition of clinker and to elucidate the mechanisms of phase transformations. The phase composition of the samples was analyzed using a Bruker D8 Advance X-ray diffractometer (Bruker, Ettlingen, Germany) under the following conditions: 4–90° scan range, Cu–Kα radiation (λ = 0.15406 nm), tube voltage of 40 kV, tube current of 40 mA, and a continuous scanning speed of 1°/min. Elemental composition was determined using an Axios 1 kW wavelength-dispersive X-ray fluorescence spectrometer (PANalytical, Almelo, the Netherlands), with data processing and interpretation performed using SuperQ5 software (Omnian 37). The surface microstructure was examined using a JXA-8230 electron probe microanalyzer (JEOL, Tokyo, Japan) operated at an accelerating voltage of 20 kV, an electron beam current below 1 nA, and aperture diaphragm No. 3. Energy-dispersive spectrometry (EDS) microanalysis (JEOL, Tokyo, Japan) was conducted with an electron beam current up to 6 nA and a dead time of up to 14%. Quantitative zinc concentrations in solutions and solid samples were measured using an Optima 8300DV inductively coupled plasma atomic emission spectrometer (PerkinElmer, Inc., Waltham, MA, USA) and an AA-7000 atomic absorption spectrometer (Shimadzu, Kyoto, Japan).

2.3. Experimental Method

The zinc leaching process from clinker consists of three key stages: grinding the clinker to achieve 90% of the fraction size below −0.071 mm, microwave phase transformation, and ultrasonic-assisted leaching. The technology involves several detailed steps, as shown in Figure 1. Initially, sample preparation is conducted, followed by roasting the clinker in a high-temperature microwave reactor of the "ENERGIA K-50" unit (Ust-Kamenogorsk, Kazakhstan) (915 MHz, 25 kW). This reactor is characterized by high power, operational stability, and enhanced efficiency [50,51]. Subsequently, the clinker undergoes leaching, as illustrated in Figure 2.

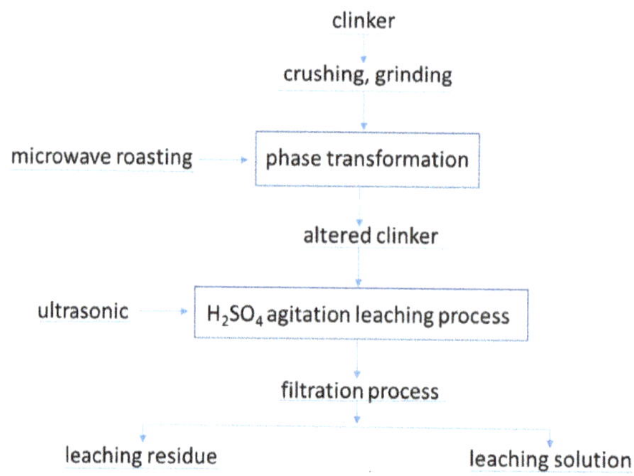

Figure 1. A flow diagram of the experiment.

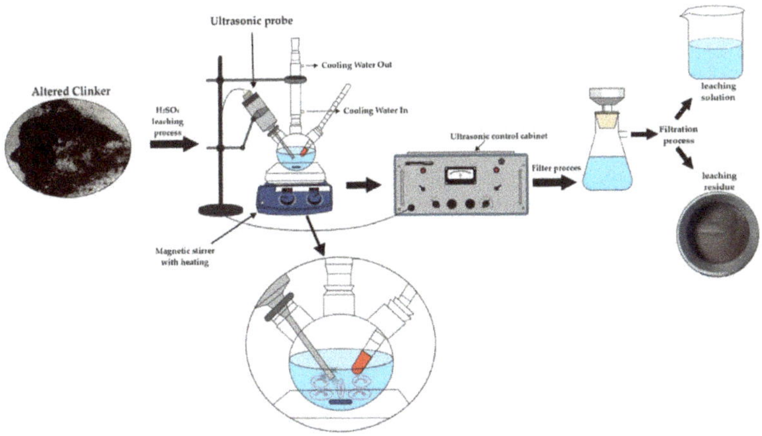

Figure 2. Schematic diagram of the leaching experimental setup. The lower part of the figure illustrates the ultrasonic probe and the cavitation effect, where microbubbles collapse, generating localized high pressure and temperature, which enhances the leaching efficiency.

For graph generation and data analysis during the experiment, Origin software (Origin Pro 9.8.0.200) was utilized. The zinc leaching rate was calculated using the following formula:

$$\eta_{Zn} = \left(\frac{(w_{clkr.} \times m_{clkr.}) - (w_{res} \times m_{res})}{w_{clkr.} \times m_{clkr.}}\right) \times 100\% \quad (1)$$

where η_{Zn}—zinc leaching rate (%); $w_{clkr.}$—zinc content in the initial clinker (%); w_{res}—zinc content in the residue after leaching (%); $m_{clkr.}$—mass of the initial clinker (g); and m_{res}—mass of the residue after leaching (g).

During the experiments, a controlled variable approach was applied to evaluate the effects of factors such as reaction temperature, sulfuric acid concentration, leaching duration, and pulp density. Four series of experiments were conducted under identical experimental conditions, as indicated by the asterisks in Table 2. To ensure the reliability of the results, each experiment was repeated at least three times, and the average values of the obtained data were used for analysis.

Table 2. Leaching parameters and ranges applied in the experiments.

Parameter	Value
Time (min)	30, 60 *, 120, 240, 300, 360
H_2SO_4 Concentration (g/dm^3)	20, 40, 60, 80, 100 *, 120, 140, 160
Pulp Density (%)	20 *, 25, 35
Temperature (°C)	20, 40, 60 *, 80

* These parameters were kept constant.

2.4. Optimal Experimental Design for Leaching Process

To determine the optimal conditions and enhance the accuracy and reproducibility of the results, this study analyzed the zinc extraction process from clinker using the response surface methodology (RSM) and central composite design (CCD). The independent factors considered included temperature, sulfuric acid concentration, leaching duration, and pulp density (solid-to-liquid ratio). The application of the RSM mathematical approach enabled the development of a second-order model describing the relationship between the response variable (zinc extraction) and the input parameters:

$$y = b_0 + \sum_{i=1}^{k} b_i x_i + \sum_{i=1}^{k} b_{ii} x_i^2 + \sum_{i=1}^{k-1} \sum_{j=i+1}^{k} b_{ij} x_i x_j \quad (2)$$

where y is the predicted zinc extraction value, b_0 is the constant coefficient, b_i represents the linear coefficients, b_{ii} denotes the quadratic coefficients, b_{ij} are the interaction coefficients between variables, and k is the number of factors.

The experiments were designed and analyzed using Design Expert 7.0 software (Stat-Ease, Inc., Minneapolis, MN, USA), which facilitated the development of a second-order regression model with high predictive accuracy.

The optimal levels and ranges of the variables—temperature, sulfuric acid concentration, leaching duration, and pulp density—identified as independent factors are presented in Table 3.

Table 3. Levels and codes of factors for CCD.

Factors	Symbol	Coding Level		
		−1	0	1
Time (min)	A	30	195	360
H_2SO_4 Concentration (g/dm^3)	B	20	90	160
Pulp Density (%)	C	20	25	35
Temperature (°C)	D	20	50	80

3. Results and Discussion

3.1. Mineralogical Characteristics of the Clinker

Mineralogical analysis of the clinker was performed using an OLYMPUS BX51 microscope (Olympus Corporation, Tokyo, Japan). The typical structure and particle size of the minerals comprising the clinker are presented in Figure 3. The mineralogical examination revealed the presence of several key minerals, including sphalerite (ZnS), which appears as rare, fine-grained anhedral particles, chalcopyrite ($CuFeS_2$), commonly associated with pyrite (FeS_2) and sphalerite, as well as hematite (Fe_2O_3) and carbonaceous inclusions of varying sizes and shapes.

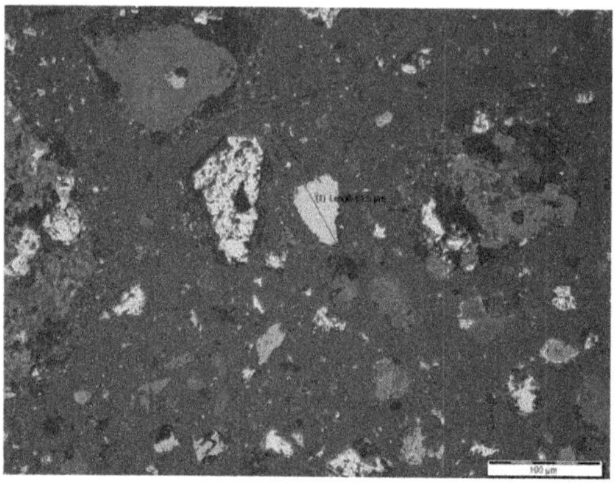

Figure 3. Sphalerite.

X-ray phase analysis identified hematite (24.2%) and magnesium iron oxide (18.2%) as the primary phase components of the clinker. Additional phase components are presented in Figure 4 and Table 4.

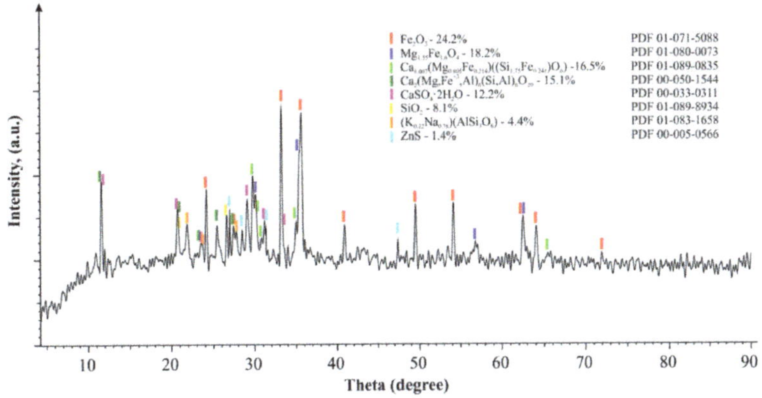

Figure 4. Diffractogram of the clinker (reprinted from Ref. [49]).

Table 4. Results of the X-ray phase analysis of the clinker (reprinted from Ref. [49]).

Compound Name	Formula	S-Q, %
Hematite	Fe_2O_3	24.2%
Magnesium iron oxide	$Mg_{1.55}Fe_{1.6}O_4$	18.2%
Diopside, ferrian	$Ca_{1.007}(Mg_{0.805}Fe_{0.214})$ $((Si_{1.75}Fe_{0.241})O_6)$	16.5%
Calcium magnesium iron aluminum silicate	$Ca_2(Mg,Fe^{+3},Al)_6(Si,Al)_6O_{20}$	15.1%
Gypsum	$CaSO_4 \cdot 2H_2O$	12.2%
Quartz	SiO_2	8.1%
Albite, potassian	$(K_{0.22}Na_{0.78})(AlSi_3O_8)$	4.4%
Sphalerite	ZnS	1.4%

3.2. *Phase Transformation via Microwave Irradiation*

Based on the data from our previous study [49], which extensively investigated the phase changes in clinker at various temperatures, the optimal conditions for microwave roasting were determined to be 600 °C for 5–7 min. Under these conditions, the best conversion of zinc-containing phases into zinc oxide (ZnO) is achieved, according to the following chemical reaction:

$$ZnS + 1.5O_2 = ZnO + SO_2 \quad (3)$$

In this study, these conditions were applied to clinker treatment. Microwave roasting at 600 °C for 5–7 min effectively transforms sphalerite into zinc oxide (Figure 5).

The microwave roasting process induces significant changes in the composition and microstructure of sphalerite. In its initial state, prior to microwave irradiation (Figure 5a), sphalerite is characterized by a high sulfur content (35.93%) and zinc content (64.07%). After irradiation at 600 °C for 3–4 min (Figure 5b), the sulfur content on the external surface significantly decreases to 1.54%, while the zinc content is reduced to 59.60%. Concurrently, oxygen appears (23.23%), indicating the onset of oxidative reactions on the mineral's surface and partial conversion of sphalerite to zinc oxide. However, analysis of the internal region of the same sample (Figure 5c) reveals a higher sulfur content (36.41%) and zinc content (63.59%), similar to the initial state (Figure 5a), suggesting incomplete material conversion.

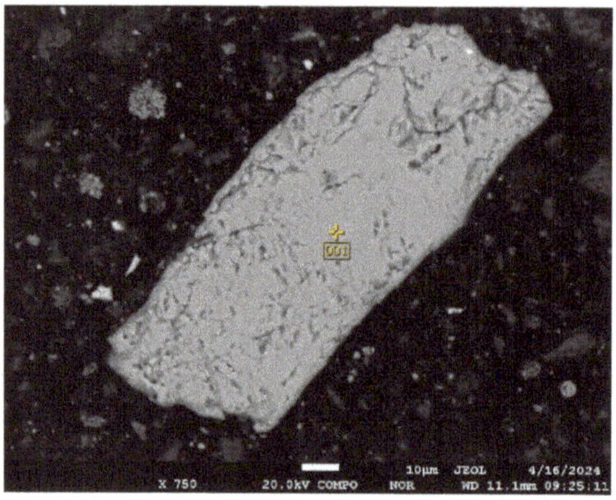

Chemical formula	ms%	mol%	Sigma	Net	K ratio	Line
S	35.93	53.34	0.26	8878621	0.2805589	K
Zn	64.07	46.66	2.11	3706249	0.6204246	K
Total	100.00	100.00				

(**a**)

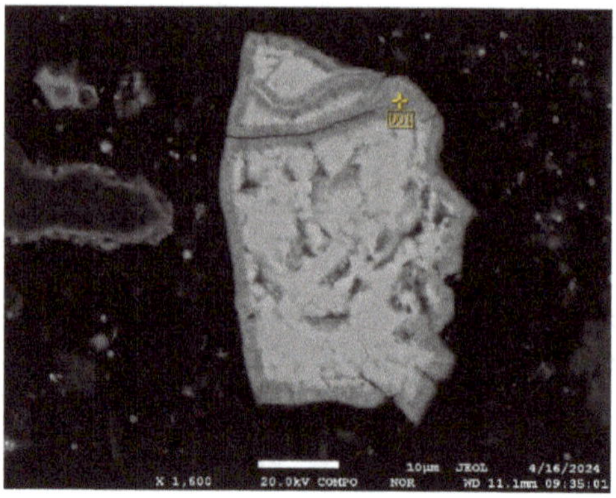

Chemical formula	ms%	mol%	Sigma	Net	K ratio	Line
C	15.64	35.06	0.44	202549	0.0051799	K
O	23.23	39.10	0.46	1045882	0.1165136	K
S	1.54	1.29	0.25	298928	0.0129774	K
Zn	59.60	24.55	2.01	2763044	0.6354559	K
Total	100.00	100.00				

(**b**)

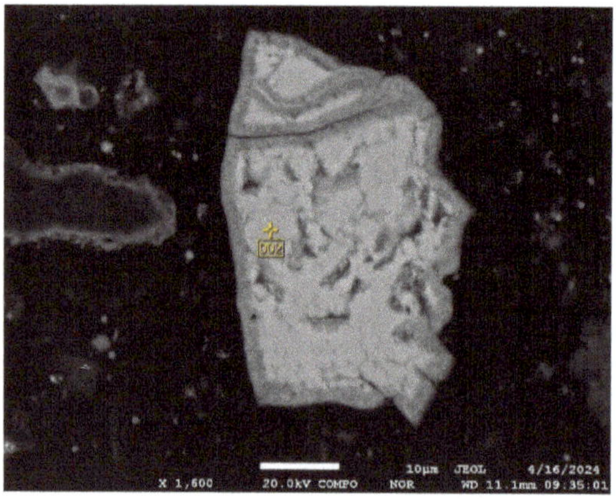

Chemical formula	ms%	mol%	Sigma	Net	K ratio	Line
S	36.41	53.86	0.27	6495800	0.2821026	K
Zn	63.59	46.14	2.22	2649945	0.6096592	K
Total	100.00	100.00				

(**c**)

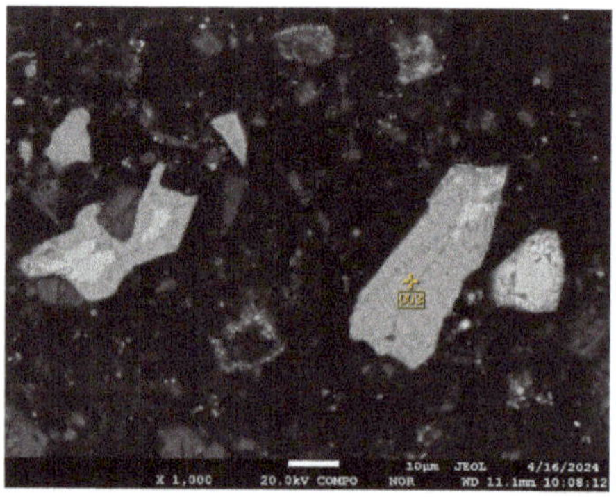

Chemical formula	ms%	mol%	Sigma	Net	K ratio	Line
C	16.15	36.51	0.40	174886	0.0054496	K
O	22.20	37.66	0.44	813701	0.1104528	K
S	0.53	0.44	0.24	82639	0.0043715	K
Zn	61.13	25.38	1.90	2312497	0.6480323	K
Total	100.00	100.00				

(**d**)

Figure 5. (**a**) Microstructure of the initial sample and energy-dispersive analysis of sphalerite prior to microwave irradiation at 25 °C; (**b**,**c**) microstructure and energy-dispersive analysis of sphalerite after microwave irradiation at 600 °C for 3–4 min; (**d**) microstructure and energy-dispersive analysis of sphalerite after microwave irradiation at 600 °C for 5–7 min.

A sample exposed to irradiation at 600 °C for 5–7 min (Figure 5d) demonstrates nearly complete transformation of sphalerite into zinc oxide, as evidenced by the presence of oxygen (22.20%). The sulfur content decreases to 0.53%, while the zinc content reaches 61.13%. These results confirm that the optimal microwave roasting conditions of 5–7 min ensure a high degree of sphalerite conversion to zinc oxide, which is a key factor for efficient material processing.

The detection of carbon in the samples is attributed to its deposition under the electron beam during microprobe analysis. The source of the carbon is the carbon tape used as a binding material during sample preparation for microprobe analysis.

3.3. *Statistical Analysis and Model Fitting*

3.3.1. Data Analysis

Table 5 displays the results of the analysis of variance (ANOVA) for the response surface model of the zinc leaching process from clinker.

Table 5. ANOVA for response surface quadratic model.

Source	Sum of Squares	df	Mean Square	F Value	p-Value Prob > F	Standard Error	95% CI (Lower–Upper)
Model	5466.47	14	390.46	4.39	0.0037	-	-
D- temperature	497.65	1	497.65	5.59	0.0319	±12.4	(470.2–525.1)
AD Interaction	1122.80	1	1122.80	12.62	0.0029	±18.7	(1085.5–1160.3)
BD Interaction	460.63	1	460.63	5.18	0.0380	±10.5	(438.2–483.1)
CD Interaction	414.24	1	414.24	4.65	0.0476	±9.8	(395.1–432.6)
A^2—Quadratic	1389.91	1	1389.91	15.62	0.0013	±22.5	(1350.4–1429.2)
D^2—Quadratic	346.96	1	346.96	3.90	0.0670	±8.6	(330.1–364.2)
Residual Error	1334.90	15	88.99	-	-	-	-
Lack of Fit	1290.90	10	129.09		0.1661		
Pure Error	44.00	5	8.80	-	-	-	-
Cor Total	6801.37	29	390.46	-	-	-	-

The F-statistic in analysis of variance (ANOVA) serves as a statistically grounded indicator of the proportion of variance explained by the model relative to the total variance. In this study, the F-statistic of the model, equal to 4.39, confirms its statistical significance according to the ANOVA results. The probability that such a high F-value could arise purely due to random factors is less than 0.1%, highlighting the reliability of the model.

The p-value of less than 0.0001 for the quadratic model further emphasizes its statistical significance. It is important to note that p-values not exceeding 0.05 are interpreted as confirming the significance of the corresponding factors. In this context, the significant components of the model include D, A, AD, and BD. To enhance the interpretation of the effects of significant factors, the model can be simplified to a regression equation focusing on the most significant variables with a 95% confidence level. The resulting regression equation is as follows:

$$E(Zn) = -3.36 - 4.38A - 6.48B - 25.70C + 144.50D + 2.80AB - 36.32AC - 33.49AD + 18.19BC + 39.53BD + 170.46CD - 28.22A^2 - 1.65B^2 + 42.43C^2 + 20.32D^2 \quad (4)$$

To improve the accuracy of assessing the adequacy of the quadratic model in approximating experimental data, several key diagnostic plots were developed and analyzed. Most experimental points are distributed along the diagonal axis, as shown in Figure 6a, indicating minimal deviations and high data reliability. The model errors demonstrate a normal distribution, as confirmed by the linear pattern of residual distribution. Further analysis of the random distribution of points along the t-axis (ranging from −3.00 to 3.00) and their proximity to zero, as seen in Figure 6b,c, supports the conclusion that the quadratic model effectively describes the relationship between the main experimental parameters and the leaching rate. This result underscores the reliability of the model and its suitability for predictive purposes.

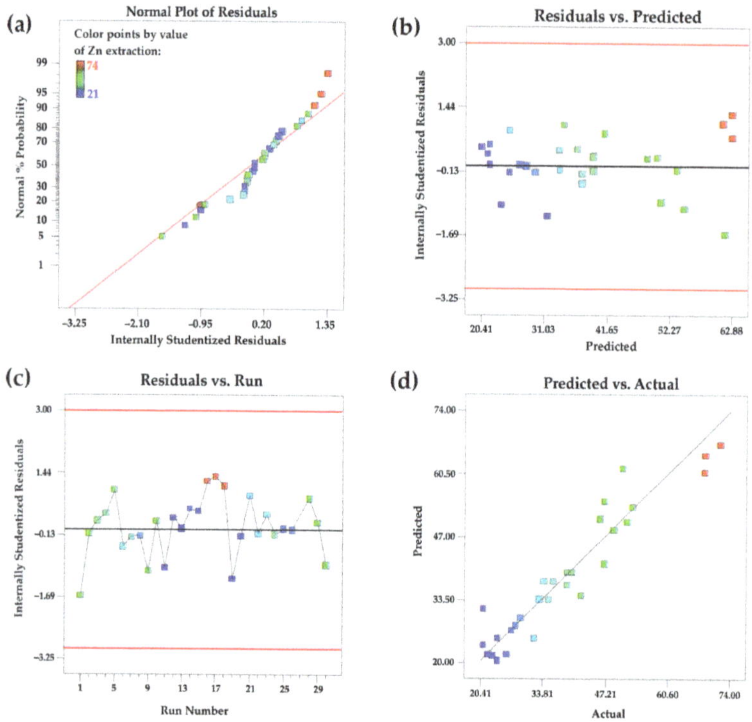

Figure 6. (**a**) A plot of normal probability vs. the internally studentized residuals, (**b**) internally studentized residuals vs. the predicted responses, (**c**) internally studentized residuals vs. run number, and (**d**) predicted responses vs. the actual values.

Figure 6d presents a diagram illustrating the relationship between predicted and actual values. The plot shows that the slope of the regression line approaches unity, with most points aligning along a straight line. This indicates a high degree of agreement between the calculated and experimental data, confirming the reliability and accuracy of the proposed quadratic model in predicting process parameters.

The analysis of the effects of experimental factors on zinc extraction efficiency (A: leaching duration; B: sulfuric acid concentration; C: pulp density; and D: temperature), as shown in Figure 7, reveals key trends. Temperature, sulfuric acid concentration, pulp density, and leaching duration exert significant influence on the zinc leaching rate from clinker.

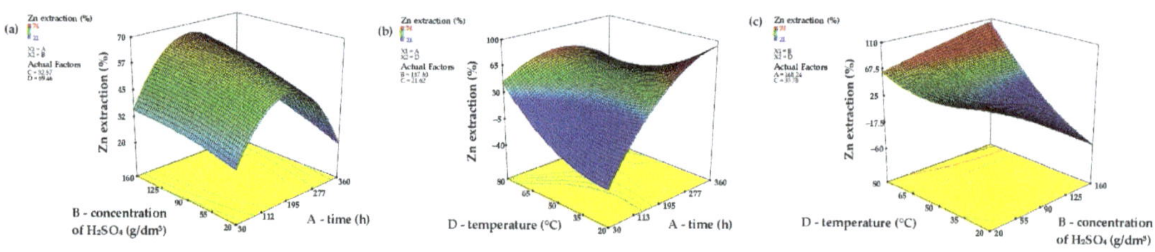

Figure 7. Three-dimensional response surfaces (with other parameters maintained at their central levels), illustrating the combined effects of B and A (**a**); D and A (**b**); D and B (**c**) (A—leaching duration, B—sulfuric acid concentration, D—temperature).

3.3.2. Internal Relationships Between Factors

The coefficients for factors A, B, C, and D are −4.38, −6.48, −25.70, and +144.50, respectively, as presented in Equation (4). These values quantitatively reflect the impact of each factor on the target response, consistent with theoretical assumptions [52,53]. Analysis of the coefficients indicates that, except for temperature, all other investigated parameters have a negative influence on the zinc leaching rate.

Additionally, the order of the factors' influence on the leaching rate is as follows: temperature (D) > leaching duration (A) > sulfuric acid concentration (B) > pulp density (C).

The three-dimensional response surfaces, derived from the quadratic model, provide a comprehensive analysis of the interrelationship between key process factors and zinc extraction efficiency. Figure 7a examines the interaction between leaching duration (A) and sulfuric acid concentration (B). The analysis reveals that increasing both parameters significantly enhances zinc extraction. However, at high H_2SO_4 concentrations, a saturation point is reached, resulting in response stabilization. Contour plots confirm the dominant role of leaching duration, particularly in the initial stages of the reaction, highlighting its substantial contribution to the process kinetics.

Figure 7b illustrates the interaction between leaching duration (A) and temperature (D). The three-dimensional response surface graphs clearly indicate the predominant effect of temperature. As the temperature increases, the steepness of the surface rises significantly, particularly at longer leaching durations. Even at relatively short time intervals, high temperatures considerably accelerate the process, confirming its critical role in achieving maximum extraction efficiency.

Figure 7c presents the analysis of the interaction between sulfuric acid concentration (B) and temperature (D). The response surface plots demonstrate a synergistic effect between these factors: their simultaneous increase leads to a notable improvement in zinc extraction. Temperature retains its dominant importance, as evidenced by the steeper gradient of the surface along the temperature axis.

The results unequivocally underscore the pivotal role of temperature (D) as the primary factor determining process efficiency. It is followed by the interaction of leaching duration (A) and sulfuric acid concentration (B). Based on the analysis of F-values and factor interactions, the hierarchy of influence is as follows: AD > AC > BD > AB.

Each interaction factor has its own points of extreme values, enabling the prediction of optimal process conditions, as shown in Figure 7. The Design Expert 7.0 software was used for modeling and optimizing the zinc leaching process. According to the calculations, the optimal parameters included a sulfuric acid concentration of 140 g/dm^3, pulp density of 20%, leaching duration of 2 h, and temperature of 80 °C. Under these conditions, the predicted zinc extraction rate was 74.0%, with a desirability level of 0.967 for the model.

Experimental validation conducted under the optimized conditions confirmed the high accuracy of the model: the actual zinc extraction rate was 72.6%, which is close to the predicted value. This agreement demonstrates the reliability and predictive capability of the response surface-based model.

3.4. Effect of Ultrasonic-Assisted Leaching of Clinker

The phenomenon of ultrasonic cavitation and its impact on the leaching process are vividly illustrated in Figure 8. Cavitation induced by ultrasonic waves generates high-pressure microjets and shock waves, causing mineral particles to move turbulently and subjecting them to intense multidirectional impact forces. This dynamic environment plays a crucial role in enhancing the efficiency of zinc leaching [51].

Ultrasound was generated using a device (UZDN-LUCH 2, Almaty, Kazakhstan) operating at 15 kHz and 200 W, which ensured the stable formation of cavitation zones.

Ultrasonic cavitation significantly reduces diffusion resistance by creating a highly active interaction zone between the leaching solution and the mineral surface [54]. The collapse of cavitation bubbles produces localized high temperatures (up to 5000 K) and pressures (up to 1000 atm), effectively removing passivation layers from particle surfaces [55]. This exposes fresh, reactive surfaces, allowing sulfuric acid to interact more effectively with zinc-containing minerals. Furthermore, the mechanical action of microjets facilitates the disintegration of agglomerated particles, further increasing the reactive surface area.

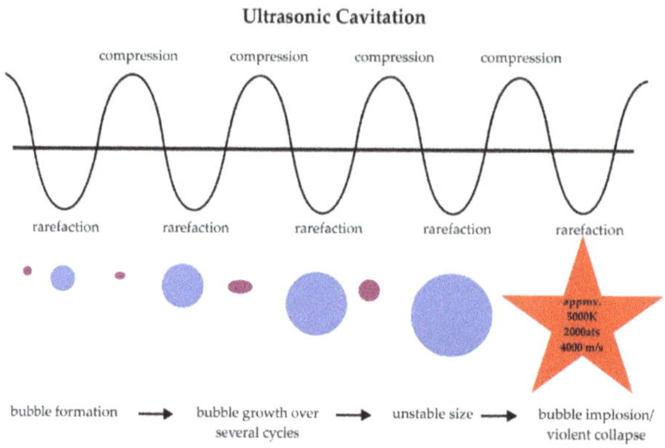

Figure 8. Ultrasonic cavitation.

The preliminary treatment of clinker using microwave irradiation, combined with leaching studies supported by mathematical modeling, enabled the determination of optimal conditions under which zinc recovery reached 72.6%. To further enhance efficiency, ultrasound was integrated into the leaching process. Figure 9 presents the results, demonstrating that the application of ultrasonic treatment increased zinc recovery by 4.1%, reaching 76.5%, compared to 72.6% achieved under identical conditions without ultrasound.

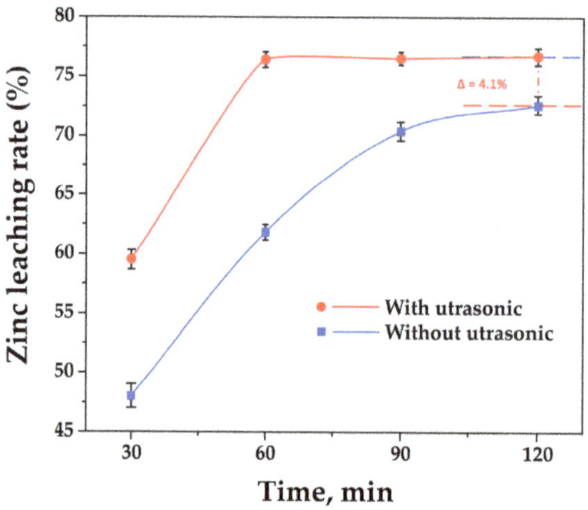

Figure 9. The effect of ultrasound on clinker leaching.

Experimental results demonstrated that applying ultrasonic treatment during leaching increased zinc recovery to 76.5% within just 1 h, compared to 72.6% achieved in 2 h without ultrasound. This highlights the ability of ultrasound not only to enhance recovery but also to significantly reduce leaching time [56], showcasing the high efficiency of the technology.

4. Conclusions

This study demonstrates the effectiveness of a combined approach involving microwave thermal treatment and ultrasonic-assisted leaching to enhance zinc recovery from refractory clinker. The experimental results enabled the optimization of key leaching parameters, including sulfuric acid concentration, temperature, pulp density, and leaching time, while also assessing the influence of microwave treatment and ultrasonic cavitation on the dissolution rate.

Statistical data processing using response surface methodology (RSM) and central composite design (CCD) allowed for the identification of key trends in the effects of process parameters. It was shown that temperature plays a critical role in leaching efficiency, while the interaction between microwave treatment and ultrasonic cavitation enhances zinc recovery. Although the model demonstrated reasonable predictive capability within the experimental dataset, further research with an expanded data set is required to improve its accuracy and validate its applicability in industrial conditions.

Microwave treatment facilitated phase transformations that improved zinc solubility, whereas ultrasonic cavitation disrupted passivating layers and intensified mass transfer, resulting in enhanced leaching efficiency and reduced process duration. Under optimized conditions, zinc recovery reached 76.5%, confirming the potential of this approach for the hydrometallurgical processing of refractory materials.

Future research should focus on evaluating the economic feasibility of this method under industrial conditions, analyzing energy efficiency, and exploring alternative solvent systems to further enhance zinc recovery. Additionally, comparative studies with conventional leaching approaches and advanced process simulations could provide deeper insights into the scalability of this technique.

Author Contributions: Conceptualization, B.K., Z.B. and Y.A.; methodology, A.B., T.S., K.S. and S.S.; software, B.K., T.S., K.S. and N.T.; validation, B.K., A.B. and Y.A.; formal analysis, A.B. and N.T.; investigation, A.B., B.K. and Z.D.; resources, B.K., K.S. and T.O.; data curation, Z.B., A.B. and T.O.; writing—original draft preparation, Z.B. and S.S.; writing—review and editing, Z.B., B.K., A.B. and Z.D.; visualization, T.S. and N.T.; supervision, Z.B. and Y.A.; project administration, B.K., Z.B. and S.S.; funding acquisition, B.K. and Z.B. All authors have read and agreed to the published version of the manuscript.

Funding: This research is funded by the Science Committee of the Ministry of Science and Higher Education of the Republic of Kazakhstan (Grant No. AR 19675985).

Institutional Review Board Statement: Not applicable.

Informed Consent Statement: Not applicable.

Data Availability Statement: The original contributions presented in this study are included in the article. Further inquiries can be directed to the corresponding authors.

Conflicts of Interest: The authors declare no conflict of interest.

Abbreviations

The following abbreviations are used in this manuscript:

MDPI	Multidisciplinary Digital Publishing Institute
DOAJ	Directory of open access journals
TLA	Three letter acronym
LD	Linear dichroism

References

1. Sharipova, A.S.; Linnik, K.A.; Zagorodnyaya, A.N.; Bakhytuly, N. Behaviour of lead and selenium during sequential sodium carbonate and nitric acid leaching of slime generated by the sulphuric acid plant of the Balkhash copper smelter. *Tsvetnye Met.* **2021**, 35–42. [CrossRef]
2. Ultarakova, A.A.; Yessengaziyev, A.M.; Kuldeyev, E.I.; Kassymzhanov, K.K.; Uldakhanov, O.K. Processing of Titanium Production Sludge with the Extraction of Titanium Dioxide. *Metalurgija* **2021**, *60*, 411–414.
3. Panichkin, A.; Kenzhegulov, A.; Mamaeva, A.; Uskenbayeva, A.; Kshibekova, B.; Imbarova, A.; Alibekov, Z. Effect of Carbon and Cooling Rate on the Structure of Hypereutectic High Chromium Cast Iron in the Cast State and after Heat Treatment. *J. Compos. Sci.* **2023**, *7*, 483. [CrossRef]
4. Ospanov, K.; Smailov, K.; Nuruly, Y. Patterns of Non-Traditional Thermodynamic Functions ΔrG0/n and ΔfG0(Averaged) Changes for Cobalt Minerals. *Chem. Bull. Kazakh Natl. Univ.* **2020**, *96*, 22–30. [CrossRef]
5. Kenzhaliyev, B.K.; Tussupbayev, N.K.; Abdykirova, G.Z.; Koizhanova, A.K.; Fischer, D.Y.; Baltabekova, Z.A.; Samenova, N.O. Evaluation of the Efficiency of Using an Oxidizer in the Leaching Process of Gold-Containing Concentrate. *Processes* **2024**, *12*, 973. [CrossRef]
6. Ju, J.; Feng, Y.; Li, H.; Ma, R.; Li, Y.; Zhao, H.; Wang, H.; Jiang, S. An Innovative Method for the Efficient and Selective Extraction of Co, Ni, Cu, and Mn from Oceanic Cobalt-Rich Crusts by Ammonium Sulfate Roasting: Behavior, Roasting Kinetics and Mechanism. *Miner. Eng.* **2024**, *207*, 108543. [CrossRef]
7. Zhao, F.; Jiang, X.; Wang, S.; Feng, L.; Li, D. The Recovery of Valuable Metals from Ocean Polymetallic Nodules Using Solid-State Metalized Reduction Technology. *Minerals* **2019**, *10*, 20. [CrossRef]
8. Sakellariadou, F.; González, F.J.; Hein, J.R.; Rincón-Tomás, B.; Arvanitidis, N.; Kuhn, T. Seabed Mining and Blue Growth: Exploring the Potential of Marine Mineral Deposits as a Sustainable Source of Rare Earth Elements (MaREEs) (IUPAC Technical Report). *Pure Appl. Chem.* **2022**, *94*, 329–351. [CrossRef]
9. Wegerer, S.; Semel, M.D.; Weixler, L. Vertical Exploration Approach for Seafloor Massive Sulfide Deposits. In Proceedings of the Offshore Technology Conference, Houston, TX, USA, 1–4 May 2023; ISBN 978-1-61399-974-5. [CrossRef]
10. Balaram, V. Deep-Sea Mineral Deposits as a Future Source of Critical Metals, and Environmental Issues—A Brief Review. *Miner. Miner. Mater.* **2023**, *2*, 5. [CrossRef]
11. Nath, S.; Singh, K.K.; Tangjang, S.; Das, S. Industrial Solid Wastes and Environment: An Overview on Global Generation, Implications, and Available Management Options. In *Springer Water*; Springer: Berlin/Heidelberg, Germany, 2024; pp. 221–246. [CrossRef]
12. Varjani, S. Trends in Mitigation of Industrial Waste: Global Health Hazards, Environmental Implications and Waste Derived Economy for Environmental Sustainability. *Sci. Total Environ.* **2022**, *811*, 152357. [CrossRef]
13. Bazarbayeva, S.M. Dataset on Industrial Waste Compositions in West Kazakhstan and Conditions for Processing Them into Construction Materials. *Data Brief* **2024**, *54*, 110265. [CrossRef] [PubMed]
14. Kenzhaliyev, B.; Surkova, T.; Berkinbayeva, A.; Amanzholova, L.; Mishra, B.; Abdikerim, B.; Yessimova, D. Modification of Natural Minerals with Technogenic Raw Materials. *Metals* **2022**, *12*, 1907. [CrossRef]
15. Dosmukhamedov, N.; Zholdasbay, E.; Argyn, A. Integrated Chlorination Technology for Producing Alumina and Silica from Ash-Slag Waste of the TPP of Kazakhstan. *J. Mater. Res. Technol.* **2022**, *23*, 1435–1446. [CrossRef]
16. Zhanikulov, N.N.; Sapargaliyeva, B.; Agabekova, A.B.; Alfereva, Y.O.; Syrlybekkyzy, S.; Nurshakhanova, L.K.; Nurbayeva, F.K.; Sabyrbaeva, G.S.; Kozlov, P.T.; Kolesnikova, O.G. Studies of Utilization of Technogenic Raw Materials in the Synthesis of Cement Clinker from It and Further Production of Portland Cement. *J. Compos. Sci.* **2023**, *7*, 226. [CrossRef]
17. Jandieri, G. Increasing the Efficiency of Secondary Resources in the Mining and Metallurgical Industry. *J. S. Afr. Inst. Min. Metall.* **2023**, *123*, 1–8. [CrossRef]
18. Krishnan, S.; Zulkapli, N.S.; Kamyab, H.; Taib, S.M.; Din, M.F.M.; Majid, Z.A.; Chaiprapat, S.; Kenzo, I.; Ichikawa, Y.; Nasrullah, M.; et al. Current Technologies for Recovery of Metals from Industrial Wastes: An Overview. *Environ. Technol. Innov.* **2021**, *22*, 101525. [CrossRef]
19. Kolesnikov, A.; Fediuk, R.; Kolesnikova, O.; Zhanikulov, N.; Zhakipbayev, B.; Kuraev, R.; Akhmetova, E.; Shal, A. Processing of Waste from Enrichment with the Production of Cement Clinker and the Extraction of Zinc. *Materials* **2022**, *15*, 324. [CrossRef]
20. Rahmati, S.; Adavodi, R.; Hosseini, M.R.; Veglio', F. Efficient Metal Extraction from Dilute Solutions: A Review of Novel Selective Separation Methods and Their Applications. *Metals* **2024**, *14*, 605. [CrossRef]
21. Ultarakova, A.; Karshyga, Z.; Lokhova, N.; Yessengaziyev, A.; Kassymzhanov, K.; Mukangaliyeva, A. Studies of Niobium Sorption from Chloride Solutions with the Use of Anion-Exchange Resins. *Processes* **2023**, *11*, 1288. [CrossRef]
22. Ultarakova, A.; Karshyga, Z.; Lokhova, N.; Yessengaziyev, A.; Kassymzhanov, K.; Mukangaliyeva, A. Studies on the Processing of Fine Dusts from the Electric Smelting of Ilmenite Concentrates to Obtain Titanium Dioxide. *Materials* **2022**, *15*, 8314. [CrossRef]

23. Abdulvaliev, R.A.; Surkova, T.Y.; Baltabekova, Z.; Yessimova, D.M.; Stachowicz, M.; Smailov, K.M.; Dossymbayeva, Z.D.; Ainur, B. Effect of Amino Acids on the Extraction of Copper from Sub-Conditional Raw Materials. *Complex Use Miner. Resour.* **2024**, *335*, 50–58. [CrossRef]
24. Walke, S.; Mandake, M.B. Optimization of Chemical Engineering Processes in the Mining and Metal Industry: A Review. *J. Mines Met. Fuels* **2024**, *71*, 240–251. [CrossRef]
25. Kuandykova, A.; Taimasov, B.; Potapova, E.; Sarsenbaev, B.; Kolesnikov, A.; Begentayev, M.; Kuldeyev, E.; Dauletiyarov, M.; Zhanikulov, N.; Amiraliyev, B.; et al. Production of Composite Cement Clinker Based on Industrial Waste. *J. Compos. Sci.* **2024**, *8*, 257. [CrossRef]
26. Toshkodirova, R.E.; Abdurakhmonov, S. Processing of Clinker—Technogenic Waste of Zinc Production. *Univ. Tech. Sci.* **2020**, *11*, 78–81.
27. Trebukhov, S.; Volodin, V.; Nitsenko, A.; Burabaeva, N.; Ruzakhunova, G. Recovery of Zinc from the Concentrate of Domestic Waste Processing by Vacuum Distillation. *Metals* **2022**, *12*, 703. [CrossRef]
28. Lobanov, V.G.; Kolmachikhina, O.B.; Polygalov, S.E.; Khabibulina, R.E.; Sokolov, L.V. Features of the Presence of Precious Metals in the Zinc Production Clinker. *Russ. J. Non-Ferr. Met.* **2022**, *63*, 594–598. [CrossRef]
29. Kiprono, N.R.; Kawalec, A.; Klis, B.; Smolinski, T.; Rogowski, M.; Kalbarczyk, P.; Samczynski, Z.; Norenberg, M.; Ostachowicz, B.; Adamowska, M.; et al. Radiation Techniques for Tracking the Progress of the Hydrometallurgical Leaching Process: A Case Study of Mn and Zn. *Metals* **2024**, *14*, 744. [CrossRef]
30. Maltrana, V.; Morales, J. The Use of Acid Leaching to Recover Metals from Tailings: A Review. *Metals* **2023**, *13*, 1862. [CrossRef]
31. Wang, S.; Gao, F.; Li, B.; Liu, Y.; Deng, T.; Zhang, Y.; Chen, W. Clinkerization of Carbonatable Belite–Melilite Clinker Using Solid Waste at Low Temperature. *Constr. Build. Mater.* **2024**, *418*, 135357. [CrossRef]
32. Zhidebekkyzy, A.; Temerbulatova, Z.; Amangeldiyeva, B.; Sakhariyeva, A. Towards a Circular Economy: An Analysis of Kazakhstani Case. *J. Econ. Res. Bus. Adm.* **2023**, *143*, 16–32. [CrossRef]
33. Qiao, Y.; Wang, G. Recent Status of Production, Administration Policies, and Low-Carbon Technology Development of China's Steel Industry. *Metals* **2024**, *14*, 480. [CrossRef]
34. Diaz, F.; Sommerfeld, M.; Hovestadt, G.; Latacz, D.; Friedrich, B. Lessons Learned from Attempts at Minimising CO_2 Emissions in Process Metallurgy—Pyrolysed Secondary Raw Materials, Bio-Coke, and Hydrogen as Alternative Reducing Agents. In Proceedings of the 12th International Conference of Molten Slags, Fluxes and Salts, Brisbane, Australia, 17–19 June 2024. [CrossRef]
35. Harvey, J.-P.; Courchesne, W.; Vo, M.D.; Oishi, K.; Robelin, C.; Mahue, U.; Leclerc, P.; Al-Haiek, A. Greener Reactants, Renewable Energies and Environmental Impact Mitigation Strategies in Pyrometallurgical Processes: A Review. *MRS Energy Sustain.* **2022**, *9*, 212–247. [CrossRef] [PubMed]
36. Hamidi, A.; Nazari, P.; Shakibania, S.; Rashchi, F. Microwave Irradiation for the Recovery Enhancement of Fly Ash Components: Thermodynamic and Kinetic Aspects. *Chem. Eng. Process.* **2023**, *191*, 109472. [CrossRef]
37. Lin, S.; Li, K.; Yang, Y.; Gao, L.; Omran, M.; Guo, S.; Chen, J.; Chen, G. Microwave-Assisted Method Investigation for the Selective and Enhanced Leaching of Manganese from Low-Grade Pyrolusite Using Pyrite as the Reducing Agent. *Chem. Eng. Process.* **2021**, *159*, 108209. [CrossRef]
38. Al-Harahsheh, M.; Kingman, S.W. Microwave-Assisted Leaching—A Review. *Hydrometallurgy* **2004**, *73*, 189–203. [CrossRef]
39. Stojković, M.; Ristić, M.; Đolić, M.; Perić Grujić, A.; Onjia, A. Recovery of Rare Earth Elements from Coal Fly and Bottom Ashes by Ultrasonic Roasting Followed by Microwave Leaching. *Metals* **2024**, *14*, 371. [CrossRef]
40. Fang, X.; Peng, Z.; Yin, T.; Rao, M.; Li, G. Microwave Treatment of Copper–Nickel Sulfide Ore for Promotion of Grinding and Flotation. *Metals* **2024**, *14*, 565. [CrossRef]
41. Ye, L.; Peng, Z.; Tian, R.; Tang, H.; Zhang, J.; Rao, M.; Li, G. A novel process for highly efficient separation of boron and iron from ludwigite ore based on low-temperature microwave roasting. *Powder Technol.* **2022**, *410*, 117848. [CrossRef]
42. Tao, F.; Zhao, G.; Chen, W.; Tao, D. 1/2 Order Subharmonic Waves of Two Cavitation Bubbles. *Ultrason. Sonoch.* **2024**, *110*, 107022. [CrossRef]
43. Moussatov, A.; Granger, C.; Dubus, B. Cone-Like Bubble Formation in Ultrasonic Cavitation Field. *Ultrason. Sonoch.* **2003**, *10*, 191–195. [CrossRef]
44. Li, H.; Hu, C.; He, X.; Wang, J.; Tian, S.; Zhu, X.; Mao, X. Mechanism and Kinetics Study of Vanadium Leaching from Landfilled Metallurgical Residues by Ultrasonic with Ozonation Enhancement in a Low-Acid Medium. *Ultrason. Sonoch.* **2024**, *109*, 106998. [CrossRef] [PubMed]
45. Liu, B.; Shi, C.; Huang, Y.; Han, G.; Sun, H.; Zhang, L. Intensifying Separation of Pb and Sn from Waste Pb-Sn Alloy by Ultrasound-Assisted Acid Leaching: Selective Dissolution and Sonochemistry Mechanism. *Ultrason. Sonoch.* **2024**, *102*, 106758. [CrossRef]
46. Zhu, R.; Wang, S.; Chen, Y.; Xiang, D.; Zhang, L.; Liu, J.; Ye, J. Ultrasound Enhanced In-Situ Chemical Oxidation for Leaching Ag from Zinc Leaching Residue and Response Surface Optimization. *Chem. Eng. J.* **2024**, *489*, 151243. [CrossRef]

47. Zhang, D.; Fu, L.; Liu, H.; Li, H.; Wang, S.; Zhang, M.; Zhu, M.; Zhang, L. High-Efficiency Leaching of Chalcopyrite by Ozone with Ultrasonic Promotion: Kinetics and Mechanism. *J. Mol. Liq.* **2024**, *401*, 124682. [CrossRef]
48. Liu, L.; Gan, M.; Fan, X.; Gao, Z.; Sun, Z.; Ji, Z.; Wei, J.; Ma, S. Mechanism of Ultrasonic Enhanced Acetic Acid Efficiently Leaching of Steel Slag and Synthesis of Calcium Carbonate Whiskers. *Sep. Purif. Technol.* **2024**, *339*, 126615. [CrossRef]
49. Kenzhaliyev, B.; Surkova, T.; Berkinbayeva, A.; Baltabekova, Z.; Smailov, K. Harnessing Microwave Technology for Enhanced Recovery of Zinc from Industrial Clinker. *Metals* **2024**, *14*, 699. [CrossRef]
50. Ma, A.; Zheng, X.; Gao, L.; Li, K.; Omran, M.; Chen, G. Enhanced Leaching of Zinc from Zinc-Containing Metallurgical Residues via Microwave Calcium Activation Pretreatment. *Metals* **2021**, *11*, 1922. [CrossRef]
51. Liu, J.; Li, S.; Zhang, L.; Yang, K. Application of the Microwave and Ultrasonic Combined Technique in the Extraction of Refractory Complex Zinc Ore. *Metals* **2023**, *13*, 356. [CrossRef]
52. Chen, H.; Wu, D.; Wang, Z. Investigation of the Leaching Kinetics of Zinc from Smithsonite in Ammonium Citrate Solution. *Metals* **2024**, *14*, 519. [CrossRef]
53. Tian, J.J.; Wu, D.D.; Li, S.Y.; Ma, W.H.; Wang, R.Z. Effect of process variables on leaching behavior and kinetics of silver element from waste photovoltaic modules. *Sep. Purif. Technol.* **2024**, *335*, 126062. [CrossRef]
54. Scruby, C.B.; Drain, L.E. *Laser Ultrasonics: Techniques and Applications*; Routledge: London, UK, 2019.
55. Vargas, S.A.; Delgado-Macuil, R.J.; Ruiz-Espinosa, H.; Rojas-López, M.; Amador-Espejo, G.G. High-intensity ultrasound pretreatment influence on whey protein isolate and its use on complex coacervation with kappa carrageenan: Evaluation of selected functional properties. *Ultrason. Sonoch.* **2021**, *70*, 105340. [CrossRef]
56. Slaczka, A.S. Effect of ultrasound on ammonium leaching of zinc from galmei ore. *Ultrasonics* **1986**, *24*, 53–55. [CrossRef]

metals

Article

Gold Leaching from an Auriferous Ore by Alkaline Thiosulfate–Glycine–Copper Solution

Alex S. Redrovan *, Ernesto de la Torre and Carlos F. Aragón-Tobar

Department of Extractive Metallurgy, Escuela Politécnica Nacional, Ladrón de Guevara E11-253, P.O. Box 17-01-2759, Quito 170525, Ecuador; ernesto.delatorre@epn.edu.ec (E.d.l.T.); carlos.aragont@epn.edu.ec (C.F.A.-T.)
* Correspondence: alex1998saul@hotmail.com; Tel.: +593-969423657

Abstract: The thiosulfate–glycine–copper system has emerged as a promising alternative for gold recovery, offering significant advantages over cyanidation and ammoniacal thiosulfate leaching. Recognizing the limitations of thiosulfate degradation in ammoniacal systems, this study focused on optimizing the thiosulfate–glycine–copper system for gold recovery using an auriferous ore with (10 g t^{-1}) of Au. The ore was associated with aluminosilicates such as grossular (64%) and clinochlore (12%). Leaching conditions were systematically varied, including thiosulfate (0.5–1 M), glycine (0.3–1.75 M), copper sulfate (2–10 mM), pH (9.3–10.5), temperature (20–60 °C), 6 h, and potassium permanganate concentrations (0.004–0.04 M), and dosing intervals were also optimized. Thus, the best conditions were thiosulfate (0.7 M), glycine (1.75 M), copper sulfate (5 mM), pH 9.3, 60 °C, and permanganate addition every 2 h. This system achieved 89.3% gold recovery in just 6 h, comparable to cyanidation (89.8% in 24 h) and ammoniacal thiosulfate (58% in 6 h), but without generating toxic effluents, such as in the cyanidation process. Additionally, a gold dissolution mechanism was proposed, highlighting glycine's role in stabilizing cupric ions and enhancing thiosulfate efficiency. This study underscores the thiosulfate–glycine–copper system as a sustainable and effective method for gold recovery.

Keywords: thiosulfate; glycine; gold; leaching; eco-friendly; adsorption; activated carbon

1. Introduction

1.1. Traditional Methods for Gold Recovery

The most widely used industrial method for the recovery of gold from auriferous ores is sodium cyanide leaching, due to its low reagent and high recovery (89%) [1,2]. However, its use poses serious environmental risks due to the high toxicity of the generated effluents, which can lead to severe ecological and public health consequences [3–5]. For this reason, alternatives, such as sodium thiosulfate, have been extensively studied for their lower cost and reduced toxicity [3,6].

Despite its advantages, the industrial scale use of thiosulfate presents two main challenges: first, the degradation of sodium thiosulfate into by-products such as tetrathionates, polythionates, and sulfates, which increases reagent consumption [2,7]; and second, the low affinity of the gold–thiosulfate complex to adsorb onto activated carbon [6,8]. To mitigate the first problem, a catalyst, such as stable cupric ions, is typically used to accelerate the dissolution rate of gold [5,9,10]. However, cupric ions can be reduced to cuprous ions by dissolved oxygen in the solution, which in turn accelerates the degradation and consumption of thiosulfate [11,12]. To prevent this, cupric ions are generally stabilized using

 https://doi.org/10.3390/met15020204

ammonia, which forms the "cuprotetramine" complex, keeping copper in its Cu (II) state, thereby facilitating the formation of the gold–thiosulfate complex [6,9]. Although ammonia improves gold recovery, its use in high concentrations poses environmental risks similar to cyanide [4,13].

On the other hand, the recovery of dissolved gold from thiosulfate solutions using activated carbon presents significant limitations [14–16]. The low affinity of the gold–thiosulfate complex, due to the difference in polarity with the carbon and the molecule's size relative to the pores of conventional activated carbon, hinders this process. In contrast, the traditional method using activated carbon achieves up to 98% adsorption of the gold–cyanide complex [15,17–19]. However, it has been demonstrated that modifying the specific surface characteristics of activated carbon, by introducing nitrogen-containing functional groups, can significantly improve the gold adsorption rate [14,20].

1.2. Thiosulfate–Glycine System: An Alternative for Gold Extraction

As mentioned in the previous Section, gold leaching with sodium thiosulfate solutions requires stable cupric ions to catalyze the dissolution of gold. Given the toxicity of ammonia, it is necessary to find an effective substitute [21].

Glycine, a non-toxic and low-cost amino acid, is commonly used in the selective recovery of copper from oxidized ores. It is primarily used as a dietary supplement or a regulator of the central nervous system [22,23]. However, [22,24,25] demonstrated that selective leaching from pure gold sheets and copper–gold concentrates, using glycine in synergy with other amino acids or cyanide, achieved 70% gold recovery within 50 h. This recovery is due to glycine's ability to form strong complexes with Cu^{2+}, Co^{2+}, and Ni^{2+} [26,27], thereby reducing reagent consumption. Particularly with Cu^{2+}, the copper–glycinate complex is more stable than the ammoniacal copper complex "cuprotetramine" [4,25,28]. Furthermore, [1,29] demonstrated glycine's ability to dissolve gold, recovering 50% of the gold and silver under alkaline conditions and in synergy with oxidizing agents, such as hydrogen peroxide or potassium permanganate. However, glycine leaching conditions—temperature (60 °C), pH (11), and leaching time (167 h)—reveal that the leaching rate is very slow [1,23].

For these reasons, the use of glycine within the sodium thiosulfate system emerges as a novel and underexplored alternative for the metallurgical treatment of gold ores. It can replace ammonia, reducing thiosulfate consumption; additionally, it acts as a co-leaching agent for gold [5,27,30–32].

The objective of this research is to develop an alternative leaching system using sodium thiosulfate. For this purpose, an Ecuadorian gold-bearing ore was used to optimize the thiosulfate–glycine–copper system by varying the concentrations of sodium thiosulfate, glycine, copper sulfate, potassium permanganate, pH, and temperature, until achieving recoveries comparable to those obtained with cyanidation and ammoniacal thiosulfate leaching.

2. Materials and Methods

2.1. Materials

The chemicals used in this study were analytical grade copper sulfate pentahydrate (AppliChem Panreac, 99%, Darmstadt, Germany), anhydrous sodium thiosulfate (LOBA Chemie, 98%, Mumbai, India), glycine (Sigma-Aldrich, 99%, St. Louis, MO, USA), sodium hydroxide (Merck, 99%, Darmstadt, Germany), and technical grade potassium permanganate. All experiments were conducted with distilled water, and pH control was achieved using a 20% v v^{-1} ammonium solution for the ammoniacal thiosulfate system and a 4 M sodium hydroxide solution for the thiosulfate–glycine–copper system.

2.2. Methods

2.2.1. Ore Characterization

The auriferous ore used in this study was obtained from the Guayzimi mining area, located in the province of Zamora Chinchipe, southern Ecuador. A 50 kg sample, with a d80 of 10 cm, was initially crushed in a closed circuit using a BICO Inc. jaw crusher (Quito, Ecuador) and an ASEA HM 4100 roller crusher (Quito, Ecuador) to achieve a d80 of 2 mm. The material was then homogenized and quartered. Finally, the ore was ground in a ball mill at 62.5% solids for 40 min.

For mineralogical characterization and leaching tests, several analyses were performed. The elemental composition was determined using X-ray Fluorescence (XRF) with a Bruker S8 Tiger (Bruker, Karlsruhe, Germany). The mineralogical composition was analyzed via X-ray Diffraction (XRD) using a Bruker AXS D8 Advance (Bruker, Karlsruhe, Germany). The copper content was measured using Atomic Absorption Spectroscopy (AAS) (Perkin Elmer AA 300, Shelton, CT, USA), and precious metals (Au, Ag) were estimated through a fire assay, followed by dore dissolution for AAS measurement.

The concentration of dissolved elements (Au, Ag, Cu, Fe, and Pb) during all leaching tests was measured using AAS (Perkin Elmer AA 300).

2.2.2. Sodium Cyanide Leaching

As a reference for gold recovery, a cyanidation test was conducted using 100 g of ore with a sodium cyanide concentration of 1 (g L^{-1}) (technical grade) for 24 h to compare the recoveries with those obtained from thiosulfate systems. The test conditions were as follows: ambient temperature (20 °C), pH 12, 750 RPM, and 15% solids. After the test, the sample was filtered using a vacuum filter, and both the pregnant solution and wash solution were analyzed to determine the gold concentration by AAS (Perkin Elmer AA 300).

2.2.3. Ammonium Thiosulfate Leaching

As a reference for gold recovery with the traditional ammoniacal thiosulfate system, tests were conducted using 100 g of ore at 15% solids for 6 h with mechanical stirring at 750 RPM. The concentration of sodium thiosulfate (sodium thiosulfate anhydrous, LOBA Chemie, 98%) was varied between 0.3 and 1 M, the pH ranged from 10 to 12, the copper sulfate concentration was adjusted between 2 and 10 mM, and the temperatures were tested between 20 and 60 °C (Figure 1a) to determine the optimal conditions for gold recovery.

At the beginning of each test, the pH was adjusted with a 20% (v v^{-1}) ammoniacal solution, increasing the ammonium ion concentration to form the cuprotetramine complex [11,13,33,34]. Upon completion of the tests, the samples were filtered through a vacuum filter; the solution was analyzed using AAS (Perkin Elmer AA 300), and the dried ore sample was subjected to fire assay to complete the gold recovery balance.

During the tests, 10 mL aliquots were taken at 1, 2, and 4 h, which were replenished with distilled water. One part of the filtered solution was used for AAS (Perkin Elmer AA 300) to analyze the leaching kinetics, while the other part was used to determine the concentration of free thiosulfate by iodometric titration. The consumed thiosulfate was weighed and replenished to maintain the initial concentration of each test.

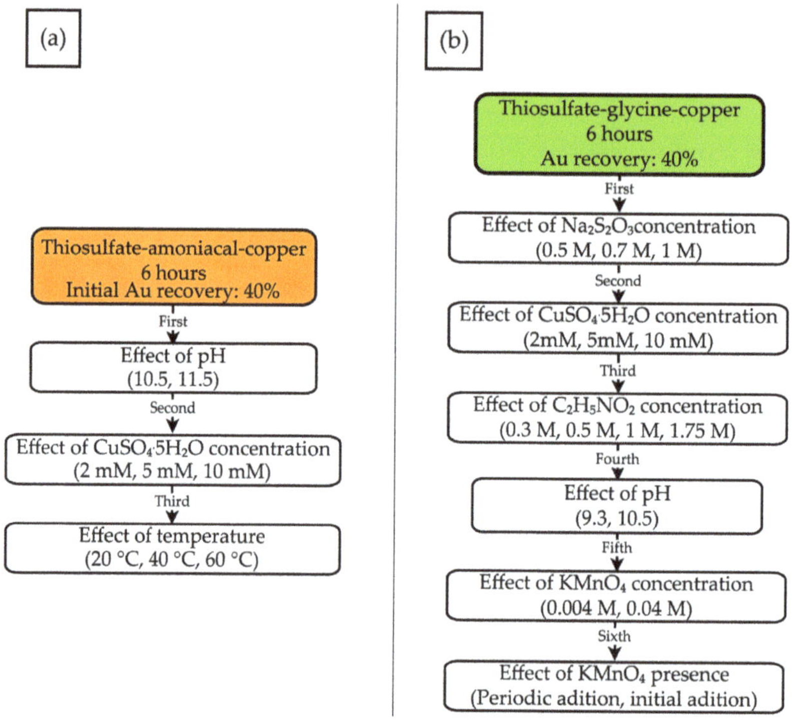

Figure 1. Comparison between the methodology followed for the (**a**) ammoniacal thiosulfate and (**b**) thiosulfate–glycine system.

2.2.4. Thiosulfate–Glycine Leaching

Leaching tests with the thiosulfate–glycine system were conducted to determine the effectiveness of glycine as a replacement for ammonia. The tests were carried out in a container covered with aluminum due to the photosensitivity of glycine in the presence of potassium permanganate. The tests used 100 g of mineral at 15% solids for 6 h under mechanical agitation at 750 RPM. The variables considered included the following: sodium thiosulfate concentration from 0.3 to 0.7 M, glycine concentration (Glycine, Sigma-Aldrich, 99%) from 0.3 to 1.75 M, copper sulfate concentration at 2, 5, and 10 mM, potassium permanganate concentration at 0.004 M and 0.04 M, pH 9.3, and a temperature of 60 °C, as shown in Figure 1b, to achieve the optimal gold dissolution. At the end of the test, the sample was filtered using a vacuum filter, the solution was sent for AAS (Perkin Elmer AA 300), and the dried sample was subjected to the fire assay to quantify gold recovery.

pH adjustment was performed at the beginning of the test by adding a 4 M NaOH solution. Additionally, potassium permanganate was tested as a catalyst in the formation of the copper–glycinate complex [33,35,36] and in the co-leaching of gold due to the action of glycine [1,29,37]. The effectiveness of potassium permanganate was evaluated by increasing its concentration to 0.04 M at the start of the test and adding lower concentrations (0.004 M) at intervals of 0, 2, and 4 h.

Finally, as in the ammoniacal thiosulfate tests, 10 mL aliquots were taken at 1, 2, and 4 h and replaced with distilled water. One part of the filtered solution was used for atomic absorption to analyze the leaching kinetics, and the second part was used to determine the thiosulfate concentration through iodometric titration.

2.2.5. Glycine Leaching

A 100 g sample with 15% solids was used for 6 h of mechanical agitation at 750 RPM, with a pH of 9.3, temperature of 60 °C, and a glycine concentration of 1.75 M. Additionally, potassium permanganate was added to reach a concentration of 0.004 M at the beginning, and at 2 and 4 h.

The pH of the glycine solution, compared to the thiosulfate–glycine system tests, showed lower NaOH consumption to reach a pH of 10.5. During the test, aliquots of the solution were taken at 1, 2, and 4 h to establish the leaching kinetics and glycine concentration.

At the end of the test, the sample was filtered using a vacuum filter, the solution was sent for atomic absorption spectroscopy analysis, and the dried mineral was subjected to the fire assay.

3. Results

3.1. Characterization of the Auriferous Ore

The ore used in this study was obtained from the Guayzimi mining area in Zamora Chinchipe. The mineralogical analysis, presented in Table 1 and Figure 2 and conducted through X-ray diffraction (XRD), revealed that the primary mineral composition consists of Grosular (64%), Clinochlore (12%), and quartz (10%).

Table 1. Mineralogical characterization of Guayzimi ore by XRD analysis.

Mineral	Formula	Content (%)
Grossular	$Ca_3Al_2(SiO_4)_3$	64
Clinochlore	$(Mg, Fe)_5Al (Si, Al)_4O_{10} (OH)_8$	12
Quartz	SiO_2	10
Calcite	$CaCO_3$	8
Kaolinite	$Al_2(Si_2O_5) (OH)_4$	4
Plagioclases	$(Na, Ca) Al (Si, Al) Si_2O_8$	2

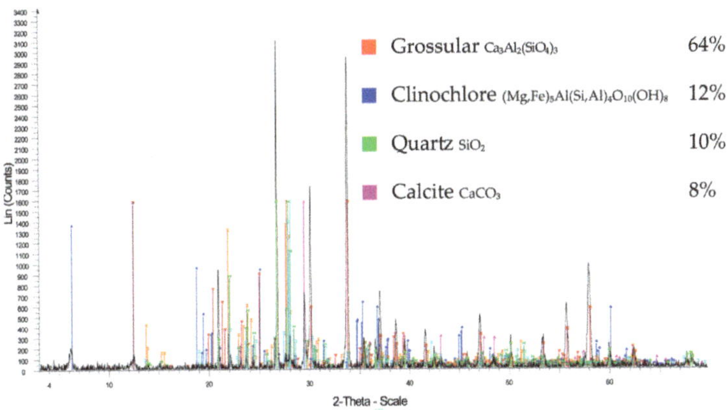

Figure 2. XRD pattern for the characterization of the auriferous ore from Guayzimi-Ecuador, identifying grossular (orange), clinochlore (blue), quartz (green), and calcite (purple) as the principal minerals. Other minerals are not highlighted as their concentrations do not exceed 5%.

Table 2 shows the concentration of valuable metals along with the ore chemical composition, determined through X-ray fluorescence (XRF), fire assay, and acid digestion. The ore has a head grade of 10 (g t^{-1}) of gold with a copper content of 0.008%. Other elements detected include iron (5%), calcium (15.1%), silicon (18.8%), and sulfur (0.1%). Chalcopyrite was also observed in hand sample geological assessments; however, its low concentration (>1%) prevents detection by XRD.

Table 2. Elemental composition, gold, and copper content of the ore determined using XRF, fire assay, and acid digestion.

Element	Concentration
Au	10 g t^{-1}
Cu	0.008%
Si	18.8%
Ca	15.1%
Fe	5%
Al	4.5%
Mg	1.6%
Mn	0.9%
K	0.8%
Na	0.4%
S	0.1%

3.2. Sodium Cyanide Leaching

A cyanidation test was conducted to establish a comparative baseline for gold recovery. As shown in Figure 3, under the conditions of 15% solids, pH 11 (adjusted with NaOH), and a sodium cyanide concentration of 1 g (NaCN) L^{-1}, a gold recovery of 89% was achieved after 24 h.

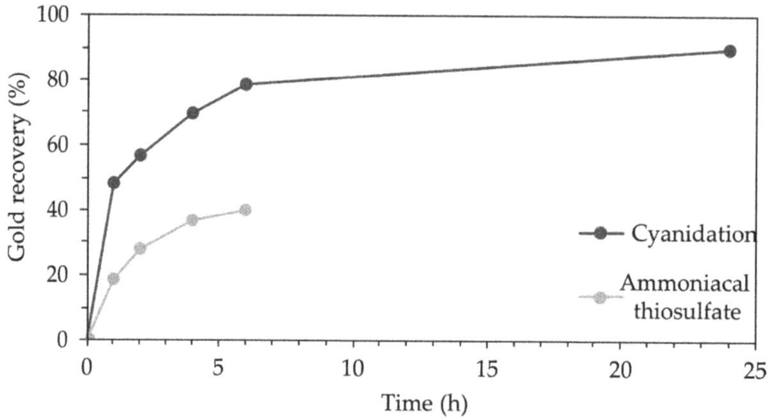

Figure 3. Comparison between baseline cyanidation tests (1 g L^{-1}; 15% solids; pH 12 for 24 h) and ammoniacal thiosulfate tests ($Na_2S_2O_3$ 0.7 M; copper sulfate 5 mM; pH 10.5; 15% solids for 6 h).

3.3. Ammoniacal Thiosulfate–Copper Leaching

Leaching tests using the ammoniacal thiosulfate–copper system were conducted to evaluate gold recovery under different leaching conditions. The optimized conditions were as follows: sodium thiosulfate concentration of 0.7 M, pH 10.5 (adjusted with a 20% v v^{-1} ammoniacal solution), copper sulfate concentration of 5 mM, 750 RPM agitation, and 15% solids for 6 h. Under these conditions, a maximum gold recovery of 40% was achieved, as shown in Figure 3.

Additionally, a reduction in thiosulfate concentration to 0.5 M was observed during the first 2 h of testing. Consequently, further tests were conducted by systematically varying the conditions, including thiosulfate concentration (0.5–1 M), copper sulfate concentration

(2–10 mM), and pH (10.5–11.5) to optimize gold recovery from the characterized ore using the ammoniacal thiosulfate–copper system.

Evaluation of Leaching Conditions for Gold Recovery Using the Ammoniacal Thiosulfate System

Leaching tests were performed by varying cupric ion concentration, pH, and temperature to assess the efficiency of the ammoniacal thiosulfate system. Results indicate that reducing the copper sulfate concentration to 2 mM increases gold recovery to 48%, as shown in Figure 4a, whereas a higher concentration of 10 mM only achieves a gold dissolution rate of 33%.

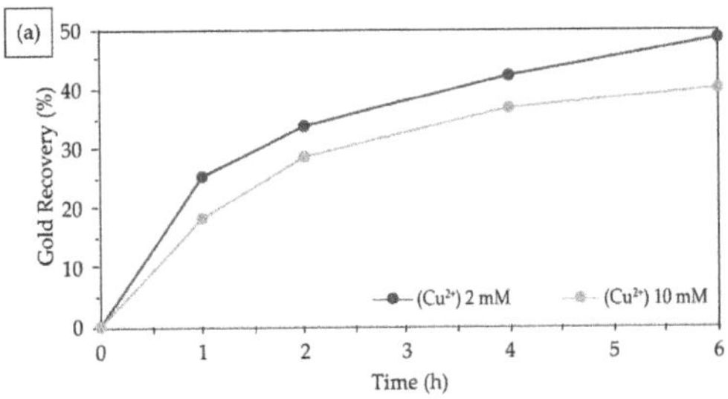

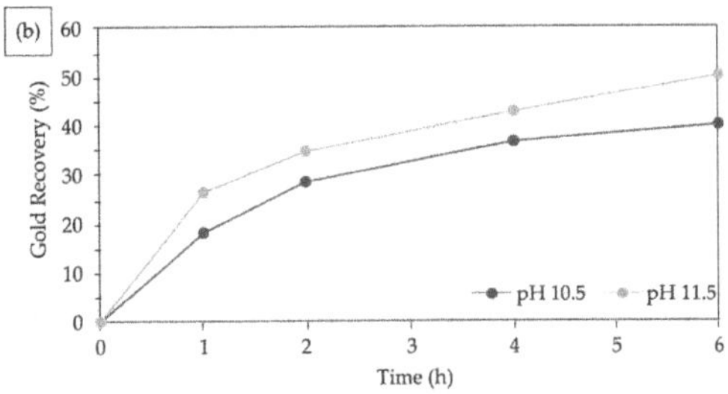

Figure 4. (**a**) Effect of copper concentration on Au recovery using the ammoniacal thiosulfate system: $Na_2S_2O_3$ (0.7 M); pH (10.5). (**b**) Effect of pH on gold recovery using the ammoniacal thiosulfate system: $Na_2S_2O_3$ (0.7 M); copper sulfate (2 mM).

Additionally, increasing the ammonia content to raise the pH from 10.5 to 11.5 resulted in an improvement in gold recovery from 40% to 50%, as depicted in Figure 4b.

Finally, tests conducted at different temperatures revealed that raising the temperature to 60 °C enhances recovery to 58% in 6 h, as presented in Figure 5.

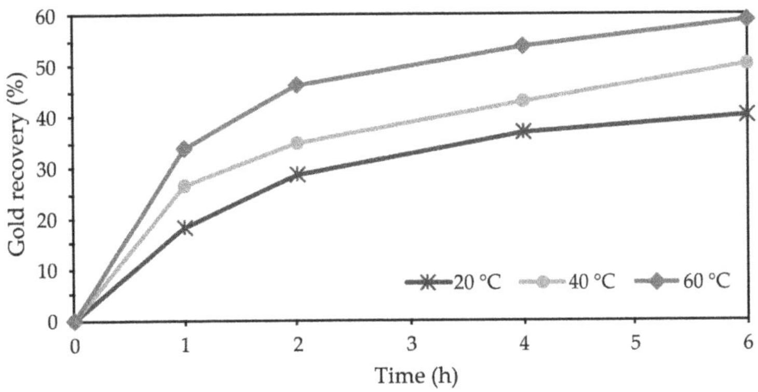

Figure 5. Effect of temperature on gold recovery using the ammoniacal thiosulfate system: $Na_2S_2O_3$ (0.7 M); copper sulfate (2 mM) in 6 h.

The optimal leaching conditions for the ammoniacal thiosulfate system were established at 0.7 M sodium thiosulfate, 2 mM copper sulfate, pH 10.5, and a temperature of 60 °C for 6 h, achieving a gold recovery of 58%.

3.4. Thiosulfate–Glycine–Copper Leaching

In the baseline test with the thiosulfate–glycine–copper system, using 100 g of mineral with a 15% solid concentration, a gold recovery of 40% was achieved under the following conditions: 0.7 M sodium thiosulfate, 0.3 M glycine, 60 °C, agitation at 750 RPM, and pH 10.5 (adjusted with NaOH) over 6 h, as shown in Figure 6a. This test served as a reference point for subsequent experimental variations, including adjustments in sodium thiosulfate concentration, glycine concentration, copper sulfate pentahydrate, potassium permanganate, and pH, aiming to enhance gold recovery efficiency.

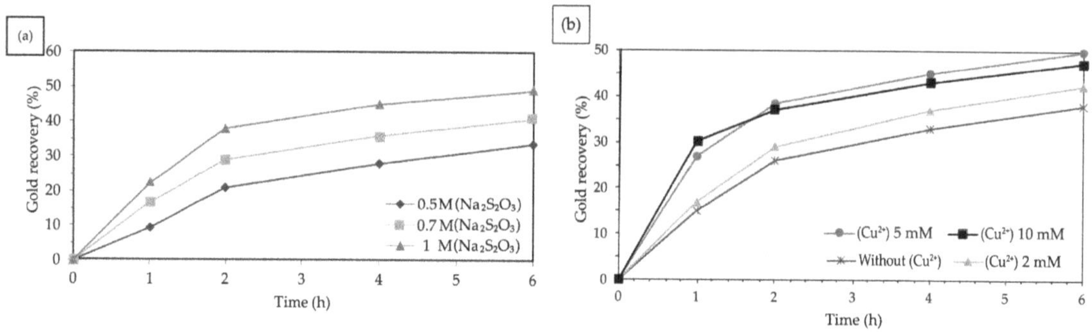

Figure 6. (**a**) Effect of sodium thiosulfate concentration on gold recovery using the thiosulfate–glycine system: $C_2H_5NO_2$ (0.3 M); 60 °C; pH (10.5); copper sulfate (2 mM). (**b**) Effect of copper concentration on gold recovery using the thiosulfate–glycine system: $Na_2S_2O_3$ (0.7 M); glycine (0.3 M); 60 °C; pH (10.5).

3.4.1. Effect of Sodium Thiosulfate Concentration

Table 3 presents the results of tests in which the sodium thiosulfate concentration was varied (0.5 M, 0.7 M, and 1 M), while all other experimental conditions remained constant. As shown in Figure 6a, gold recovery increased proportionally with the sodium thiosulfate concentration. The highest recovery, 49%, was achieved with a concentration of 1 M.

Table 3. Effect of sodium thiosulfate concentration on gold recovery using the thiosulfate–glycine system after 6 h. Glycine 0.3 M, 60 °C, pH 10.5.

Sodium Thiosulfate Concentration	Gold Recovery (%)
0.5 M	33.7
0.7 M	40.8
1 M	49.6

However, the difference in gold recovery between 1 M (49%) and 0.7 M (40.8%) was not substantial enough to justify the increase in reagent concentration. Thus, a 0.7 M concentration was selected for subsequent tests.

Throughout the tests, iodometric titration of the thiosulfate concentration showed no consumption of reagent over the 6 h of agitation.

3.4.2. Effect of Copper Ion Concentration

Figure 6b illustrates gold recovery in the absence of added copper sulfate, indicating that dissolved copper from the mineral in the presence of glycine was insufficient, yielding a recovery of less than 30%. To improve this outcome, tests were conducted with varying copper sulfate concentrations of 2 mM, 5 mM, and 10 mM, as detailed in Figure 6b.

Increasing the copper sulfate concentration to 5 mM enhanced gold dissolution, achieving a recovery of 48%. However, raising the concentration to 10 mM did not further improve recovery efficiency. The excess of cupric ions accelerated thiosulfate degradation, as evidenced by a reduction in thiosulfate concentration to 0.65 M when using 10 mM copper sulfate during the first 2 h of testing, quantified through iodometry.

In the thiosulfate–glycine system, no thiosulfate consumption was detected throughout the test when copper sulfate concentrations were kept below 5 mM.

3.4.3. Effect of Glycine Concentration

Leaching tests were conducted with a sodium thiosulfate concentration of 0.7 M and copper sulfate at 5 mM, varying the glycine concentration between 0.3 M, 0.5 M, 1 M, and 1.75 M. The results, shown in Figure 7, demonstrate that an increase in glycine concentration enhances gold recovery, achieving a dissolution of 69% of the gold with a glycine concentration of 1.75 M after 6 h.

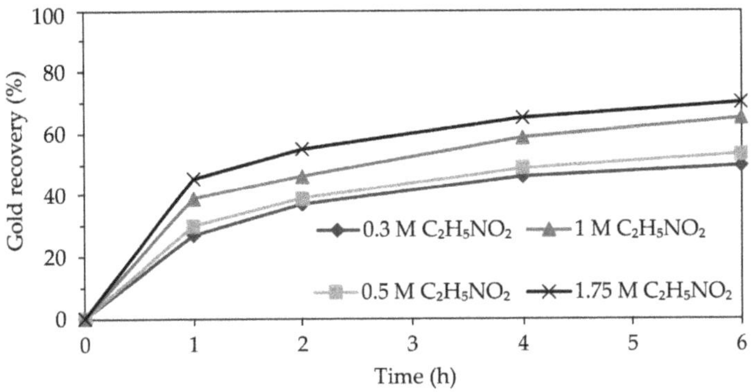

Figure 7. Effect of glycine ($C_2H_5NO_2$) concentration on Au recovery using the thiosulfate–glycine system: $Na_2S_2O_3$ (0.7 M); copper sulfate (5 mM); 60 °C; pH (10.5); 15% solids for 6 h.

3.4.4. Effect of pH

Although previous studies [5,13] suggest that optimal gold recovery in a thiosulfate system is achieved at pH levels above 10.5, the results in Figure 8 indicate that gold recovery is equivalent at pH 9.3 and 10.5, reaching 65% in both cases. These tests were conducted with the same reagent concentrations. It was observed that glycine concentration is directly related to NaOH consumption, requiring larger volumes of NaOH to maintain a pH of 10.5 when the glycine concentration exceeds 1 M.

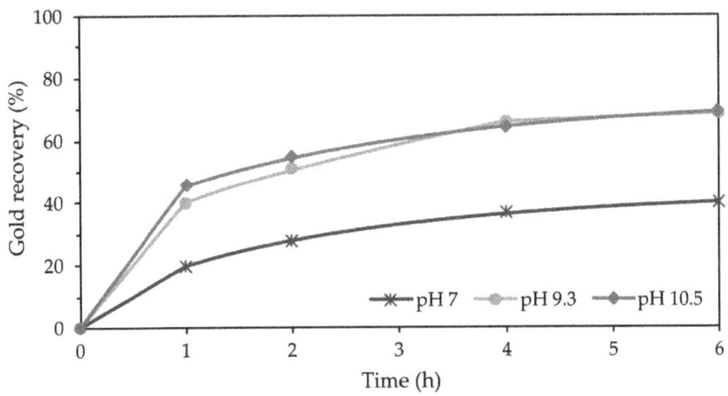

Figure 8. Effect of pH on gold recovery using the thiosulfate–glycine system: $Na_2S_2O_3$ (0.7 M); $C_2H_5NO_2$ (1 M); copper sulfate (5 mM); 60 °C; 15% solids for 6 h.

3.4.5. Effect of Oxidizing Agents: Potassium Permanganate

Tests were conducted to evaluate the effect of potassium permanganate as an oxidizing agent in the gold-leaching process using the thiosulfate–glycine system. As shown in Figure 9a, the addition of potassium permanganate at a concentration of 0.004 M improved the gold recovery, achieving 65% recovery after 6 h.

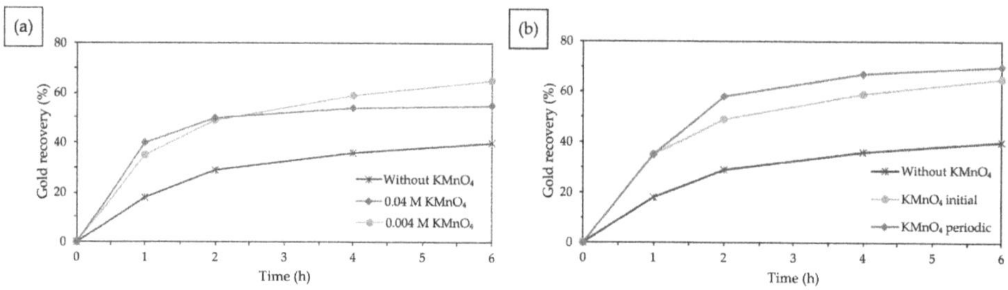

Figure 9. (**a**) Effect of potassium permanganate concentration on gold recovery using the thiosulfate–glycine system: $Na_2S_2O_3$ (0.7 M); $C_2H_5NO_2$ (1 M); copper sulfate (5 mM); 60 °C; pH (9.3); 15% solids for 6 h. (**b**) Effect of periodic addition of potassium permanganate on gold recovery using the thiosulfate–glycine system: $Na_2S_2O_3$ (0.7 M); $C_2H_5NO_2$ (1 M); copper sulfate (5 mM); 60 °C; pH (9.3); 15% solids for 6 h.

Effect of Potassium Permanganate Concentration

Tests with an initial potassium permanganate concentration of 0.004 M resulted in a 65% gold recovery, as shown in Figure 9a. However, when the concentration was increased to 0.04 M, gold recovery decreased to 55%. Additionally, thiosulfate consumption was observed during the first 2 h.

Effect of Periodic Addition of Potassium Permanganate

Figure 9b shows the results of tests in which potassium permanganate at 0.004 M was periodically added at the start, at 2 h, and at 4 h. This approach achieved a 70% gold recovery after 6 h, maintaining a constant sodium thiosulfate concentration throughout the test.

Additionally, to assess the efficiency of glycine alone in dissolving gold, tests were conducted using only glycine at a concentration of 1.75 M, which resulted in a 6% gold dissolution. When potassium permanganate was added to the glycine, the gold recovery increased to 22%, as illustrated in Figure 10.

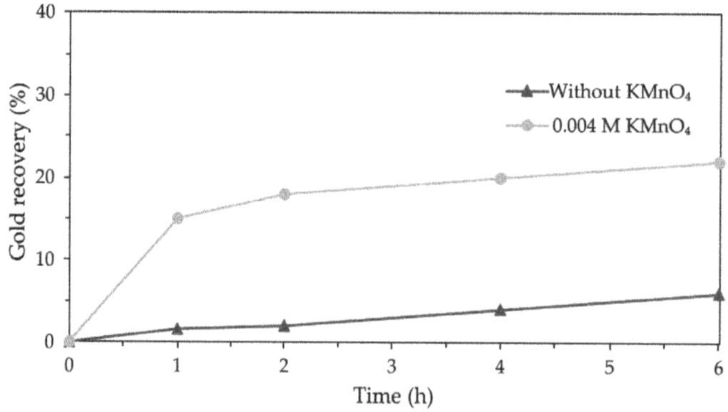

Figure 10. Gold recovery using glycine and potassium permanganate solutions (PP): $C_2H_5NO_2$ (1.75 M); potassium permanganate (PP) (0.004 M); 60 °C; pH (9.3); 15% solids for 6 h.

3.4.6. Synergy of Conditions for Gold Recovery with a Thiosulfate–Glycine–Copper-Permanganate System

Tests conducted under optimal conditions 0.7 M sodium thiosulfate, 1.75 M glycine, 5 mM copper sulfate, 0.004 M potassium permanganate added periodically, pH 9.3, 60 °C, and 750 RPM achieved an 89% gold recovery after 6 h, as shown in Figure 11. In the initial four control tests, gold recovery reached 85% within the first 4 h of leaching.

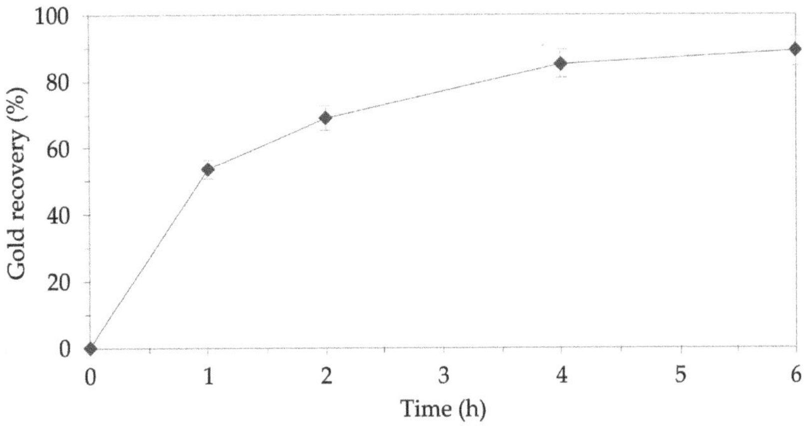

Figure 11. Gold recovery using the thiosulfate–glycine system: $Na_2S_2O_3$ (0.7 M); $C_2H_5NO_2$ (1.75 M); copper sulfate (5 mM); potassium permanganate 0.004 M added every two hours; 60 °C; pH (9.3); 15% solids for 6 h.

4. Discussion

4.1. Characterization of Auriferous Ore

The mineralogical characterization revealed that the ore is composed of 78% aluminosilicates (Table 1), suggesting that the gold is trapped within this matrix, particularly associated with clinochlore and grossular.

The copper content, at 0.008% (Table 2), acts as a catalyst in the dissolution of gold in ammoniacal-thiosulfate-leaching systems [3,5]. These findings highlight the importance of leaching systems that leverage the interaction between copper and thiosulfate to improve gold recovery. Furthermore, the presence of chalcopyrite, albeit at low concentration, can influence the kinetics of gold leaching, interferes with the dissolution process, and reduces the ability to form the thiosulfate gold complex.

4.2. Gold Recovery by Cyanidation Leaching

Cyanidation, although effective, achieved an 89% recovery for this mineral, which differs from the 98% recoveries reported in previous studies [1,2]. This recovery is explained by the mineralogical association of gold with aluminosilicates, specifically clinochlore and grossular, which limits the complete release of the metal. This result will be used as a reference to compare the recovery achieved with other leaching systems.

Additionally, the formation of copper cyanide complexes competes with gold, decreasing the concentration of dissolved sodium cyanide [1,2]. These results demonstrate the need to consider alternative methods that utilize copper as a catalytic agent, such as sodium thiosulfate.

4.3. *Ammoniacal Thiosulfate-Copper Leaching System*

Leaching with ammoniacal thiosulfate and copper resulted in only a 40% gold recovery, compared to the expected 85% under optimal conditions [13]. This low efficiency can be explained by the rapid degradation of thiosulfate into products such as polythionates, reducing its availability to form a stable complex with gold. This phenomenon was observed within the first 2 h of testing, as noted in other studies [38,39].

Furthermore, factors such as cupric ion concentration above 5 mM, pH, temperature, and the mineralogical composition of the ore accelerate the degradation of thiosulfate. The copper in the ore acts as a catalyst in forming the cuprotetramine complex, which impacts thiosulfate stability. These results suggest that, while the ammoniacal thiosulfate–copper system is a viable alternative, optimizing operational conditions is essential to minimize thiosulfate degradation and improve gold recovery.

Evaluation of Leaching Conditions for Gold Recovery Using the Ammoniacal Thiosulfate System

Tests with ammoniacal thiosulfate demonstrated that a lower copper sulfate concentration (2 mM) increases gold recovery from 38% to 48%. In contrast, higher copper concentrations (>5 mM) reduce gold recovery to 33%. This is because concentrations exceeding 5 mM accelerate thiosulfate degradation, limiting its effectiveness as a leaching agent. These findings align with reports by Aylmore and Muir (2001) [38], who observed reduced gold recovery with excessive copper concentrations in the leaching system.

Moreover, increasing the solution pH due to higher ammonia concentration favors the formation of the cupric–tetramine complex, stabilizing cupric ions, as noted by Senanayake (2007) [40] and Xu et al. (2017) [41]. This effect was observed when increasing the pH from 10.5 to 11.5 (Figure 12), where recovery increased from 40% to 50%. However, it is important to highlight that higher ammonia concentrations present environmental risks [42].

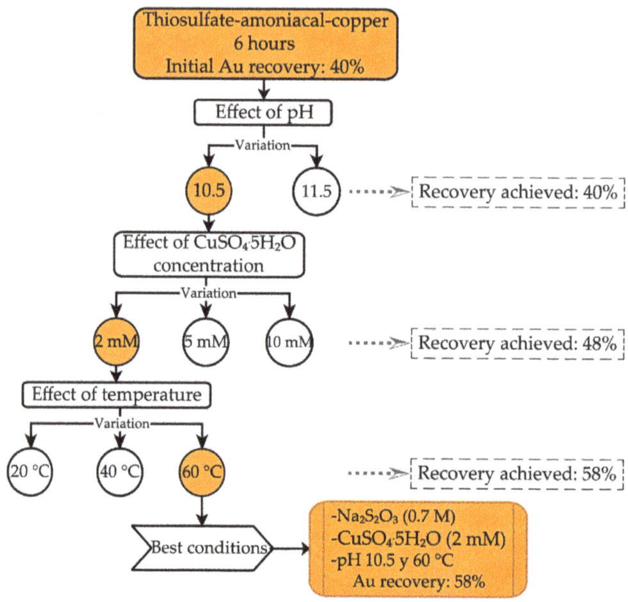

Figure 12. Stepwise optimization of pH, copper sulfate concentration, and temperature for improved gold recovery (40% to 58%).

Finally, tests conducted at 60 °C and 2 mM of copper sulfate showed a maximum recovery of 58% (Figure 12), confirming that temperature has a direct effect on enhancing gold dissolution in the ammoniacal thiosulfate system, consistent with previous studies by Senanayake (2005) [31] and Aylmore and Muir (2001) [38]. These results suggest that temperature (60 °C) is a critical factor for optimizing the leaching process and achieving higher gold recovery.

4.4. Thiosulfate–Glycine–Copper-Leaching System

The baseline test using the thiosulfate–glycine–copper system achieved an initial gold recovery of 38%, which is promising to consider that it was conducted at 60 °C and with a low glycine concentration (0.3 M). Glycine, as mentioned in previous research [24], has the potential to form a stable cupric complex, facilitating gold leaching. However, process efficiency needs improvement.

Variations in the concentration of thiosulfate, glycine, copper sulfate, potassium permanganate, and pH are expected to accelerate gold dissolution (Figure 13). Higher concentrations of these reagents may improve the stability of cupric ions by forming the copper glycinate complex, preventing thiosulfate degradation and increasing gold dissolution, as noted by Barani et al. (2022) [11] and Oraby and Eksteen (2015) [29]. Moreover, the addition of potassium permanganate as an oxidizing agent could accelerate gold leaching in this system.

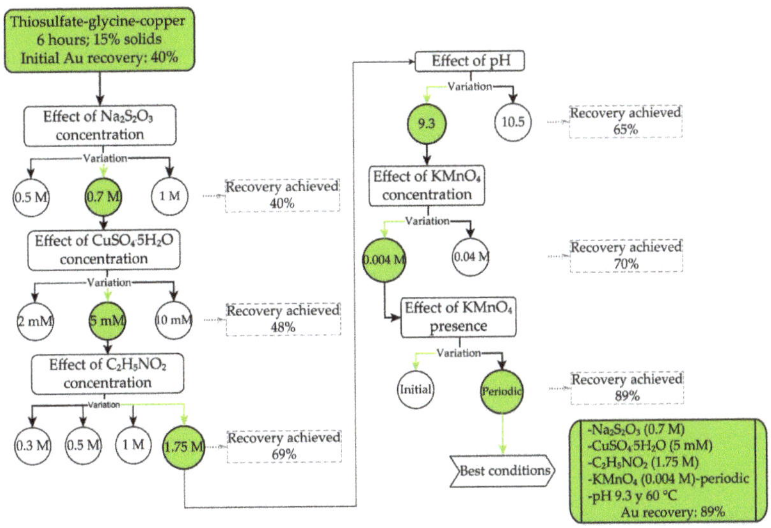

Figure 13. Sequential optimization of thiosulfate, copper sulfate, glycine, pH, potassium permanganate concentration, and dosing frequency to achieve 89% gold recovery.

The temperature for all experiments was maintained at 60 °C, based on the results obtained with the ammoniacal system at this temperature and to prevent glycine degradation at temperatures above 60 °C [43].

4.4.1. Effect of Sodium Thiosulfate Concentration

Increasing the sodium thiosulfate concentration results in improved gold recovery, as expected, since a higher thiosulfate concentration facilitates the formation of the gold–thiosulfate complex, in line with the studies by Gámez et al. (2019) [7]. However, the difference in gold recovery between 0.7 M (41%) and 1 M (49%) is not significant enough to justify the use of a higher reagent concentration.

Additionally, the absence of thiosulfate consumption during tests indicates that glycine forms more stable cupric complexes than the cuprotetramine complex. This is consistent with stability studies presented by Wang (2022) [12].

4.4.2. Effect of Copper Ion Concentration

The concentration of dissolved cupric ions is a key factor in gold recovery in thiosulfate systems. The results show that a copper sulfate concentration of 5 mM is optimal for promoting gold dissolution, increasing efficiency without compromising thiosulfate stability. However, an excess copper sulfate concentration (10 mM) accelerated thiosulfate degradation, reducing its concentration to 0.65 M during the first two hours of the test.

In comparison to the ammoniacal thiosulfate system, which showed significant thiosulfate consumption even at low copper concentrations, the thiosulfate–glycine system demonstrates greater stability. This is attributed to glycine's ability to form more stable complexes, such as the copper glycinate, as also reported in previous studies [1,29,44].

4.4.3. Effect of Glycine Concentration

A gold recovery of 69% obtained with a glycine concentration of 1.75 M demonstrates that glycine plays a key role in improving the gold recovery process. The observed trend, where gold recovery increases proportionally with glycine concentration, can be explained by its dual capability: forming the copper glycinate complex, which catalyzes the formation of the gold–thiosulfate complex, and acting directly as a co-leaching agent, contributing to gold dissolution.

These results are consistent with previous studies suggesting that glycine, by stabilizing cupric ions through the formation of the copper glycinate complex, enhances thiosulfate stability and additionally facilitates gold lixiviation, thereby improving the overall efficiency of the gold recovery process [1,22,29].

4.4.4. Effect of pH on Thiosulfate–Glycine System

Although previous studies indicate that a pH close to 10.5 optimizes gold recovery in sodium-thiosulfate-leaching systems [45,46], this study demonstrated that maintaining a pH of 9.3 is sufficient to achieve similar results (Figure 8). This is because glycine has the ability to stabilize cupric ions over a broader pH range, reducing the need to operate under highly alkaline conditions.

The formation of a 'buffer' at pH 9.4, attributed to the properties of glycine [47–49], explains the increased NaOH consumption observed at glycine concentrations above 1 M when adjusting the pH to values above 10. Nevertheless, for the thiosulfate–glycine–copper system, maintaining a pH of 9.3 is sufficient to achieve 65% gold recovery.

4.4.5. Effect of Oxidizing Agents: Potassium Permanganate

The presence of potassium permanganate accelerates gold dissolution in the thiosulfate–glycine system, consistent with findings by Eksteen and Oraby (2015) [24] and Senanayake (2012) [45], who reported significant improvements in gold recovery when using additional oxidants. Potassium permanganate's oxidizing effect creates a more favorable environment for glycine's role as a co-leaching agent, promoting the formation of stable complexes with copper and gold and enhancing the gold recovery.

Effect of Potassium Permanganate Concentration

The results indicate that increasing the potassium permanganate concentration does not improve gold recovery. In fact, increasing the concentration from 0.004 M to 0.04 M resulted in a decrease in final gold recovery from 65% to 55%. This demonstrates that an excessively oxidizing environment accelerates thiosulfate degradation.

This phenomenon aligns with previous studies indicating that excessive oxidant concentrations can break down thiosulfate into unwanted products, reducing the availability of thiosulfate [24,45]. Thiosulfate consumption observed during the first 2 h of testing with a concentration of 0.04 M confirmed this.

Effect of Periodic Addition of Potassium Permanganate

The results confirmed that periodic addition of potassium permanganate improves gold recovery in the thiosulfate–glycine system. The improvement in gold recovery to 70%, compared to the 65% obtained with a single addition of permanganate at the start of the test, suggests that the effective action time of permanganate as an oxidizing agent is approximately 2 h. This aligns with previous research indicating that a continuous and controlled oxidant concentration prevents premature thiosulfate degradation [38,45,50].

Although glycine with permanganate increases the gold dissolution rate, Figure 10 shows that glycine alone is inefficient (6% recovery), suggesting that its role is more effective in stabilizing copper complexes within the thiosulfate system rather than acting as an independent leaching agent. Even with potassium permanganate, glycine alone does not achieve acceptable recovery (22%), reinforcing that glycine's primary function is to stabilize cupric ions, accelerating gold dissolution, rather than acting as an independent lixiviant.

4.4.6. Synergy of Conditions for Gold Recovery with a Thiosulfate–Glycine–Copper–Permanganate System

The results demonstrate that the synergy between thiosulfate, glycine, copper, and permanganate improves gold dissolution, achieving an 89% recovery in 6 h. The efficiency observed in the tests suggests that this system can recover most of the gold within the first 4 h, these results are consistent with previous studies [27] that propose the combination of thiosulfate and glycine to improve leaching kinetics.

Using glycine to stabilize cupric ions, along with periodic addition of potassium permanganate, is key to accelerating the process without significant thiosulfate degradation. Additionally, the results indicate that reducing leaching time does not affect system efficiency, suggesting that this combination of reagents is both effective and economical in terms of time and reagent consumption [27].

If the leaching system is to be scaled up to an industrial level, the primary challenges include maintaining precise temperature control and protecting the solution from sunlight during the process. Tests with permanganate revealed the photosensitivity of the solution, leading to the generation of byproducts that significantly reduced gold recovery. To mitigate this issue, aluminum covers were required to shield the containers during agitation.

4.4.7. Mechanism of Gold Dissolution in the Thiosulfate–Glycine–Copper–Permanganate System

As shown in Figure 14, the mechanism of gold dissolution in the thiosulfate–glycine–copper–permanganate system relies on their synergistic interaction in a moderately alkaline environment (pH 9.3) and at a temperature of 60 °C. The general reaction involved in gold dissolution is presented in Equation (1). Based on experimental assays and analyzed thermodynamic diagrams [27], for the dissolution mechanism, the presence of two electrochemical areas between the gold surface and the leaching solution was assumed: the cathodic area and the anodic area.

$$Au + 2(S_2O_3)^{2-} + KMnO_4 + Cu(C_2H_2NO_2)^{3+}H_2O \rightarrow Au(S_2O_3)_2{}^{3-} + Cu(C_2H_2NO_2)_3 + MnO_2 + O_2 + OH^- \quad (1)$$

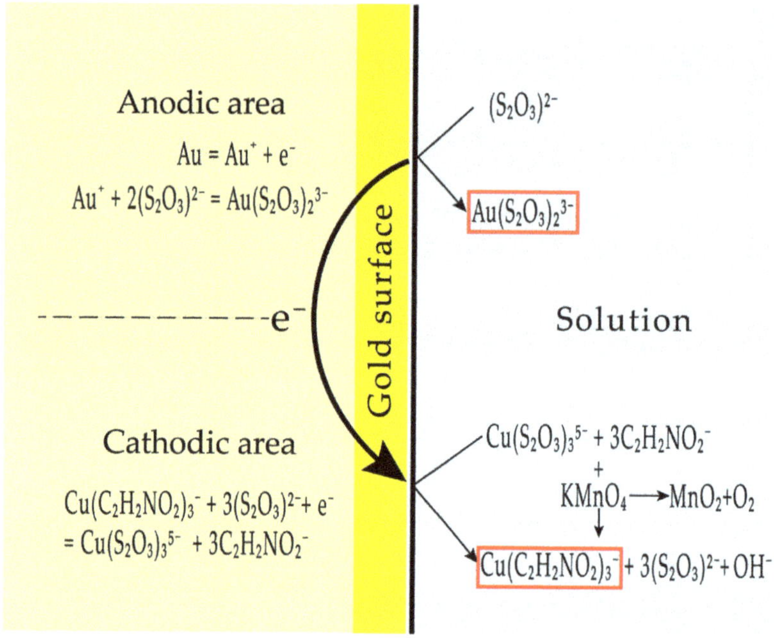

Figure 14. Proposed electrochemical mechanism of gold dissolution in the thiosulfate–glycine–copper–permanganate system. The central arrow represents the electron exchange between the anodic and cathodic areas, while the highlighted boxes indicate the key chemical species, cupric glycinate and gold thiosulfate, targeted for formation within the system.

In the cathodic area, glycine plays a dual role. First, it forms the cupric glycinate complex ($[Cu(C_2H_4NO_2)_2]^0$), which remains stable due to the high affinity between glycine and copper [27]. Through a REDOX reaction, this cupric glycinate complex interacts with dissolved thiosulfate ions to form the $Cu(S_2O_3)_3{}^{5-}$ complex via electron transfer, reducing copper (II) to copper (I). The continuous regeneration of the cupric glycinate complex is facilitated by the oxidizing environment created by dissolved oxygen and potassium permanganate. This ensures the constant oxidation of Cu(I) back to Cu(II) [27]. This process not only stabilizes the cupric ions in the glycinate complex but also prevents the accumulation of unstable species that could degrade thiosulfate into polythionates. Additionally, free glycine produced during the REDOX regeneration process can form a gold–glycinate complex.

In the anodic area, cupric ions stabilized as cupric glycinate facilitate the direct action of thiosulfate on metallic gold, resulting in the formation of the gold–thiosulfate complex $Au(S_2O_3)_2{}^{3-}$. The stability of the cupric glycinate in the medium prevents thiosulfate decomposition, as confirmed experimentally by the absence of reagent consumption. Furthermore, although the formation of a gold–glycinate complex is possible, this complex is thermodynamically less stable than the gold–thiosulfate complex [22,27]. Therefore, in the presence of thiosulfate, gold initially associated with glycine transitions to the gold–thiosulfate complex $Au(S_2O_3)_2{}^{3-}$, regenerating glycine for further complexation in the cathodic area during the REDOX cycle.

Finally, experimental assays demonstrated that in the presence of glycine and permanganate, no thiosulfate consumption was recorded. This supports the conclusion that in an oxidizing environment, glycine's primary role is to stabilize copper in its Cu(II) state (Figure 14). This stabilization prevents the formation of unstable cuprous complexes that accelerate thiosulfate degradation. Although glycine has the capacity to form complexes with gold, experimental results show that its primary interaction is with copper. Glycine stabilizes copper in its cupric state, maintaining an optimal environment for thiosulfate to dissolve gold (Figure 14). This reaffirms that glycine's primary role is associated with copper rather than gold.

5. Conclusions

The auriferous ore from the mining area in Ecuador exhibited a head grade of 10 (g t^{-1}) of gold and 0.008% Cu. The main mineralogy obtained through the XRD analysis revealed aluminosilicates, such as grossular (64%) and clinochlore (12%), to which the gold is associated. Additionally, the chemical analysis performed via XRF indicated a composition of 18.8% Si, 15.1% Ca, 5% Fe, and 4.5% Al, with trace amounts of manganese (1%), sulfur (0.1%), and titanium (0.2%).

Baseline cyanidation tests achieved a gold recovery of 89.8% at room temperature in 24 h, highlighting the process's efficiency. In contrast, ammoniacal thiosulfate leaching reached a maximum recovery of 53.1% under optimal conditions (0.7 M thiosulfate, 2 mM copper sulfate, pH 10.5, and 60 °C). However, the degradation of thiosulfate accelerated with copper sulfate concentrations above 2 mM due to the increased availability of cupric ions generated by copper dissolution from the ore in the presence of ammonia.

The thiosulfate–glycine system exhibited an initial recovery of 40%, and it was determined that a sodium thiosulfate concentration of 0.7 M is sufficient, as higher concentrations do not significantly improve recovery. Regarding copper requirements, the glycine system showed greater stability compared to the ammoniacal system, achieving a gold recovery of 47% with 5 mM copper sulfate and no significant consumption of sodium thiosulfate. This can be attributed to the ability of glycine to form more stable cupric complexes compared to cupric–tetramine, although its capacity to dissolve copper from the ore is lower than that of ammonia. Copper concentrations above 10 mM were counterproductive, yielding the same recoveries as at 5 mM and accelerating the degradation and consumption of thiosulfate during the first four hours of agitation.

The glycine concentration also proved crucial for stabilizing dissolved cupric ions, preventing thiosulfate degradation, and acting as a co-leaching agent for gold. A glycine concentration of 1.75 M resulted in the highest recovery of 60%. Additionally, the system maintained stable recoveries of approximately 65% at both pH 10.5 and pH 9.3, due to glycine's ability to form stable cupric complexes over a broader pH range compared to ammonia. This behavior minimizes the need to adjust the pH to values above 10, given the buffering effect observed at pH 9.4.

The use of potassium permanganate as an oxidizing agent at a concentration of 0.004 M accelerated the co-leaching action of glycine (1 M), improving gold recovery to 65% in 6 h. However, higher concentrations (0.04 M) decreased recovery to 55%, as an excessively oxidizing environment, while initially enhancing gold dissolution, also accelerated thiosulfate degradation, resulting in reagent consumption after the second hour of agitation. The periodic addition of potassium permanganate (0.004 M) every two hours maintained a lower and more stable oxidizing environment, enabling a gold recovery of 70% in 6 h. In contrast, tests conducted solely with glycine (1.75 M) and periodically added permanganate (0.004 M) achieved a limited recovery of 22%, underscoring the importance of the synergy between thiosulfate, glycine, copper, and permanganate.

Finally, the optimized thiosulfate–glycine–copper–permanganate system achieved a gold recovery of 89.3% under the following conditions: 0.7 M sodium thiosulfate, 1.75 M glycine, 5 mM copper sulfate, 0.004 M potassium permanganate added periodically every 2 h, pH 9.3, and 60 °C. Although the total leaching time was 6 h, 85% of the gold was recovered within the first 4 h, demonstrating the efficiency of the system. This performance is attributed to glycine's ability to stabilize cupric ions, preventing thiosulfate degradation, and its role as a co-leaching agent in an oxidizing environment controlled by potassium permanganate.

Author Contributions: Conceptualization, E.d.l.T.; methodology, E.d.l.T. and A.S.R.; validation, A.S.R. and E.d.l.T.; formal analysis, A.S.R.; investigation, A.S.R.; resources, A.S.R.; data curation, A.S.R.; writing—original draft preparation, A.S.R. and C.F.A.-T.; writing—review and editing, A.S.R. and C.F.A.-T.; visualization, A.S.R.; supervision, E.d.l.T.; project administration, E.d.l.T.; funding acquisition, E.d.l.T. All authors have read and agreed to the published version of the manuscript.

Funding: The research presented in this study was made possible by the financing of the Department of Extractive Metallurgy (DEMEX) of the Escuela Politécnica Nacional thanks to the research project PIGR-2208. Tel.: +593-2-297-6300 (ext. 5806).

Data Availability Statement: The original contributions presented in this study are included in the article. Further inquiries can be directed to the corresponding author.

Conflicts of Interest: The authors declare no conflicts of interest.

References

1. Altinkaya, P.; Wang, Z.; Korolev, I.; Hamuyuni, J.; Haapalainen, M.; Kolehmainen, E.; Yliniemi, K.; Lundström, M. Leaching and recovery of gold from ore in cyanide-free glycine media. *Miner. Eng.* **2020**, *158*, 106610. [CrossRef]
2. Barani, K.; Dehghani, M.; Azadi, M.R.; Karrech, A. Leaching of a polymetal gold ore and reducing cyanide consumption using cyanide-glycine solutions. *Miner. Eng.* **2021**, *163*, 106802. [CrossRef]
3. Wang, Q.; Hu, X.; Zi, F.; Yang, P.; Chen, Y.; Chen, S. Environmentally friendly extraction of gold from refractory concentrate using a copper—Ethylenediamine—Thiosulfate solution. *J. Clean. Prod.* **2019**, *214*, 860–872. [CrossRef]
4. Wu, H.; Feng, Y.; Huang, W.; Li, H.; Liao, S. The role of glycine in the ammonium thiocyanate leaching of gold. *Hydrometallurgy* **2019**, *185*, 111–116. [CrossRef]
5. Xie, F.; Chen, J.; Wang, J.; Wang, W. Review of gold leaching in thiosulfate-based solutions. *Trans. Nonferr. Met. Soc. China* **2021**, *31*, 3506–3529. [CrossRef]
6. Ospina-Correa, J.D.; Mejía-Restrepo, E.; Serna-Zuluaga, C.M.; Posada-Montoya, A.; Osorio-Cachaya, J.G.; Ta-mayo-Sepúlveda, J.A.; Calderón-Gutiérrez, J.A. Process mineralogy of refractory gold ore in thiosulfate solutions. *Hydrometallurgy* **2018**, *182*, 104–113. [CrossRef]
7. Gámez, S.; Garcés, K.; de la Torre, E.; Guevara, A. Precious metals recovery from waste printed circuit boards using thiosulfate leaching and ion exchange resin. *Hydrometallurgy* **2019**, *186*, 1–11. [CrossRef]
8. Wu, J.; Ahn, J.; Lee, J. Characterization of gold deportment and thiosulfate extraction for a copper gold concentrate treated by pressure oxidation. *Hydrometallurgy* **2022**, *207*, 105771. [CrossRef]
9. Celep, O.; Altinkaya, P.; Yazici, E.Y.; Deveci, H. Thiosulphate leaching of silver from an arsenical refractory ore. *Miner. Eng.* **2018**, *122*, 285–295. [CrossRef]
10. Sitando, O.; Dai, X.; Senanayake, G.; Nikoloski, A.N.; Breuer, P. A fundamental study of gold leaching in a thiosulfate oxygen copper system in the presence of activated carbon. *Hydrometallurgy* **2020**, *192*, 105232. [CrossRef]
11. Barani, K.; Kogani, Y.; Nazarian, F. Leaching of complex gold ore using a cyanide-glycine solution. *Miner. Eng.* **2022**, *180*, 107475. [CrossRef]
12. Wang, J.; Xie, F.; Pan, Y.; Wang, W. Leaching of Gold with Copper-citrate-thiosulfate Solutions. *Miner. Process. Extr. Metall. Rev.* **2022**, *43*, 916–925. [CrossRef]
13. Gámez, S.A.; De la Torre, E. Recuperación de oro a partir de minerales polisulfurados con soluciones amoniacales de tiosulfato de sodio. *Rev. Politéc.* **2015**, *36*, 42.
14. Escobar-Ledesma, F.R.; Aragón-Tobar, C.F.; Espinoza-Montero, P.J.; De La Torre-Chauvin, E. In-creased recovery of gold thiosulfate alkaline solutions by adding thiol groups in the porous structure of activated carbon. *Molecules* **2020**, *25*, 2902. [CrossRef]
15. Olvera, O.G.; Domanski, D.F.R. Effect of activated carbon on the thiosulfate leaching of gold. *Hydrometallurgy* **2019**, *188*, 47–53. [CrossRef]
16. Sitando, O.; Senanayake, G.; Dai, X.; Breuer, P. The adsorption of gold(I) on minerals and activated carbon (preg-robbing) in non-ammoniacal thiosulfate solutions—Effect of calcium thiosulfate, silver(I), copper(I) and polythionate ions. *Hydrometallurgy* **2019**, *184*, 206–217. [CrossRef]

17. Chen, Y.; Zi, F.; Hu, X.; Yang, P.; Ma, Y.; Cheng, H.; Wang, Q.; Qin, X.; Liu, Y.; Chen, S.; et al. The use of new modified activated carbon in thiosulfate solution: A green gold recovery technology. *Sep. Purif. Technol.* **2020**, *230*, 115834. [CrossRef]
18. He, P.; Hu, X.; Zi, F.; Wang, C. A Method of Recovery Au(I) from Thiosulfate Solution by Novel Functionalised Activated Carbon. *Russ. J. Non-Ferr. Met.* **2022**, *63*, 46–56. [CrossRef]
19. Yu, H.; Zi, F.; Hu, X.; Nie, Y.; Chen, Y.; Cheng, H. Adsorption of gold from thiosulfate solutions with chemically modified activated carbon. *Adsorpt. Sci. Technol.* **2017**, *36*, 408–428. [CrossRef]
20. Chen, Y.; Zi, F.; Hu, X.; Yu, H.; Nie, Y.; Yang, P.; Cheng, H.; Wang, Q.; Qin, X.; Chen, S.; et al. Grafting of organic sulfur-containing functional groups on activated carbon for gold(I) adsorption from thiosulfate solution. *Hydrometallurgy* **2019**, *185*, 102–110. [CrossRef]
21. Muir, D.M.; Aylmore, M.G. Thiosulphate as an Alternative to Cyanide for Gold Processing—Issues and Impediments. In Proceedings of the Green Processing 2002—Proceedings: International Conference on the Sustainable Proceeing of Minerals, Cairns, QLD, Australia, 29–31 May 2002; pp. 125–133. Available online: https://www.scopus.com/record/display.uri?eid=2-s2.0-2342642835&origin=resultslist&sort=plf-f&src=s&sot=b&sdt=b&s=TITLE-ABS-KEY(Thiosulphate+as+an+Alternative+to+Cyanide+for+Gold+Processing+-+Issues+and+Impediments)&sessionSearchId=ec33dcdea3835817912a4605780cdc08&relpos=1 (accessed on 1 October 2024).
22. Oraby, E.A.; Eksteen, J.J.; Tanda, B.C. Gold and copper leaching from gold-copper ores and concentrates using a synergistic lixiviant mixture of glycine and cyanide. *Hydrometallurgy* **2017**, *169*, 339–345. [CrossRef]
23. Tauetsile, P.J.; Oraby, E.A.; Eksteen, J.J. Activated carbon adsorption of gold from cyanide-starved glycine solutions containing copper. Part 1: Isotherms. *Sep. Purif. Technol.* **2019**, *211*, 594–601. [CrossRef]
24. Eksteen, J.J.; Oraby, E.A. The leaching and adsorption of gold using low concentration amino acids and hydrogen peroxide: Effect of catalytic ions, sulphide minerals and amino acid type. *Miner. Eng.* **2015**, *70*, 36–42. [CrossRef]
25. Li, H.; Deng, Z.; Oraby, E.; Eksteen, J. Amino acids as lixiviants for metals extraction from natural and secondary resources with emphasis on glycine: A literature review. *Hydrometallurgy* **2023**, *216*, 106008. [CrossRef]
26. Feng, D.; van Deventer, J.S.J. The role of amino acids in the thiosulphate leaching of gold. *Miner. Eng.* **2011**, *24*, 1022–1024. [CrossRef]
27. Wang, J.; Wang, R.; Pan, Y.; Liu, F.; Xu, Z. Thermodynamic analysis of gold leaching by copper-glycine-thiosulfate solutions using Eh-pH and species distribution diagrams. *Miner. Eng.* **2022**, *179*, 107438. [CrossRef]
28. Li, J.; Kou, J.; Sun, C.; Zhang, N.; Zhang, H. A review of environmentally friendly gold lixiviants: Fundamentals, applications, and commonalities. *Miner. Eng.* **2023**, *197*, 108074. [CrossRef]
29. Oraby, E.A.; Eksteen, J.J. The leaching of gold, silver and their alloys in alkaline glycine–peroxide solutions and their adsorption on carbon. *Hydrometallurgy* **2015**, *152*, 199–203. [CrossRef]
30. Feng, D.; van Deventer, J.S.J. Thiosulphate leaching of gold in the presence of carboxymethyl cellulose (CMC). *Miner. Eng.* **2011**, *24*, 115–121. [CrossRef]
31. Senanayake, G. Gold leaching by thiosulphate solutions: A critical review on copper(II)-thiosulphate-oxygen interactions. *Miner. Eng.* **2005**, *18*, 995–1009. [CrossRef]
32. Zhang, X.M.; Senanayake, G. A Review of Ammoniacal Thiosulfate Leaching of Gold: An Update Useful for Further Research in Non-cyanide Gold Lixiviants. *Miner. Process. Extr. Metall. Rev.* **2016**, *37*, 385–411. [CrossRef]
33. Oraby, E.A.; Eksteen, J.J.; O'Connor, G.M. Gold leaching from oxide ores in alkaline glycine solutions in the presence of permanganate. *Hydrometallurgy* **2020**, *198*, 105527. [CrossRef]
34. Zhang, Y.; Li, Q.; Liu, X.; Jiang, T. A thermodynamic analysis on thiosulfate leaching of gold under the catalysis of Fe^{3+}/Fe^{2+} complexes. *Miner. Eng.* **2022**, *180*, 107511. [CrossRef]
35. Oraby, E.A.; Eksteen, J.J. The selective leaching of copper from a gold–copper concentrate in glycine solutions. *Hydrometallurgy* **2014**, *150*, 14–19. [CrossRef]
36. Zhang, Y.; Cui, M.; Wang, J.; Liu, X.; Lyu, X. A review of gold extraction using alternatives to cyanide: Focus on current status and future prospects of the novel eco-friendly synthetic gold lixiviants. *Miner. Eng.* **2022**, *176*, 107336. [CrossRef]
37. Liedberg, B.; Lundström, I.; Wu, C.R.; Salaneck, W.R. Adsorption of glycine on hydrophilic gold. *J. Colloid Interface Sci.* **1985**, *108*, 123–132. [CrossRef]
38. Aylmore, M.G.; Muir, D.M. Thiosulfate leaching of gold—A review. *Miner. Eng.* **2001**, *14*, 135–174. [CrossRef]
39. Grosse, A.; Dicinoski, G.; Shaw, M.; Haddad, P. Leaching and recovery of gold using ammoniacal thiosulfate leach liquors (a review). *Hydrometallurgy* **2003**, *69*, 1–21. [CrossRef]
40. Senanayake, G. Review of rate constants for thiosulphate leaching of gold from ores, concentrates and flat surfaces: Effect of host minerals and pH. *Miner. Eng.* **2007**, *20*, 1–15. [CrossRef]
41. Xu, B.; Kong, W.; Li, Q.; Yang, Y.; Jiang, T.; Liu, X. A review of thiosulfate leaching of gold: Focus on thiosulfate consumption and gold recovery from pregnant solution. *Metals* **2017**, *7*, 222. [CrossRef]

42. Xing, H.; Peng, M.; Li, Z.; Chen, J.; Zhang, H.; Teng, X. Ammonia inhalation-mediated mir-202-5p leads to cardiac autophagy through PTEN/AKT/mTOR pathway. *Chemosphere* **2019**, *235*, 858–866. [CrossRef] [PubMed]
43. Khezri, M.; Rezai, B.; Akbar Abdollahzadeh, A.; Molaeinasab, M.; Wilson, B.P.; Lundström, M. Glycine leaching of Sarcheshmeh chalcopyrite concentrate at high pulp densities in a stirred tank reactor. *Miner. Eng.* **2020**, *157*, 106555. [CrossRef]
44. Liu, Z.; Guo, X.; Tian, Q.; Zhang, L. A systematic review of gold extraction: Fundamentals, advancements, and challenges toward alternative lixiviants. *J. Hazard. Mater.* **2022**, *440*, 129778. [CrossRef]
45. Senanayake, G. Gold leaching by copper(II) in ammoniacal thiosulphate solutions in the presence of additives. Part I: A review of the effect of hard-soft and Lewis acid-base properties and interactions of ions. *Hydrometallurgy* **2012**, *115–116*, 1–20. [CrossRef]
46. Sun, C.; Zhang, X.; Kou, J.; Xing, Y. A review of gold extraction using noncyanide lixiviants: Fundamentals, advancements, and challenges toward alkaline sulfur-containing leaching agents. *Int. J. Miner. Metall. Mater.* **2020**, *27*, 417–431. [CrossRef]
47. Ferreira, C.M.H.; Pinto, I.S.S.; Soares, E.V.; Soares, H.M.V.M. (Un) suitability of the use of pH buffers in biological, biochemical and environmental studies and their interaction with metal ions—A review. *RSC Adv.* **2015**, *5*, 30989–31003. [CrossRef]
48. Knight, J.; Assimos, D.G.; Callahan, M.F.; Holmes, R.P. Metabolism of primed, constant infusions of [1,2-13C2] glycine and [1-13C1] phenylalanine to urinary oxalate. *Metabolism* **2011**, *60*, 950–956. [CrossRef]
49. Yoo, K.; Kim, H. Development of Ammoniacal leaching processes; a review. *Resour. Recycl.* **2012**, *21*, 3–17. [CrossRef]
50. Sitando, O.; Senanayake, G.; Dai, X.; Nikoloski, A.N.; Breuer, P. A review of factors affecting gold leaching in non-ammoniacal thiosulfate solutions including degradation and in-situ generation of thiosulfate. *Hydrometallurgy* **2018**, *178*, 151–175. [CrossRef]

metals

MDPI

Communication

Innovative Process for Strategic Metal Recovery from Electric Arc Furnace Slag by Alkaline Leaching

Nour-Eddine Menad *, Alain Seron and Sara Bensamdi †

BRGM—Bureau de Géologie et Minière, 45060 Orléans, France; a.seron@brgm.fr (A.S.); sara.bensamdi@etfr.eurofins.com (S.B.)

* Correspondence: n.menad@brgm.fr

† Current address: Eurofins, 20 rue du Kochersberg, 67700 Saverne, France.

Abstract: Currently, Electric Arc Furnace Slag (EAFS) is undervalued and is therefore only used in road construction, while blast furnace slag (BFS) is used as an interesting alternative in construction materials to replace natural aggregates in the manufacture of concrete. Steel slag (SS) represents a promising secondary resource due to its high content of critical metals, such as chromium (Cr) and vanadium (V). These metals are essential for various strategic industries, making it crucial to consider slag as a resource rather than waste. However, the primary challenge lies in selectively recovering these valuable metals. In this work, we explore the development of a hydrometallurgical process aimed at efficiently extracting Cr and V from Electric Arc Furnace Slag (EAFS). The characterization of the investigated EAFS shows that the main crystalline phases contained in this heterogeneous material are srebrodolskite, larnite, hematite, and spinel such as probably magnesio-chromite. The targeted metals seem to be dispersed in various mineral species contained in the SS. An innovative hydrometallurgical method has been explored, involving physical preparation by co-grinding slag with alkaline reagents followed by treatment in a microwave furnace to modify the metal-bearing species to facilitate metal processing dissolution. The results obtained showed that the leaching rates of Cr and V were, respectively, 100% and 65% after 15 min of treatment in the microwave furnace, while, after 2 h of conventional heat treatment, as explored in a previous study, 98% and 63% of the Cr and V were, respectively, leached.

Keywords: slags; strategic metals; recovery; microwaves; alkaline reagents; leaching

1. Introduction

The generic term steel slag (SS) encompasses several materials with different physico-chemical characteristics. These are artificial mineral matrices, classified as waste produced in very large quantities (several million tons per year in France) simultaneously with cast iron or steel in liquid form at temperatures close to 1500 °C by the steel industry. They are particularly rich in strategic metals [1]. These materials are nevertheless mostly stored in a "slag heap" in the steel industry and constitute a potential impact on the environment and public health. Long considered waste from steel manufacturing, iron and steel slag are today recognized as truly precious and valuable materials. Historically, steelmakers disposed of these materials by storing them in landfills near production sites. The progressive awareness of the intrinsic quality of these materials (and therefore of their market values) has allowed for the emergence of organized value chains since 1930 [2]. Contrarily to their previous status as waste, slag is basically valorized as by-products of steel and cast iron manufacturing and used in road or construction applications, for example. The fields of application are, today, diverse and varied. Blast slag can be safely used along with clinker and gypsum for the production of cement, in the preparation of materials such as ceramic glass, silica gel, ceramic tiles, bricks, etc. LD slag can be used in soil conditioners, fertilizers, the recovery of metal values, etc. EAF oxidizing slag is used as a concrete aggregate. Steel slag is also used in road and hydraulic construction [3–5].

Because of the scarcity of raw materials and increasingly drastic environmental laws, industries have been forced to create more efficient and sustainable models of socio-economic development. The recovery of certain wastes and co-products such as steel slag has become a solution to limit the storage of these materials. Each slag is considered a unique product because it comes from the manufacture of a specific type of steel [6,7]. Currently, Electric Arc Furnace Slag (EAFS) is poorly valorized and is mainly used in road construction, while blast furnace slag (BFS) could be used as an interesting alternative in building materials to replace natural aggregates in concrete manufacturing, for example [8]. Slags could be a very good secondary resource, as they contain large amounts of different metals such as chromium, vanadium etc. Therefore, they should be considered as a resource, not as waste. The challenge is to be able to selectively recover the strategic metals (Cr and V) they contain. There is a wide variety of hydrometallurgical processes described in the literature depending on the nature of the metal to be extracted, the composition of the gangue, or even the extraction context chosen according to the physicochemical conditions of the environment (temperature, pressure, pH, etc.). X. Wang et al. [9] developed a process to separate and recover vanadium and chromium from V-Cr-bearing reducing slag. The investigated process included four procedures: low-temperature sulfate roasting, reduction leaching, selective oxidation, and preparation, by, respectively, calcination or precipitation of vanadium and chromium oxides. They were able to obtain Cr and V leaching efficiencies reaching 99.7% and 99.6%, respectively, by sulphating roasting at low temperature and leaching reduction. The products of V_2O_5 with purity of 99.1% and Cr_2O_3 with purity of 98.9% were obtained by treating ammonium poly-vanadate and Cr^{III} precipitate, respectively.

The steel industries are working to develop new alternatives to valorize these slags in order to achieve a double objective: the preservation of natural virgin material resources and a reduction in slag treatment and storage costs. In this work, the development of an EAFS hydrometallurgical process to recover strategic metals by microwave-assisted leaching is considered. This method allows for the rapid heating of slag and alkaline reagent mixtures to improve the kinetics of the extraction of the valuable metals contained in them and reduce the environmental impact [10,11]. Moreover, such a process is known to allow for the formation of highly reactive entities named radicals, which induce high reaction kinetics. This process is established in order to study the operating parameters and conditions favoring thermal treatment as well as the leaching of metals. Among all the contained elements, this study was specifically focused on chromium and vanadium due to their respective contents in the targeted slag and the increasingly strong industrial demand inducing an increase in their value.

Kokko M. et al. [12] developed an innovative two-stage direct leaching process for selectively recovering calcium (Ca) and vanadium (V) from Electric Arc Furnace (EAF) Slag, which is obtained after furnace hydrogen direct reduction of iron process (HRI). The process is designed to optimize the recovery of these valuable elements through selective leaching in different stages using specific chemical solutions. For the first stage, the leaching of calcium (Ca) and magnesium (Mg), they used a mixture of nitric acid, ammonium nitrate, and water (HNO_3-NH_4NO_3-H_2O) to leach calcium and magnesium selectively. This solution works efficiently to extract these alkaline earth metals, as they are more readily soluble in the acidic environment provided by nitric acid, while other elements in the slag remain mostly unaffected. For the second stage: the leaching of vanadium (V), two different solutions were used to extract vanadium, as follows.

1. **Sodium carbonate and sodium hydroxide in water (Na_2CO_3-NaOH-H_2O)**: This solution creates an alkaline medium that facilitates the selective dissolution of vanadium, particularly vanadium in its higher oxidation states, as vanadium compounds tend to form soluble complexes in alkaline conditions.
2. **Ammonium carbonate in water ($(NH_4)_2CO_3$-H_2O)**: Ammonium carbonate also creates an alkaline environment but offers a slightly different chemical interaction, which

can further aid in vanadium dissolution, depending on the slag's composition and the specific vanadium phases present.

This two-stage process effectively separates calcium and magnesium from vanadium, minimizing cross-contamination between the elements and enhancing the purity of the recovered materials. This approach represents an efficient method for the valorization of EAF slag, reducing waste and recovering valuable resources, particularly in processes where furnace hydrogen reduction of iron is utilized.

Mantang Ding [13] proposed that submerged arc furnace carbothermic reduction method, sub-molten salt method, vanadium alloy production, acid roasting–leaching calcification and two-part acidification to extract vanadium are areas for future development.

2. Materials and Methods

A representative sample of Electric Arc Furnace Slag (EAFS) was provided by a French steel company. Due to its high content of metals, this slag cannot be used in road construction or the cement industry without pretreatment. This slag was generated using the EAF process and sampled after extracting the iron scrap by low-intensity dry magnetic separation. First, the representative slag sample was characterized to adapt its preparation pathway before the implementation of chemical modification of metal-bearing species and the leaching process. Since the EAFS sample is heterogeneous, an appropriate sampling method is crucial to ensure that the samples used for this study accurately represent the entire slag batch. Heterogeneity in slag can arise from variations in the particle size, metal content, and mineral composition, which can affect data consistency if not properly accounted for. The protocol consists of size reduction using crushing, grinding, and then sieving steps to obtain a particle size less than 125 μm, while minimizing the overproduction of fine particles, to obtain a homogenous and representative powder. After the grinding and sampling steps, fine characterization was carried out on a sample with a smaller than 125 μm particle size using conventional analytical techniques. The results of the chemical analysis were obtained by using Atomic Absorption Spectroscopy (AAS) and Inductively Coupled Plasma Atomic Emission Spectroscopy (ICP-AES) (Agilent 7800, Agilent Technologies, Santa Clara, CA, USA). X-ray Diffraction (XRD) measurements were performed at room temperature using a BRUKER D8 Advance $\theta/2\theta$ diffractometer equipped (BRUKER, Billerica, MA, USA) with a Lynxeye energy dispersive one-dimensional detector (Cu-Kα radiation, 40 kV and 40 mA; 2θ from 10° to 80° with a step of 0.02° and counting time of 0.5 s per step) This technique allows for crystalline phase identification, completed through the determination of the chemical element distribution through Scanning Electron Microscopy and Energy Dispersion Spectroscopy (FE-SEM, Tescan MIRA3, Brno, Czech Republic) (SEM-EDS) mapping, performed on a polished representative sample.

The representative samples of EAFS were then mixed and co-grinded for 15 min with different alkaline reagents by taking a slag-versus-reagent-weight ratio equal to 1. Regarding the preliminary chemical modification of the EAFS sample, various parameters have been optimized, including the nature of the alkaline reagents and the volume of water added (1 mL) before heating in a microwave furnace, to improve the reactivity of the EAFS with the alkaline reagents. Different alkaline reagents with 99.8% purity and purchased from Sigma Aldrich (St. Louis, MO, USA), have been used, including KOH; K_2CO_3; KOH + K_2CO_3; NaOH; Na_2CO_3; and NaOH + Na_2CO_3.

It is important to underline that preliminary experiments have shown that without conventional thermal or microwave treatment, co-grinding EAF/reagent mixtures did not allow any chemical reactivity likely to induce the formation of water leachable species containing metals (Cr, V) from the slag. Therefore, it is essential to implement either a thermal or a microwave pretreatment favoring the modification of the speciation of the targeted metals to obtain phases soluble in an aqueous medium. Some experiments carried out prior to this study involved the implementation of conventional thermal pretreatments [11]. The results obtained were conclusive regarding the leaching yields, but the treatment durations required to achieve the desired objective were too high. It therefore seemed necessary to pretreat the slag/alkaline reagent mixtures under microwave radiation in order to reduce the treatment duration and consequently the expended energy. The co-grinded slag mixture and alkaline reagent was treated in microwave furnace according to the following method.

Heat Treatment and Activation Under Microwave Radiation

The tests were performed in a microwave furnace made by SAIREM company (Décines-Charpieu, France), allowing power modulation from 0.6 to 6 KWh. The sample temperature measurements in the microwave furnace were carried out using a portable infrared thermometer (AMPROBE, Everett, WA, USA, IR-750-EUR). As previously performed in the conventional heat treatment method [11], 2 g of the co-grinded mixture was introduced into an alumina crucible. Considering such treatment, quantities of activation energy depending on the microwave power and residence times have been studied. These treatments may induce the formation of species (chromates and vanadates), which are easily leachable in water. An important part of this work aims at optimizing the parameters and the operating conditions necessary for the recovery of metals (V and Cr) from the slag. The modified mixtures were then introduced in water for metal leaching at an S/L value of 40 g/L. After solid/liquid separation, the obtained filtrates were analyzed by AAS to allow for the calculation of the leaching rate and, consequently, to determine the effectiveness and potential selectiveness of each pathway treatment.

The experimental protocol includes two steps (Figure 1).

(a) Physical preparation of materials including crushing, magnetic separation, grinding, and, finally, co-grinding involving slag and alkaline reagents.
(b) Modification of metal speciation by microwave treatment and then leaching.

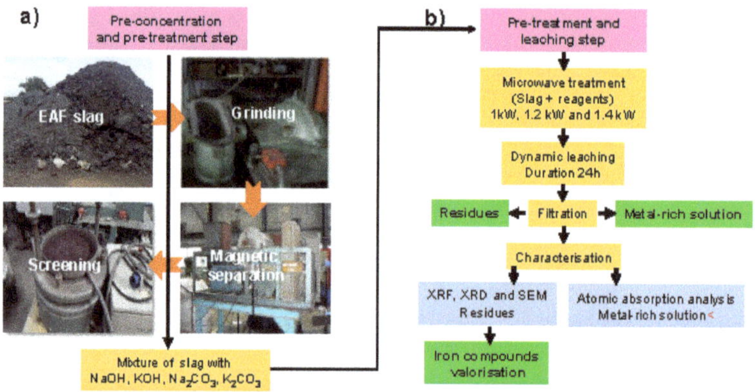

Figure 1. Experimental protocol for the recovery of Cr and V from EAFS. (**a**) physical treatment and (**b**) modification metal speciation.

3. Results

3.1. *Physicochemical Characterization*

An in-depth characterization of the microstructure of a few grains of the iron scrap-free slag by SEM clearly showed the great morphological variability of the grains, which should be correlated with their chemical heterogeneity [1]. The chemical analysis results of the investigated EAFS sample show that this slag contains 48.5% Fe, 29.5% Ca, 5.7% Si, 5.7% Mn, 5.2% Al, and 4.3% Mg, expressed, respectively, as Fe_2O_3, CaO, SiO_2, MnO, Al_2O_3, and MgO, and other elements, expressed as metals, such as Cr (3.4%), V(1450 ppm), and Mo (154 ppm) [1]. The XRD analysis of the investigated slag sample [1] revealed the following crystallized phases: larnite (Ca_2SiO_4), srebrodolskite ($Ca_2Fe_2O_5$), magnetite (Fe_3O_4), maghemite ($Fe_{2.6}O_3$), wustite (FeO), and magnesio-chromite ($MgCr_2O_4$), as reported in the literature [14–18]. Chromium-bearing compounds such as chromite were detected, but this element is probably dispersed in various other mineral phases. The most common crystalline phases were srebrodolskite, wustite, magnetite, and probably magnesio-chromite. The EDS-SEM mapping (Figure 2) made it possible to establish a partial correlation between Cr and Mn concerning certain grains moderately loaded with Mn but heavily loaded with Cr. Furthermore, grains lightly loaded with Cr were present. This clearly indicates the presence, in the sample, of at least two types of Cr carriers. The particles lightly loaded with chromium appeared poor in iron. Conversely, the iron-rich particles do not contain chromium.

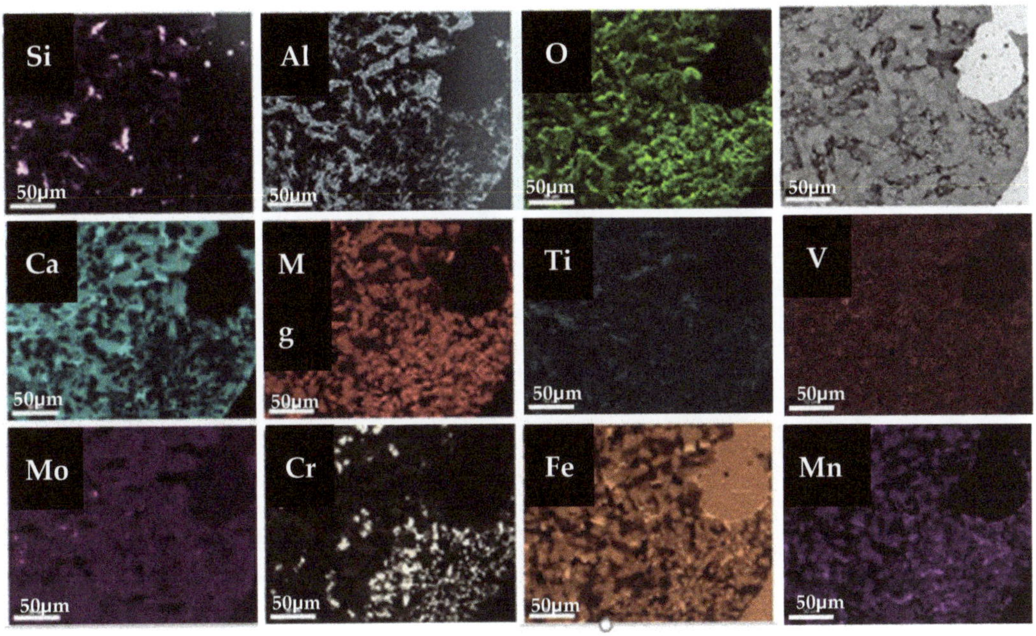

Figure 2. Mapping of different elements in EAFS by SEM-EDS.

Likewise, chromium rich particles contain, or not, little amount of iron. There is a correlation between calcium and aluminum, which most likely indicates the presence of larnite, which does not appear to contain chromium, as shown by the element mapping.

3.2. Thermodynamic Considerations

The thermodynamic calculations, necessary for establishing the phase diagrams, were carried out using the HSC software version 9 in previous work [11]. As mentioned, such diagrams are necessary to select the alkaline reagents and the temperatures allowing for high reactivity between the crystalline slag phases and reagents in order to form chromates and vanadates leachable in water solution. These diagrams clearly show that soluble chromates and vanadates of sodium or potassium can be obtained in a large range of oxygen and reagent partial pressures at different temperatures. Similar chemical behavior is expected to be exhibited under microwave radiation, which can also increase the slag/reagent mixture temperature. In such a context, this study deals with the recovery of Cr and V through the following method.

3.2.1. Leaching After Microwave Treatment

The EAFS/reagent mixture co-grinded for one hour was treated with microwaves at different powers and for residence times. After this, the EAFS/alkaline reagent mixture in deionized water was stirred and the obtained solutions were analyzed by AAS. Figure 3 shows the leaching rates of chromium (a) and vanadium (b) when the EAFS is treated using different alkaline reagents as dry (blue) and humid (red) mixtures for 15 min under 1.4 kW microwave power. Both the dry and humid mixtures exhibit similar reactivity even if leaching is a little higher when using humid mixtures, with the exception of NaOH. Consequently, most of this work has been performed by introducing small amounts of water (1 mL) into the mixtures to promote a temperature increase.

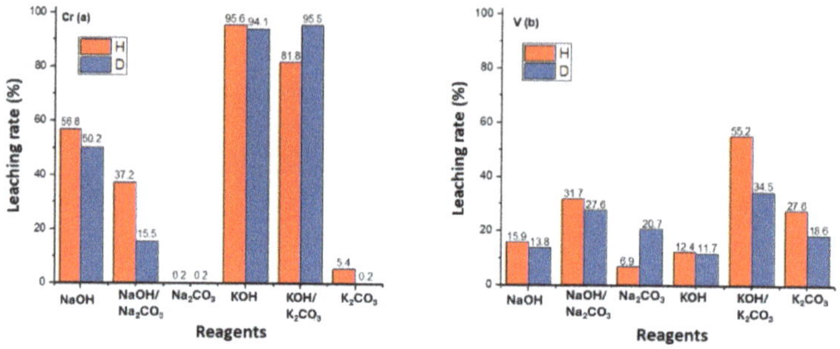

Figure 3. Leaching rates of Cr (**a**) and V (**b**) obtained using dry and humid methods through 1.4 kW microwave-assisted leaching for 15 min.

This figure clearly demonstrates that KOH-based reagents, whether used alone or in combination with K_2CO_3, are the most effective in promoting high chromium leaching from the EAF slag. This suggests that the chemical properties of these reagents play a significant role in enhancing the leaching process. In both configurations, wet and dry mixtures, KOH allows for more than 95% chromium leaching, while the mixture including KOH and K_2CO_3 allows for 95% (dry method) and 82% (wet method). Considering vanadium, through the dry and humid methods, the leaching rate is below 30%, except for the KOH + K_2CO_3 mixture, which allows for a leaching rate close to 35 and 55% considering, respectively, the dry and humid methods.

3.2.2. Effect of Microwave-Induced Temperature on Chromium Leaching

Figure 4 illustrates the results of different chromium leaching tests performed in a microwave furnace at 1.4 kWh for 15 min, inducing different temperatures for mixtures of alkaline reagents and EAF slag. The key trends observed from these tests highlight the impact of the temperature and reagent type on the efficiency of chromium extraction. Based on the results, the key findings are as follows.

1. EAF Slag with KOH: the highest chromium leaching efficiency is achieved at a roasting temperature of 670 °C.
2. EAF Slag with KOH + K_2CO_3: the optimal temperature for maximum chromium leaching is slightly lower, at 640 °C.
3. EAF Slag with Na_2CO_3 or K_2CO_3: the lowest leaching rate for each is obtained at the lowest roasting temperatures, respectively, 150 °C and 200 °C.

This suggests that the specific alkaline reagents used in the mixture, along with the precise control of the roasting temperatures, significantly impact the efficiency of the leaching process. The use of KOH alone requires a higher temperature, whereas the combination of KOH and K_2CO_3 lowers the optimal temperature needed for effective chromium recovery. These differences could be due to the varying reactivity and melting points of the alkaline reagents. These results depend also on how the reagents interact with the chromium in the EAFS during the MW treatment.

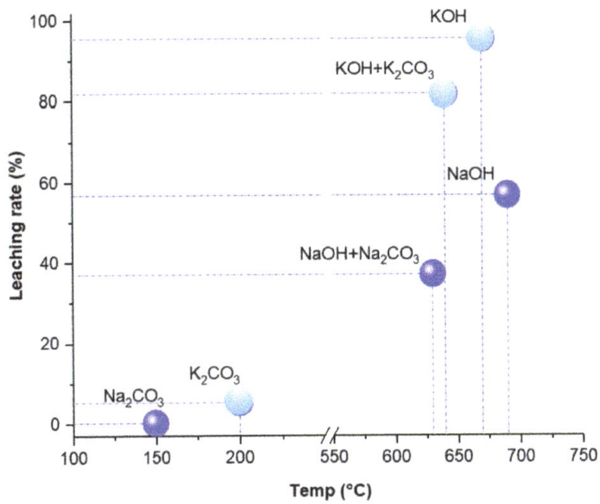

Figure 4. Leaching rates of Cr as function of MW (1.4 kWh)-induced temperature of different mixtures.

3.2.3. Effect of Microwave Power on Chromium and Vanadium Leaching

The left and right sides of Figure 5 show the leaching rates, respectively, of chromium and vanadium, for different microwave powers and different reagents, from the slag samples that were microwave-treated for 15 min. From this figure, it can be observed that when using KOH or KOH + K_2CO_3, chromium is dissolves more when increasing the MW power and is completely dissolved after 15 min under 1.4 kW of microwave power. However, when using NaOH or NaOH + Na_2CO_3 reagents, the leaching rates of chromium are close to, respectively, 60% and 20% at 1 kW, and these values are inverted when increasing the microwave power: the leaching rate increased up to 40% when using NaOH + Na_2CO_3 while it decreased close to 10% when using pure NaOH. This is likely due to the high reactivity of sodium carbonate with chromium upon increasing the microwave power, which promotes the formation of highly soluble chromate species in aqueous media. Potassium and sodium carbonates alone do not allow for the formation of great amounts of water-soluble chromium species no matter the microwave power used, as confirmed by the leaching rate of chromium being lower than 20%. This is probably due to the lower pH value induced by such reagents, which are less hard bases.

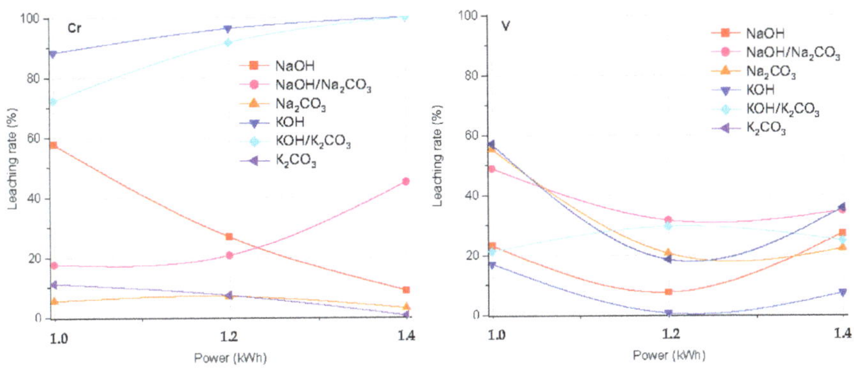

Figure 5. Leaching rates of Cr (**left side**) and V (**right side**) after slag treatment with the different alkaline reagents for 15 min using different microwave powers.

Considering vanadium, the highest leaching rates of 56 and 57% were obtained by treating the slag using, respectively, a Na_2CO_3 or K_2CO_3 mixture for 15 min with 1 kW of microwave power and these decreased when increasing the microwave power up to 1.4 kW (Figure 5, right). On the other hand, increasing the microwave power induces a decrease in vanadium leaching no matter the reagent, except for KOH + K_2CO_3. The reactivity of vanadium seems to be quite different to that of chromium when using KOH or KOH + K_2CO_3 reagents, which induce a lower leaching rate of vanadium (lower than 30%) compared to chromium (more than 70%; see Figure 5, left side).

3.2.4. Effect of Microwave Residence Time on Chromium and Vanadium Leaching

Figure 6 shows the evolution of the leaching rate of chromium (left side) and vanadium (right side) as a function of the residence time of different slag/alkaline reagent mixtures treated at 1.4 kW.

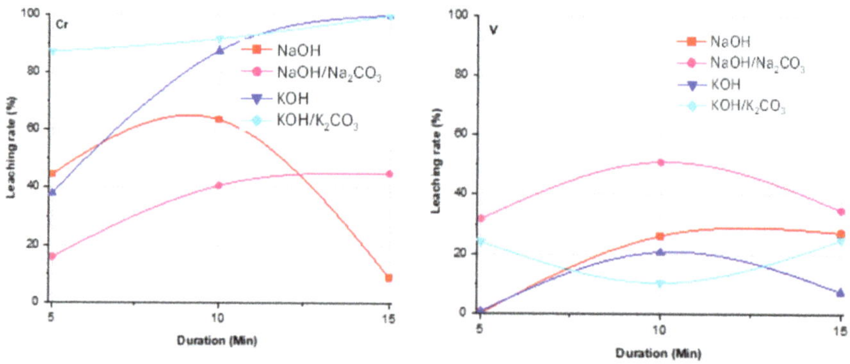

Figure 6. Leaching rates of Cr (**left side**) and V (**right side**) after slag treatment by most reactive alkaline reagents under 1.4 kW microwave power for different residence times.

The leaching rate of chromium increases from 38% to 100% in the case of using only KOH in the mixture. When using KOH + K_2CO_3, the leaching rate is very high (87%) after only 5 min of MW treatment. Then, the increase in the leaching rate up to 100% is slow when the residence time increases up to 15 min. This is due to K_2CO_3's high reactivity with chromium species, which facilitates the formation of very soluble chromate species in water.

Considering the results of the vanadium leaching, shown on the right side of Figure 6, it is clear that vanadium-bearing minerals are less reactive with alkaline reagents than chromium-bearing ones. Indeed, whatever the reagent, the leaching rate of vanadium is less than 50%.

Therefore, considering the preparation of slag/alkaline reagent mixtures using microwaves, the optimal parameters to promote the leaching of valuable metals (Cr and V) are as follows.

1. Slag/KOH or KOH + K_2CO_3 = 1, power = 1.4 kW, residence time = 15 min, V_{H2O} = 1 mL to leach 100% Cr.
2. Slag/K_2CO_3 or Na_2CO_3 = 1, P = 1 kW, residence time = 15 min, V_{H2O} = 1 mL to leach more than 55% V.

Table 1 groups together the best leaching rates of V and Cr from EAFS obtained for the two conditions of conventional and microwave treatments. It can be seen that the conventional roasting method allowed the obtainment of a maximum leaching rate of 97.5% for Cr after one hour of co-grinding the mixture of EAFS/NaOH + Na_2CO_3, and then two hours of conventional roasting at 800 °C. A leaching rate of 62.5% was obtained for V after three hours of co-grinding the slag/NaOH + Na_2CO_3 mixture followed by two hours of conventional roasting at 600 °C. On the other hand, after only 15 min of treatment under microwave conditions, the leaching rates of chromium and Vanadium are, respectively, 100% and 55%. Such results demonstrate the great benefit of using microwaves to optimize the modification of metal speciation to be leached. Consequently, the microwave-assisted leaching process optimized in this study can be considered an economical solution with less impact on the environment compared to the conventional leaching method.

Table 1. Best results obtained from leaching after conventional or MW furnace treatment.

Parameters	Microwave Roasting		Conventional Roasting	
	Cr	V	Cr	V
Reagents	KOH KOH + K_2CO_3	Na_2CO_3 K_2CO_3	NaOH + Na_2CO_3	NaOH + Na_2CO_3
Co-grinding residence time	1 h	1 h	1 h	3 h
Temperature (power)	900 °C (1.4 kW)	500 °C (1.0 kW)	800 °C	600 °C
Residence time	15 min	15 min	2 h	2 h
Leaching rate	100%	56–57%	98%	63%

At the end of the leaching experiments, the Eh and pH values of the leachates were measured. Figure 7 brings together the Eh-pH diagrams of the following systems: Cr-Na-H_2O, V-Na-H_2O, Cr-K-H_2O, and V-K-H_2O at 25 °C.

At the end of the experiments, the pH values were around 13 for all the leaching solutions. However, the redox potential (Eh) values were between 0.2 and 0.3 V/NHE in the sodium medium and between 0.3 and 0.4 V/NHE in the potassium medium. In such Eh-pH conditions (shown by the red star on the graph in Figure 7 and in the presence of KOH or NaOH, the Eh-pH diagrams show that chromium is in the form of chromate anions, CrO_4^{2-}. Such thermodynamic results for chromium speciation seem to be in perfect agreement with the experimental results obtained when KOH is used. Indeed, during the experiments, the concentration of chromium reached the maximum when using KOH and the microwave treatment. When using NaOH, the solubility of chromium is lower than 50% no matter the duration, probably due to the low reaction kinetics. These results indicate a lower reactivity of chromium bearing-minerals with such reagents. On the other hand, when potassium hydroxide is used as an alkaline reagent, VO_4^{3-} anions are supposed to be formed, as shown by the diagram in Figure 7. Yet, the leaching rate of this metal is lower than 20%.

Considering reagents including sodium hydroxide for the leaching of vanadium, taking into account the measured Eh/pH values, sodium vanadates $Na_2V_2O_7$ should be formed, which are very water-soluble phases. This explains the higher rate of V leaching when using NaOH than that when using reagents containing KOH.

It is important to notice that chromium leaching is strongly dependent on acid–base and redox conditions. The study by Bäverman et al. [19] highlighted the difficulty of establishing similarities between simulation and experimentation for this metal. It is also possible that the variety of aqueous complexes considered by the authors is insufficient to calculate correctly the speciation of chemical compounds in solution under pH or redox conditions, which can evolve significantly. Therefore, considering the solubility of chromium and vanadium species, the evaluation of saturation indices can be distorted as the system evolves.

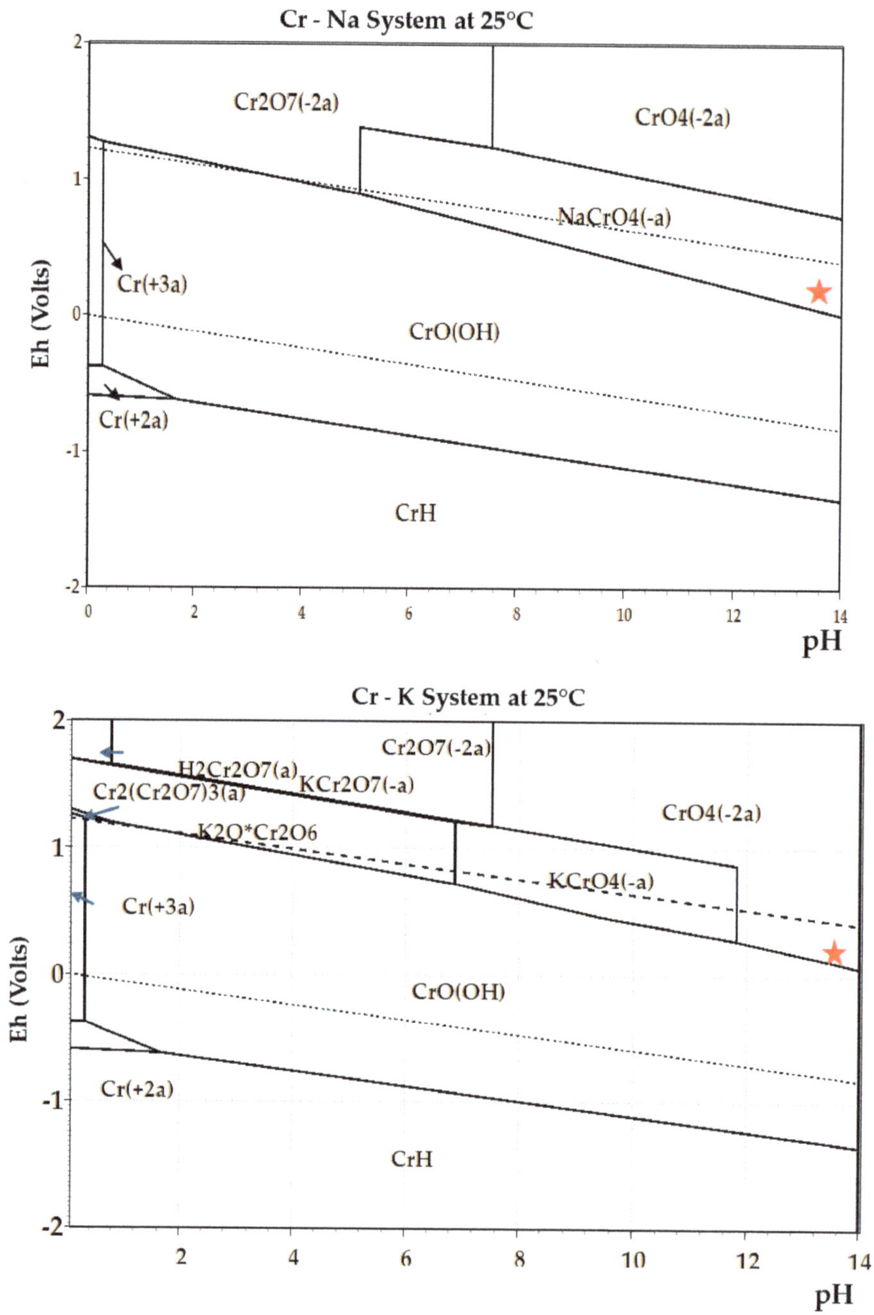

Figure 7. *Cont.*

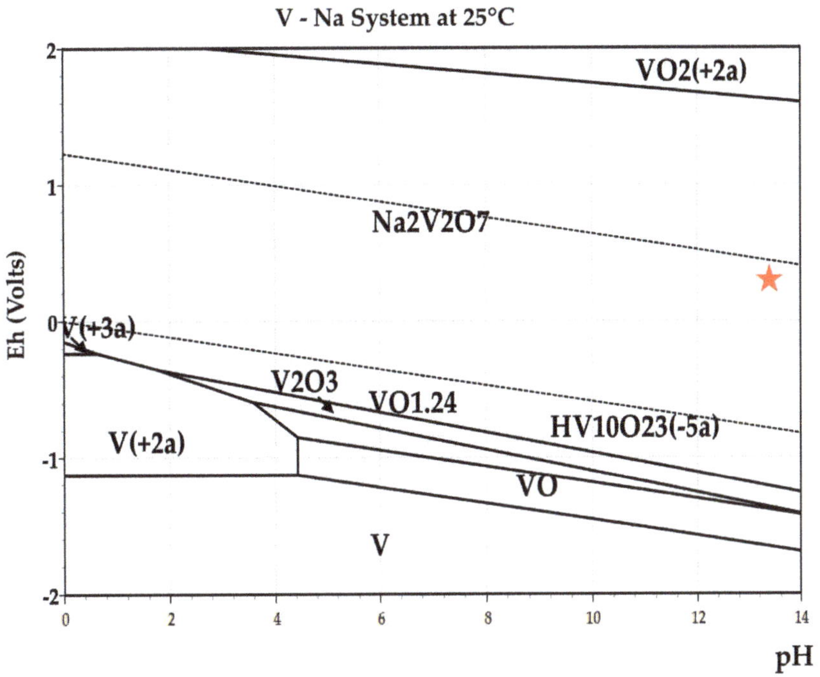

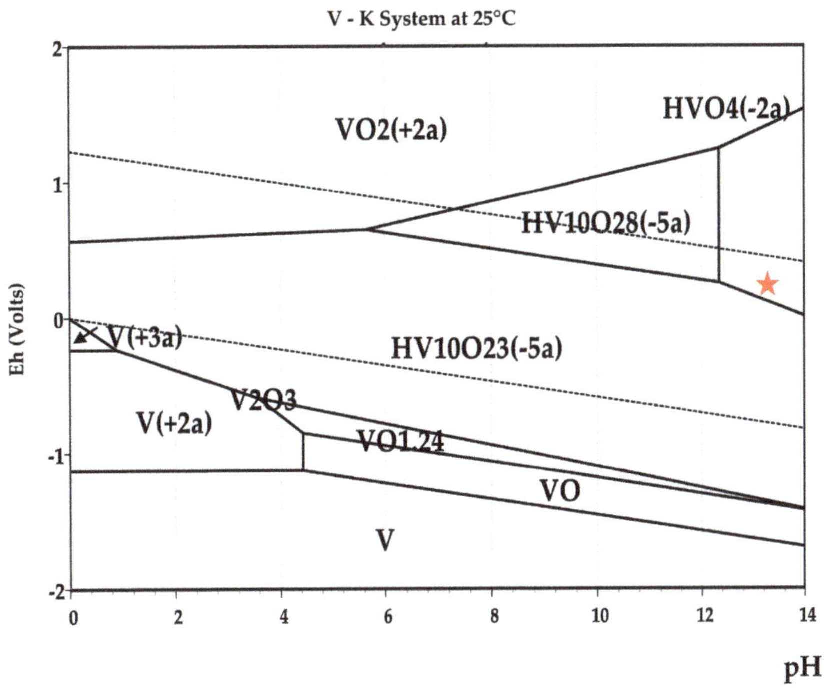

Figure 7. Eh-pH diagrams of Cr-Na-H_2O, V-Na-H_2O, V-K-H_2O, and Cr-K-H_2O systems at 25 °C.

Considering chromium species, the reaction pathway during MW treatment is expected to be as described in reaction (1). As shown by the phase diagrams of the O-Cr-K and O-Cr-Na systems at 400 and 600 °C (Figure 3), chromate species could be directly obtained by the reaction of chromite with potassium hydroxide.

$$MgCr_2O_4(s) + 4KOH(s) + 3/2O_2(g) \rightarrow MgO(s) + 2K_2CrO_4(s) + 2H_2O(g) \quad (1)$$

But the chemical reaction mechanism is probably more complex; this is due to the potential formation of $OH^\bullet$ free radicals (reaction (2)). These radicals, which mainly come from radiation–matter interactions, are considered major chemical species able to react with phases within slag due to their single electron on the peripheral layer, which gives them a high reactivity. It can be hypothesized that the reaction mechanism is consequently identical to that of reaction (3), where the hydroxyl radical $OH^\bullet$ reacts as a strong oxidant ($OH^\bullet/H_2O$: 2.34 V/ENH) with respect to chromite ore, which could be present in the form of $FeCr_2O_4$ in our slag sample. Chromates are obtained according to reaction (3) and then could be oxidized into dichromate according to reaction (4).

$$OH^- \rightarrow OH^\bullet \quad (2)$$

$$MgCr_2O_4 + 9OH^\bullet \rightarrow MgO + 2CrO_4^{2-} + 4H_2O + 4H^+ \quad (3)$$

$$2CrO_4^{2-} + 2K^+ + OH^\bullet + 3H^+ \rightarrow Cr_2O_7^{2-} + 2K^+ + 2H_2O + \quad (4)$$

Similar mechanisms are expected concerning the reactivity of slag minerals with other alkaline reagents (NaOH, Na_2CO_3, K_2CO_3, or mixtures), inducing the formation of leachable species such as chromates and vanadates in aqueous media. However, the results obtained with these other reagents are less efficient and therefore seem to indicate significantly different reaction patterns or reaction kinetics.

Considering vanadium, the best reactivities were observed when using K_2CO_3 under microwave radiation or NaOH + Na_2CO_3 using conventional heat treatment. This can be explained through Equations (5) and (6), which highlight the formation of soluble sodium and potassium vanadates.

$$Ca_3V_2O_8 + 3Na_2CO_3 \rightarrow 2Na_3VO_4 + 3CaCO_3 \quad (5)$$

$$Ca_3V_2O_8 + 3K_2CO_3 \rightarrow 2K_3VO_4 + 3CaCO_3 \quad (6)$$

4. Conclusions

Steel slag (SS) represents a promising secondary resource due to its high critical metal content, such as that of chromium (Cr) and vanadium (V). This study was conducted with the aim of developing a methodology to recover chromium and vanadium from EAF slags by using alkaline reagents. The slag characterization showed that the main crystalline phases contained in this heterogeneous material were srebrodolskite, larnite, hematite, and others, some of these minerals being expected to be Cr or V carriers. To explore this two-element leaching, an innovative technique was carried out involving physical preparation by co-grinding slag with alkaline reagents and then treatment in microwave furnaces to modify the metal speciation and consequently generate water-soluble species. Such a process allows for the leaching of a high proportion of metals, particularly chromium and vanadium. The leaching rate of chromium and vanadium were, respectively, 98% and 63% after 2 h of conventional heat treatment while MW treatment allowed for leaching close to 100% and 56–57% of the chromium and vanadium, respectively, after only 15 min of treatment in a microwave furnace. This study clearly justifies the great interest in metal-bearing species modification by microwave treatments, which allow for the enhancement of reaction kinetics and consequently have a lower energy demand.

Author Contributions: N.-E.M.: conceptualization, methodology, supervision, formal analysis, visualization, writing—review and editing; S.B.: investigation, methodology, review; A.S.: conceptualization, methodology, supervision, writing—review and editing. All authors have read and agreed to the published version of the manuscript.

Funding: This study was funded by the French National Research Agency (ANR 21 CE04 0018).

Data Availability Statement: The original contributions presented in this study are included in the article. Further inquiries can be directed to the corresponding author.

Acknowledgments: We gratefully acknowledge Cedric Duée from BRGM for the help with the scanning electron microscopy analysis.

Conflicts of Interest: Authors Nour-Eddine Menad, Alain Seron, and Sara Bensamdi were employed by the company BRGM. All authors declare that the research was conducted in the absence of any commercial or financial relationships that could be construed as a potential conflict of interest. The funders had no role in the design of the study; in the collection, analyses, or interpretation of the data; in the writing of the manuscript; or in the decision to publish the results.

References

1. Menad, N.-E.; Kana, N. Alain Seron and Ndue Kanari, New EAF Slag Characterization Methodology for Strategic Metal Recovery. *Materials* **2021**, *14*, 1513. [CrossRef] [PubMed]
2. Galkin, M.P.; Larionov, V.S.; Stepanov, A.V.; Nebolsin, V.A.; Milovanov, I.F.; Nikitin, G.S. Use of steelmaking slags. *Metallurgist* **1998**, *42*, 341–342. [CrossRef]
3. Das, B.; Prakash, S.; Reddy, P.S.R.; Misra, V.N. An overview of utilization of slag and sludge from steel industries. *Resour. Conserv. Recycl. J.* **2007**, *50*, 40–57. [CrossRef]
4. Yi, H.; Guoping, X.; Huigao, C.; Junshi, W.; Yinfeng, W.; Hui, C. An overview of utilization of steel slag. *Procedia Environ. Sci.* **2012**, *16*, 791–801. [CrossRef]
5. Wei, Y.-L.; Lin, C.-Y.; Cheng, S.-H.; Wang, P. Recycling steel-manufacturing slag and harbor sediment into construction materials. *J. Hazard. Mater.* **2014**, *265*, 253–260. [CrossRef] [PubMed]
6. Dao, P.L. Valorisation des Laitiers LWS dans les Mélanges Granulaires. Ph.D. Thesis, Lorriane University, Nancy, France, 2010. Available online: https://hal.univ-lorraine.fr/tel-01748655 (accessed on 30 October 2024).
7. Houze, C. Etude de la Valorisation des Laitiers de l'Industrie Sidérurgique et de Production des Alliages Silicomanganèse. Ph.D Thesis, Paris-Est University, Paris, France, 2014; p. 286. Available online: https://theses.hal.science/tel-00978351 (accessed on 30 October 2024).
8. Durinck, D.; Mertens, G.; Jones, P.T.; Elsen, J.; Blanpain, B.; Wollants, P. Slag Solidification Modeling Using the Scheil–Gulliver Assumptions. *J Am. Ceram. Soc.* **2007**, *90*, 1177–1185. [CrossRef]
9. Wang, X.; Gao, D.; Chen, B.; Meng, Y.; Fu, Z.; Wang, M. A clean metallurgical process for separation and recovery of vanadium and chromium from V-Cr-bearing reducing slag. *Hydrometallurgy* **2018**, *181*, 1–6. [CrossRef]
10. Bensamdi, S.; Seron, A.; Pereira, F. Nour-eddine Menad, Recovery of Valuable Metals from EAF Slag by Microwave Assisted Leaching. In Proceedings of the 7th International Slag Valorisation Symposium, Ku Leuven, Belgium, 27–29 April 2021; Proceeding 142–146. Available online: https://slag-valorisation-symposium.eu/2021/downloads.html (accessed on 30 October 2024).
11. Menad, N.; Kana, N.; Kanari, N.; Pereira, F.; Seron, A. Process for Enhancing the Valuable Metal Recovery from "Electric Arc Furnace" (EAF) Slags. *Waste Biomass Valoris.* **2021**, *12*, 5187–5200. [CrossRef]
12. Kokko, M.; Manninen, M.; Hu, T.; Lassi, U.; Vielma, T.; Pesonen, J. Hydrometallurgical recovery of vanadium and calcium from electric arc furnace slag in hydrogen-based steelmaking. *Miner. Eng.* **2024**, *217*, 108966. [CrossRef]
13. Ding, M. Utilization of Vanadium Extraction from Vanadium-Bearing Steel Slag. *IOP Conf. Ser. Earth Environ. Sci.* **2021**, *631*, 012056. [CrossRef]
14. National Academies of Sciences, Engineering, and Medicine; Division on Earth and Life Studies; Board on Environmental Studies and Toxicology; Committee on Electric Arc Furnace Slag: Understanding Human Health Risks from Unencapsulated Uses. In *Health Risk Considerations for the Use of Unencapsulated Steel Slag, Consensus Study Report*; National Academies Press: Washington, DC, USA, 2023.
15. Gu, F.; Zhang, Y.; Su, Z.; Tu, Y.; Liu, S.; Jiang, T. Recovery of chromium from chromium-bearing slags produced in the stainless-steel smelting: A review. *J. Clean. Prod.* **2021**, *296*, 126467. [CrossRef]
16. Xu, J.; Liu, M.; Ma, G.; Zheng, D.; Zhang, X.; Hou, Y. Valuable Recovery Technology and Resource Utilization of Chromium-Containing Metallurgical Dust and Slag: A Review. *Metals* **2023**, *13*, 1768. [CrossRef]
17. Bru, K.; Seron, A.; Morillon, A.; Algermissen, D.; Lerouge, C.; Menad, N. Characterization of a chromium-bearing carbon steel electric arc furnace slag after magnetic separation to determine the potential for iron and chromium recovery. *Minerals* **2021**, *12*, 47. [CrossRef]

18. Xu, J.; Liu, M.; Ma, G.; Zheng, D. Xiang Zhang and Yanglai Hou, Vanadium extraction from steel slag: Generation, recycling and management. *Metals* **2023**, *13*, 1768. [CrossRef]
19. Bäverman, C.; Spiej, A.; Moreno, L.; Neretnieks, I. Serial batch tests performed on municipal solid waste incineration bottom ash and electric arc furnace slag, in combination with computer modelling. *Waste Manag. Res.* **1997**, *15*, 55–71. [CrossRef]

metals

MDPI

Article

Effect of Nickel Impurities in Pyrite on Catalytic Degradation of Thiosulfate

Xuecong Qin, Tao Zhang, Wenhua Qin and Hongbo Zhang *

School of Resource and Environmental Engineering, Inner Mongolia University of Technology, Hohhot 010020, China; xuecong1319@imut.edu.cn (X.Q.); zhangtao@imut.edu.cn (T.Z.); qinwenhua@imut.edu.cn (W.Q.)
* Correspondence: hongbz77@imut.edu.cn

Abstract: The effects of nickel content in nickel-bearing pyrite on photocatalytic properties, light absorption properties, and oxidative decomposition of thiosulfate were studied. The leaching experiments show that the consumption of thiosulfate in the Cu^{2+}-ethylenediamine (en)-$S_2O_3{}^{2-}$ system increases with an increase in nickel content in nickel-bearing pyrite. The consumption of $Cu(en)_2{}^{2+}$ initially increases and then decreases with an increase in leaching time. There is a clear correlation between the change trend in its consumption and the doping amount of nickel in pyrite. The XPS results show that in the Cu^{2+}-ethylenediamine (en)-$S_2O_3{}^{2-}$ leaching gold system (temperature 25 °C, time 35 h, solution: 0.1 mol/L $S_2O_3{}^{2-}$, 5 mmol/L $Cu(en)_2{}^{2+}$, 200 mL solution), the nickel of pyrite-containing nickel can be transferred to the leaching solution and becomes nickel ion. In this leaching system, Cu(II), which was originally complexed with en, is reduced to Cu(I) in a short time. The consumption of $Cu(en)_2{}^{2+}$ increased rapidly in the 5 h period and then decreased gradually after 5 h. The results showed that the presence of free Ni^{2+} in the solution facilitated the conversion of bivalent copper ions to monovalent copper ions. Free Ni^{2+} ions can compete with Cu^{2+} ions for en ligands. When ethylenediamine complexes with Ni^{2+}, the decomposition of $Cu(en)_2{}^{2+}$ into $Cu(en)^+$ and en occurs more rapidly. And the en, which was originally to be oxidized with $Cu(en)^+$ to form $Cu(en)_2{}^{2+}$, forms $Ni(en)_2{}^{2+}$. As a result, the concentration of $Cu(en)_2{}^{2+}$ continues to decrease in a short period of time.

Keywords: gold extraction; thiosulfate; nickel-bearing pyrite; catalytic degradation

1. Introduction

As the reserves of easily processed gold deposits continue to decrease and environmental protection requirements become increasingly stringent, industrial research has turned towards finding nontoxic leaching agents that can replace cyanide [1]. Among the various gold leaching reagents, such as thiourea, polysulfides, halides, and thiosulfate, thiosulfate is considered the most promising green leaching agent with significant industrial application prospects because of its nontoxicity, high selectivity, and low cost [2–5]. However, problems such as high reagent consumption and difficulties in recovering gold complex ions from leach pulp have hindered the widespread industrialization of thiosulfate gold extraction technology [6,7].

To realize the industrial application of thiosulfate gold leaching, the high consumption of thiosulfate is an urgent technical bottleneck that must be solved. Sulfide minerals such as arsenopyrite, chalcopyrite, and galena have been reported to promote the composition of thiosulfate [8,9] because of their strong affinity for sulfur-containing substances and their semiconductor properties [10,11]. Moreover, most metal sulfides exhibit semiconducting properties which significantly affect the leaching of gold-bearing sulfides. The presence of impurities in the mineral lattice affects its composition, electronic structure, semiconductor properties, and the interaction between the leaching agent and mineral surface [9]. Additionally, different vacancy defects, impurity doping, and defect locations in the lattice affect both the sulfide ore surface and leaching characteristics [12–14].

Pyrite, the most common semiconductor-associated mineral in gold mines, often contains isomorphic impurities such as nickel, cobalt, and arsenic [15,16]. These impurities not only alter the semiconductor properties of pyrite but also impact the lattice defects that can affect its oxidation decomposition and surface adsorption. Although natural pyrite is widely distributed with large reserves, it typically possesses various impurities (such as nickel, silicon, arsenic.) [17,18]. Consequently, these factors introduce numerous uncertainties during experimental exploration processes.

To investigate the influence of nickel impurities in pyrite on thiosulfate gold leaching, leaching experiments were conducted using synthetic pure-phase pyrite and pyrite containing specific nickel elements. This was performed to eliminate interference from other metal impurities on the semiconductor properties of pyrite and the catalytic degradation of thiosulfate ions. Therefore, this study first synthesized pure-phase pyrite through a hydrothermal method and then synthesized pyrite containing only nickel impurities by adding a specific amount of nickel; pure-phase pyrite and nickel-containing pyrite were compared in terms of their semiconductor properties, including phase, photocatalysis, and light absorption. Additionally, the effect of nickel impurities in pyrite on the catalytic degradation of thiosulfate in a Cu^{2+}-en-$S_2O_3^{2-}$ leaching system was investigated.

2. Materials and Methods

2.1. Materials and Reagents

The reagents used in this study, including sodium thiosulfate, copper sulfate, sodium hydroxide, ferrous chloride, sodium sulfide, nickel chloride, and en, were analytically pure. All experimental solutions were prepared using ultrapure water.

2.2. Test Methods

2.2.1. Synthesis of Pure-Phase Pyrite

The basic raw materials used were 17.90 g of ferrous chloride and 43.23 g of sodium sulfide. The pH of the mixed solution was adjusted to approximately 6.6 using dilute NaOH and HCl solutions. After stirring the solution with a magnetic stirrer (MYP11-2A) for 30 min, the mixture was transferred to a high-pressure reactor. The reactor was placed in a constant-temperature vacuum drying oven (DZF-6020) at 200 °C for 24 h and then cooled to room temperature. The products in the reactor were washed, filtered, separated, and dried in a constant-temperature drying oven at 60 °C for 12 h, and the final product was synthetic pure-phase pyrite. The product was then stored in a sealed reagent bag filled with nitrogen.

2.2.2. Synthesis of Nickel-Bearing Pyrite

The synthetic procedure was the same as that for pure-phase pyrite, but the amounts of reagents used were different. The basic raw materials used were 17.90 g of ferrous chloride and 43.23 g of sodium sulfide. By adding 0.43, 1.13, 1.61, and 2.38 g of nickel chloride to the above raw materials, nickel-bearing pyrite samples with 2%, 5%, 7%, and 10% nickel contents were obtained, respectively.

2.2.3. Photocatalytic Degradation Experiment

The photocatalytic reactions were performed in a BL-GHX-V (Shanghai, China) photocatalytic reactor. The central light source of the photoreactor was a 300 W deuterium lamp. In the photocatalytic reaction, the initial concentration of methylene blue solution was 25 mg/L, and 50 mg of solid powder was added. The specific photocatalytic degradation steps have been described in the literature [19]. The methylene blue degradation rate was calculated as follows:

$$R = \frac{C_0 - C}{C_0} \times 100\% \tag{1}$$

where C_0 represents the initial concentration of methylene blue in the solution and C represents the concentration after the photocatalytic reaction.

2.2.4. Leaching Experiment and Detection Method

The leaching system was a mixed solution composed of 0.1 mol/L $Na_2S_2O_3$, 0.01 mol/L en, and 0.005 mol/L $CuSO_4$. The specific leaching experimental method has been described in detail in [20]. The analytical method for determining the gold concentration in this solution system has been extensively described in previous studies [20]. The thiosulfate concentration detection method for the leaching solution was based on previous literature [21]. The operation steps of the experiment, the consumption of thiosulfate, and the detection method of the gold concentration in the leaching solution are described in detail in the Supplementary Materials.

2.2.5. Determination of the $Cu(en)_2^{2+}$ Complex Concentration

First, $Cu(en)_2^{2+}$ standard solutions were prepared at concentrations of 2.5, 5, 7.5, 10, and 12.5 mmol/L. The absorbance of the standard solution was measured at 550 nm ($Cu(en)_2^{2+}$ has an absorption peak at 550 nm) using deionized water as a blank. The concentration–absorbance relationship was determined by plotting the concentration of the $Cu(en)_2^{2+}$ complex on the horizontal axis and the absorbance on the vertical axis to obtain a standard curve for the $Cu(en)_2^{2+}$ solution (Figure 1) [22]. An aliquot of the sample liquid was periodically removed from the leaching solution using a pipette and transferred to a glass bottle. After some time elapsed, deionized water was used as the reference sample, and the absorbance of the $Cu(en)_2^{2+}$ complex at 550 nm was measured three times in parallel using a UV-VIS spectrophotometer to determine its average value. The concentration of the $Cu(en)_2^{2+}$ complex was calculated using the following equation:

$$X = \frac{Y + 0.0113}{0.06308} \tag{2}$$

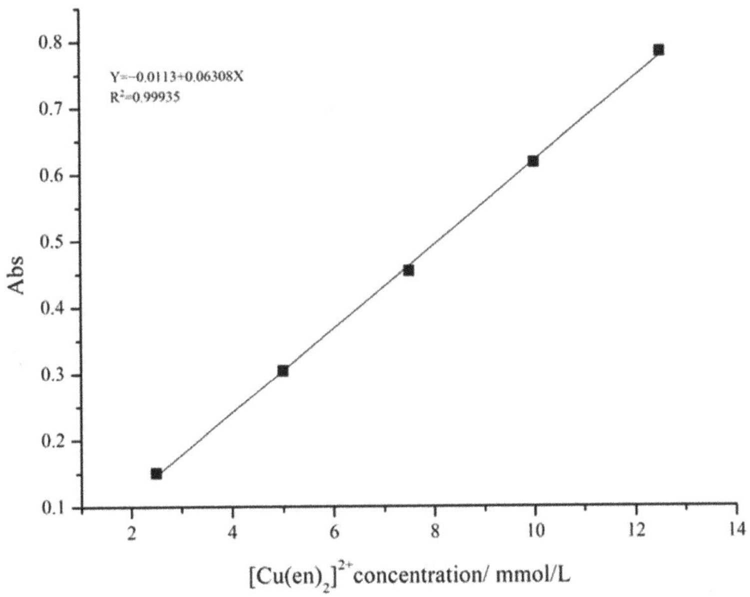

Figure 1. Relationship between the absorbance and $Cu(en)_2^{2+}$ concentration (The specific parameters are shown in Table 1).

Table 1. Data parameters obtained by linear fitting in Figure 1.

	Intercept		Slope		Statistics
	Value	Standard Error	Value	Standard Error	Adj. R-Square
C	−0.0113	0.00669	0.06308	8.06308×10^{-4}	0.99935

2.3. Sample Characterization

The full name corresponding to the abbreviation in the text: Copper–ethylenediamine–thiosulfate (Cu^{2+}-en-$S_2O_3{}^{2-}$), X-ray fluorescence spectroscopy (XRF), X-ray diffraction (XRD), ultraviolet/visible/near-infrared (UV-VIS-NIR), scanning electron microscopy (SEM), ultraviolet–visible diffuse reflection spectroscopy (UV-VIS DRS), X-ray photoelectron spectroscopy (XPS), energy-dispersive spectroscopy (EDS).

The phases of the synthetic pure-phase pyrite and nickel-bearing pyrite were detected using an XRD, and the data were analyzed using JADE software. The micromorphology and elemental composition of the samples were measured by SEM. Quantitative and qualitative analysis of leached gold foil and pyrite were performed using an XPS (X-ray photoelectron spectrometer, PHI5000 Versa ProbeII, ULVAC-PHI, Japan) [23,24]. In addition, the absorption spectra of synthetic pure-phase pyrite and nickel-bearing pyrite in the wavelength range 200–2000 nm were measured using a UV-VIS-NIR (UV-3600, Shimadzu, Japan).

3. Results and Discussion

3.1. Phase and Morphology of Synthesized Pure-Phase Pyrite

Figure 2 shows the XRD pattern of the synthetic pure-phase pyrite. The diffraction peaks of the sample in the figure correspond to (111), (200), (210), (211), (220), and (311) in the pyrite-type FeS_2 standard card (No.42-1340). The crystal half-peak widths of the sample were small, the diffraction peaks were sharp, and the diffraction intensities were high, indicating that the synthetic pyrite had good crystallinity, a well-developed crystal face, a complete crystal shape, and few impurities. The XRD pattern of the synthesized pyrite had almost no miscellaneous peaks, indicating that the obtained product was of high purity.

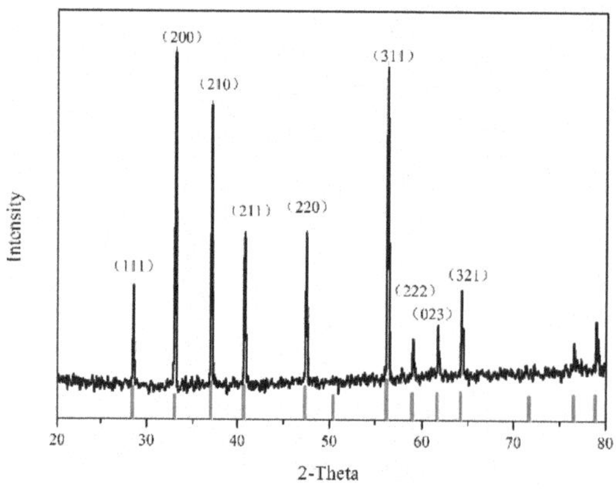

Figure 2. XRD pattern of synthetic pyrite.

After complete quantitative phase analysis of the synthesized pyrite sample in JADE, a column chart of the content of each substance was obtained. As shown in Figure 3, the purity of pyrite in the synthetic sample was higher than that of the natural pyrite used in the experiment.

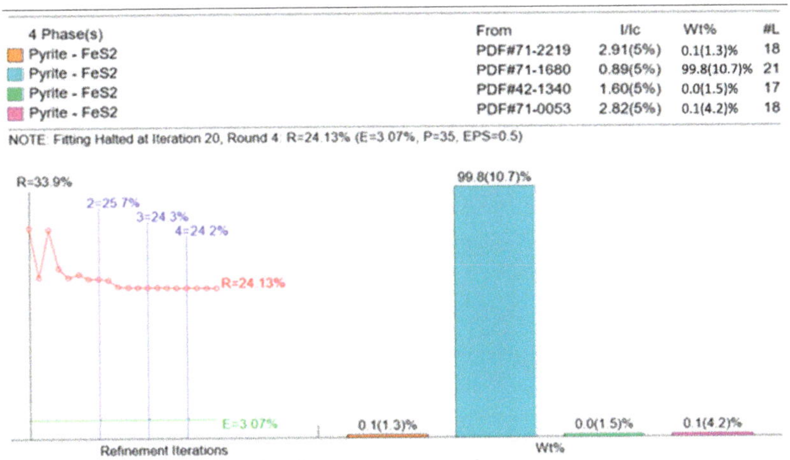

Figure 3. Phase and purity of synthetic pyrite.

Figure 4a shows SEM images of synthetic pure-phase pyrite, demonstrating smooth and complete cubic particles with well-developed boundaries and high integrity. The particle size of large particles was approximately 3 μm, and the particle size of small particles was in the range of tens to hundreds of nanometers. Figure 4b shows the spectral analysis results and element content corresponding to point 1 in Figure 4a. It can be seen from the figure that the surface of the cube particle contains only two elements, iron and sulfur, and the specific gravity of iron and sulfur is 45.65% and 54.35%, respectively, which is very close to the theoretical ideal iron and sulfur content of pyrite.

Figure 4. SEM of synthetic pure-phase pyrite; (**a**) morphology and (**b**) spectral diagram and weight percentage at corresponding point 1.

3.2. Influence of the Nickel Doping Content on the Properties of Pyrite

3.2.1. Phase and Morphology of Nickel-Bearing Pyrite

The XRD patterns of the synthesized nickel-bearing pyrite are shown in Figure 5, which indicate that Ni-doped pyrite contains more pyrite-type FeS_2 and forms $(Fe,Ni)S_2$ (pyrite, Ni-rich) and pentlandite (such as $FeNiS_2$ and $Fe_5Ni_4S_8$) phases. This indicates that in the synthesis of nickel-bearing pyrite, a significant portion of nickel exists in a homogeneous form within the nickel-rich pyrite, or as part of pentlandite formed from nickel sulfide and iron sulfide. In this study, pyrite containing a small amount of pentlandite and the lattice substitution of nickel are collectively referred to as nickel-bearing pyrite. By referencing the literature [25], it becomes evident that $(Fe,Ni)S_2$ exhibits remarkable catalytic performance and high activity.

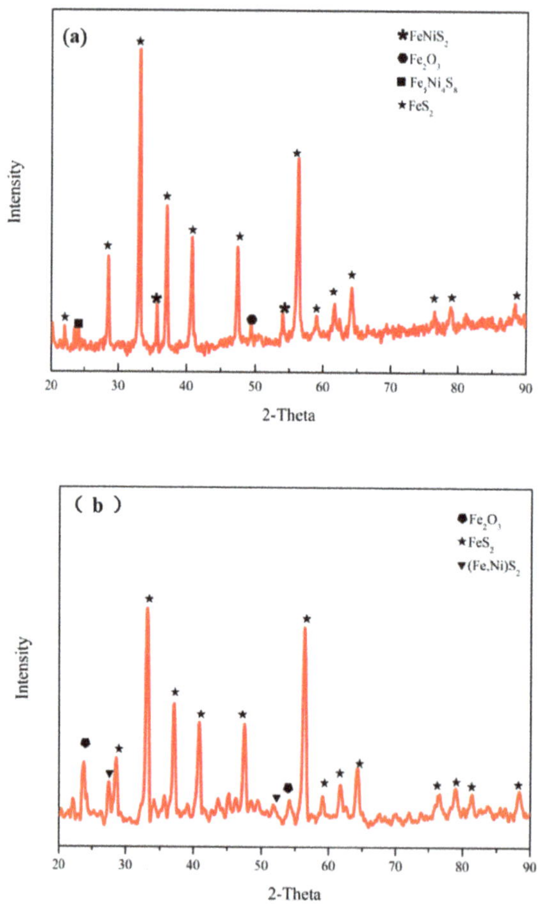

Figure 5. XRD patterns of nickel-doped pyrite: (**a**) 2% nickel-doped pyrite and (**b**) 10% nickel-doped pyrite.

The morphological changes in the synthetic nickel-bearing pyrite are illustrated in Figure 6. Compared with the synthesized pure phase pyrite samples, the morphology and particle size of nickel-bearing pyrite changed significantly. Specifically, a large number of small particles are present in nickel-bearing pyrite mixed with 2% and 5% nickel. When 7% and 10% nickel are added, these small particles aggregate to form larger diameter clumps. Therefore, the morphology and particle size of nickel-containing pyrite samples were significantly changed by increasing the amount of nickel doping. This suggests that the introduction of Ni^{2+} inhibits FeS_2 grain growth and thus reduces the particle size of the FeS_2 sample. Due to the larger specific surface area and superior catalytic activity of small particle pyrite, the addition of nickel ions will help improve the performance of FeS_2 in the catalytic degradation process.

Figure 6. *Cont.*

Figure 6. Synthetic nickel-doped pyrite surface topography: (**a**) 2% nickel, (**b**) 5% nickel, (**c**) 7% nickel, and (**d**) 10% nickel.

3.2.2. Influence of the Nickel Doping Amount on the Semiconductor Properties of Synthetic Products

Figure 7 illustrates the UV-VIS DRS spectrum of nickel-doped pyrite, which revealed that the light absorption coefficients of natural pyrite and synthetic pure-phase pyrite were similar, whereas those of nickel-bearing pyrite exhibited distinct trends depending on the doping amount. The light absorption coefficient gradually increased with the doping amount. This phenomenon can be attributed to the formation of a small quantity of pentlandite owing to the increased doping, resulting in an increase in the absorption coefficient and ultimately leading to enhanced light absorption.

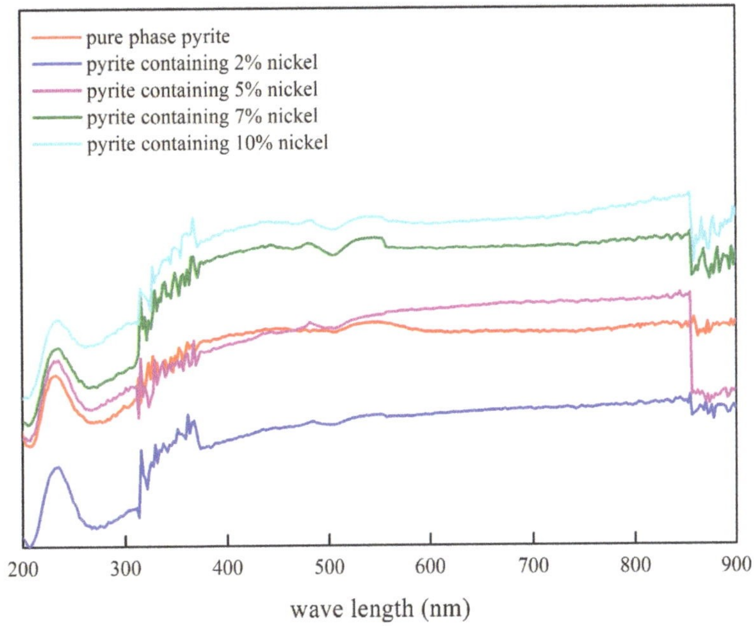

Figure 7. UV-VIS DRS spectra of undoped pyrite, synthesized nickel-doped pyrite, and natural pyrite.

3.2.3. Influence of the Nickel Doping Amount on the Photocatalytic Degradation Properties of Synthetic Products

The degradation effects of natural pyrite, synthetic pure-phase pyrite, and nickel-bearing pyrites (with Ni-doping contents of 2%, 5%, 7%, and 10%, respectively) on methylene blue samples are illustrated in Figure 8. The degradation rate of methylene blue in the solution with nickel-bearing pyrites added increased with increasing Ni-doping content. At a Ni content of 2%, the catalytic degradation rate of nickel-bearing pyrites for methylene blue was 86%, indicating superior visible-light catalytic activity compared to that of pure-phase pyrite. The photocatalytic activity of the nickel-bearing pyrite was significantly enhanced at a nickel content of 7%. After an illumination period of 4 h, the methylene blue degradation rate reached 96%, demonstrating a noticeably better catalytic effect than that achieved using natural pyrite alone. When the nickel doping amount reached 10%, the photocatalytic degradation rate of the nickel-bearing pyrite achieved the complete removal (100%) of methylene blue. Therefore, the photocatalytic effect of nickel-bearing pyrite also increased with increasing nickel doping. The results showed that nickel doping can effectively regulate the photocatalytic properties of pyrite at different doping amounts.

The reasons for the improvement in the visible-light catalytic activity of nickel-doped pyrite are as follows. After the inclusion of metal ions in the semiconductor photocatalyst, the metal can be used as an effective acceptor of electrons, trapping photoelectrons from the valence band in the conduction band and preventing the combination of semiconductor photoelectrons and holes, thus improving the visible-light catalytic activity of the semiconductor [11].

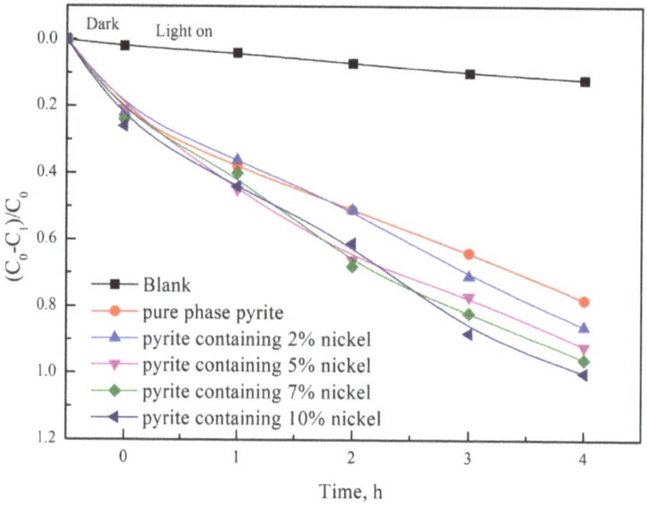

Figure 8. Changes in the photocatalytic performance of natural pyrite, undoped pyrite, and nickel-doped pyrite.

3.3. *Influence of Nickel-Containing Pyrite on the Consumption of Thiosulfate and $Cu(en)_2^{2+}$ in the System*

The catalytic degradation of thiosulfate and the consumption of the copper complex by the nickel content in pyrite were investigated by introducing different nickel-doped pyrites into a Cu^{2+}-en-$S_2O_3^{2-}$ gold leaching system.

The impact of nickel-containing pyrite on thiosulfate consumption in the Cu^{2+}-en-$S_2O_3^{2-}$ gold leaching system was also investigated (Figure 9a). The nickel contents of the pyrite samples were 2%, 5%, 7%, and 10%, with the synthesized pure-phase pyrite used as the control group. When pure-phase pyrite was introduced into the system, thiosulfate consumption increased slowly within 24 h but accelerated rapidly beyond this duration. Upon the addition of nickel-containing pyrite, a significant increase in thiosulfate consumption occurred, which correlated with higher levels of nickel doping (as shown Figure 9a). This is mainly due to the catalytic action of pyrite itself on the decomposition of thiosulfate, and the introduction of nickel inhibits the growth of FeS_2 grains, thus reducing the particle size of FeS_2 samples. Compared with large-particle pyrite, small-particle pyrite has a larger specific surface area and superior catalytic activity. Therefore, the incorporation of nickel is helpful to improve the catalytic degradation performance of pyrite for thiosulfate. In addition, some scholars have found that the doping of nickel inside pyrite will transform it into N-type semiconductor [11,12]. With the increase in doping amount, the band gap width decreases gradually, making it easier for electrons bound to the surface to transition to the conduction band and transform into free electrons or holes. Therefore, there are more impurities in the nickel-containing pyrite that can accept electrons. When electrons in the valence band are excited into the conduction band, more holes are created [11,26], thus promoting the electron-giving process of thiosulfate. Therefore, nickel-containing pyrite significantly accelerates the oxidative decomposition of thiosulfate [27].

As shown in Figure 9b, in the Cu^{2+}-en-$S_2O_3^{2-}$ gold leaching system, the addition of pure-phase pyrite had little effect on $Cu(en)_2^{2+}$ consumption compared to nickel-bearing pyrite with 2%, 5%, 7%, and 10% nickel contents. The consumption of $Cu(en)_2^{2+}$ gradually increased with leaching time when pure-phase pyrite was added to the leaching system. However, when nickeliferous pyrite was added to the gold leaching system, the consumption of $Cu(en)_2^{2+}$ initially sharply increased. This was particularly evident in the pyrite system with a large amount of nickel doping, where the consumption of the complex reached approximately 75% within the first 5 h. Subsequently, a slow increase in the content of $Cu(en)_2^{2+}$ occurred during the following 5–10 h. When the leaching time exceeded 10 h, a gradual increase in the consumption of $Cu(en)_2^{2+}$ occurred as the leaching time increased. In particular, the consumption of $Cu(en)_2^{2+}$ also increases with an increase in nickel doping, and there is a clear correlation between the two.

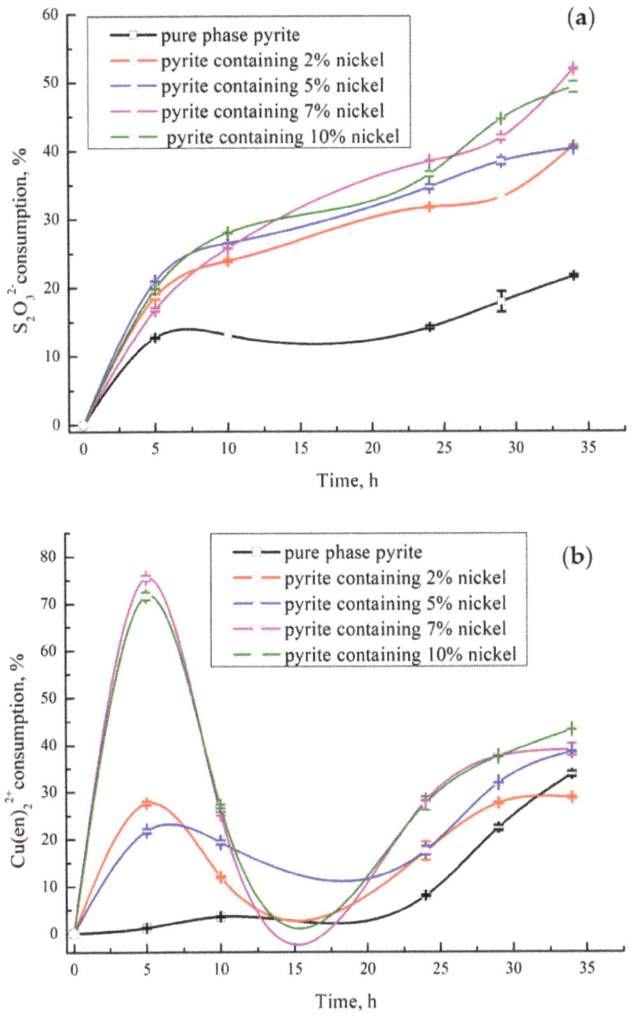

Figure 9. (**a**) Effects of nickel doping in pyrite on thiosulfate consumption; (**b**) Effects of nickel dop-ing in pyrite on $Cu(en)_2^{2+}$ consumption (in the Cu^{2+}-en-$S_2O_3^{2-}$ system, pH 9.8–10, temperature 25 °C, time 35 h, solution: 0.1 mol/L $S_2O_3^{2-}$, 5 mmol/L $Cu(en)_2^{2+}$, 200 mL solution).

3.4. *Mechanistic Analysis of Cu^{2+}-en-$S_2O_3^{2-}$ Gold Leaching with Nickel-Bearing Pyrite*

3.4.1. XPS Analysis of Gold Foil After Leaching

As shown in Figure 10a, when 7% Ni-doped pyrite was added to the Cu^{2+}-en-$S_2O_3^{2-}$ system and after fitting the Ni 2p peak of the leached gold foil surface, the peaks at 854.88, 855.43, and 856.36 eV corresponded to NiO, NiS, and $NiSO_4$, respectively [28]. The nickel in the pentlandite was present as free NiS, and the presence of $NiSO_4$ indicated that the nickel in the pentlandite was converted to Ni (II) during the degradation of thiosulfate. Based on the magnitude of the peak values, the primary nickel-containing compounds present on the surface of the leached gold foil were inferred to be NiO and NiS. Zhou's [29] study revealed the chemical equation of Ni release during the dissolution and reprecipitation of pyrite, as shown in Equation (3) below:

$$(Fe, Ni)S_2 + 3Cl^- \rightarrow FeS_2 + NiCl_3^- + 2H_2S + O_2 \quad (3)$$

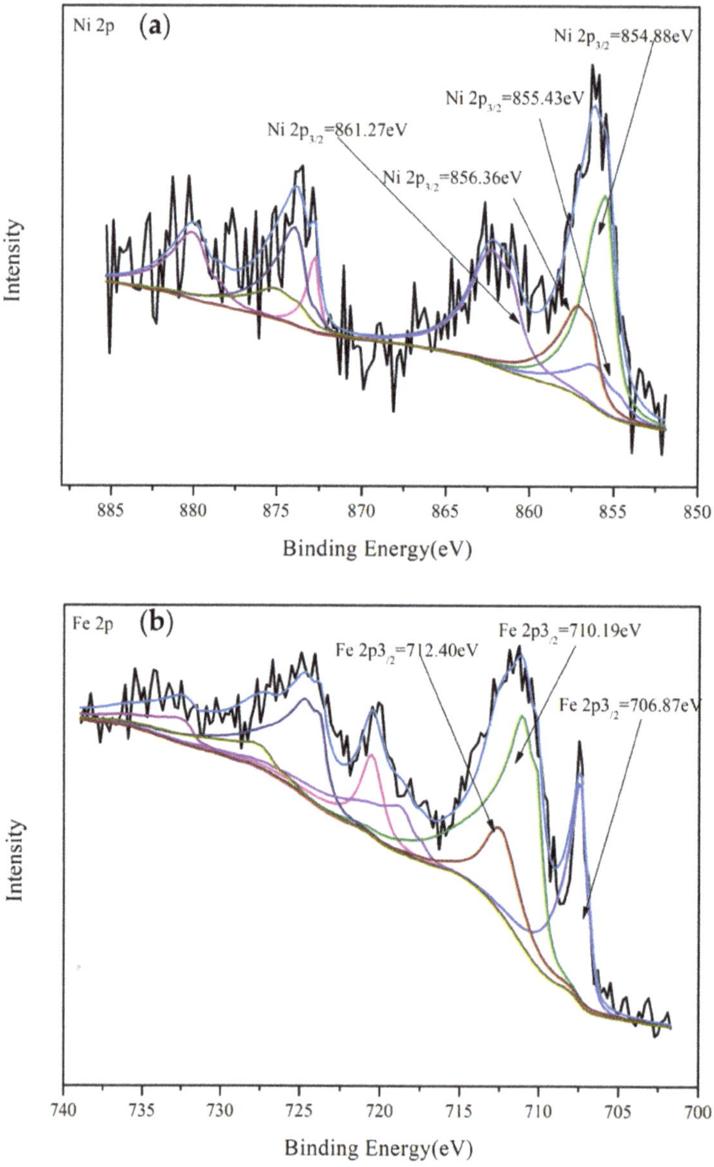

Figure 10. *Cont.*

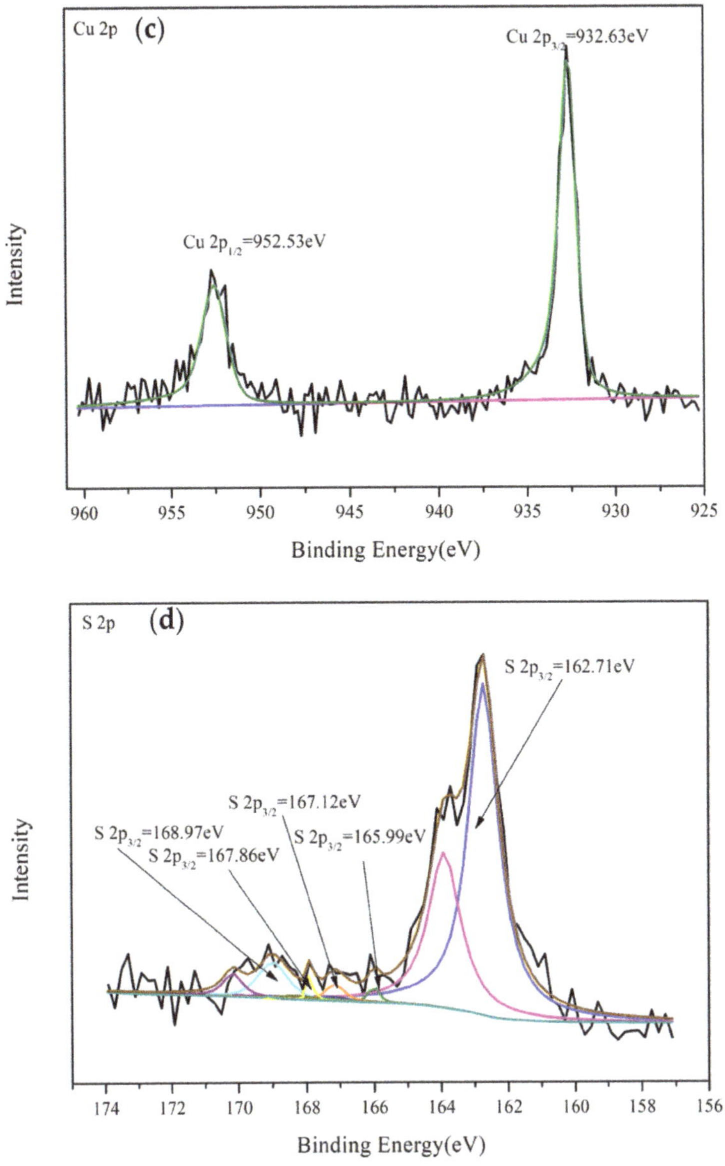

Figure 10. XPS spectrum of the gold foil surface after leaching with nickel-doped pyrite in the Cu^{2+}-en-$S_2O_3{}^{2-}$ gold leaching system: (**a**) Ni 2p, (**b**) Fe 2p, (**c**) Cu 2p, and (**d**) S 2p spectra (pH 9.8–10, temperature 25 °C, time 35 h, solution: 0.1 mol/L $S_2O_3{}^{2-}$, 5 mmol/L $Cu(en)_2{}^{2+}$, 200 mL solution).

Due to the presence of ferrous chloride in the synthetic raw material, it is possible that the nickel in the nickel pyrite is released during the leaching process and thus dissolved in the gold leaching system.

Figure 10b displays the Fe 2p spectrum observed on the surface of the leached gold foil subsequent to the introduction of nickel-bearing pyrite into the leaching system, wherein Fe 2p3/2 exhibits three primary peaks at 706.87 eV, 710.19 eV, and 712.40 eV. The peak at 706.87 eV predominantly corresponds to FeS_2, while the peak at 710.19 eV is attributed to the formation of FeS, and the peak at 712.40 eV arises from adsorption of $FeSO_4$ onto the gold foil surface [30].

In Figure 10c, the Cu 2p3/2 peak at 932.63 eV corresponds to the binding energy of Cu_2S [30,31]. This suggests that when nickel-bearing pyrite was introduced into the Cu^{2+}-en-$S_2O_3^{2-}$ gold leaching system, the only copper-containing substance present on the surface of the leached gold foil was Cu_2S. This is completely different from the situation in which CuS forms the main copper passivation layer on the surface of the gold foil after the addition of natural pyrite. Therefore, the above results indicate that Cu(II), which was originally complexed with en in the system, was reduced to cuprous ions as the leaching time increased.

In Figure 10d, the S 2p3/2 peak at 162.71 eV corresponds to the binding energy of FeS_2, while the peaks at 165.99 eV and 167.12 eV correspond to the binding energies of Na_2SO_3. The S 2p3/2 peaks at 167.86 eV and 168.97 eV correspond to the binding energies of $Na_2S_2O_3$ and $FeSO_4$, respectively [32]. The presence of SO_3^{2-} and SO_4^{2-} indicates significant decomposition and consumption of thiosulfate in the system.

3.4.2. XPS Analysis of Nickel-Bearing Pyrite After Leaching

Figure 11a–c shows the Fe 2p, Cu 2p, and S 2p spectra, respectively, on the surface of the nickel-bearing pyrite after its addition to the Cu^{2+}-en-$S_2O_3^{2-}$ gold leaching system. In Figure 11a, the peak of Cu 2p3/2 (932.91 eV) was determined to be Cu_2S [33], and no Cu^{2+} was present, indicating that the copper ions in the original system were more easily reduced to Cu^+ under the promotion of nickel ions after the addition of nickel-containing pyrite. In Figure 11b, the peak at 706.92 eV represents the binding energy of Fe 2p3/2 in the disulfide found in pyrite, while the binding energy peak of Fe 2p3/2 in FeS was observed at 710.06 eV [30].

In Figure 11c, the S 2p3/2 peak at 162.91 eV corresponds to FeS_2, and the S 2p3/2 peak at 168.89 eV corresponds to $FeSO_4$ [10]. Compared with the spectral diagram of S 2p3/2 on the surface of the leached gold foil and nickel-containing pyrite, thiosulfate adsorption was not observed on the surface of the leached nickel-containing pyrite. Owing to the addition of nickel-bearing pyrite, a significant amount of thiosulfate in the system was catalytically decomposed, resulting in only a small quantity remaining. Consequently, a substantial amount of sulfate, which is the final decomposition product of thiosulfate, was adsorbed onto the surface of the leached nickel-bearing pyrite.

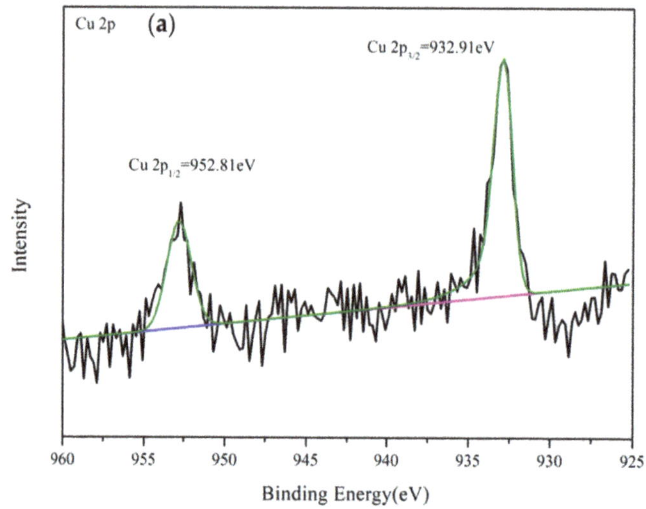

Figure 11. *Cont.*

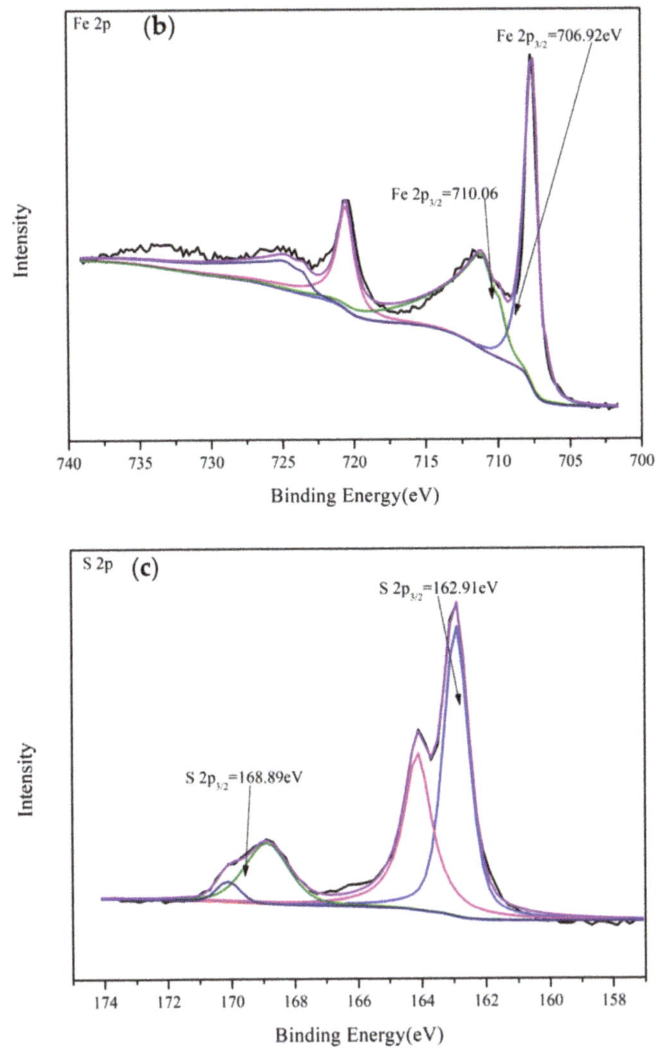

Figure 11. XPS spectra of the leached pyrite with nickel-doped pyrite in the Cu^{2+}-en-$S_2O_3{}^{2-}$ leaching system: (**a**) Cu 2p, (**b**) Fe 2p, and (**c**) S 2p spectra (pH 9.8–10, temperature 25 °C, time 35 h, solution: 0.1 mol/L $S_2O_3{}^{2-}$, 5 mmol/L $Cu(en)_2{}^{2+}$, 200 mL solution).

3.4.3. SEM Analysis of Nickel-Bearing Pyrite After Leaching

After introducing nickel-bearing pyrite into the Cu^{2+}-en-$S_2O_3{}^{2-}$ gold leaching system for the gold leaching test, the nickel-bearing pyrite was analyzed using SEM-EDS(as shown in Figure 12). A comparison of the surface topography of the nickel-bearing pyrite before and after the test revealed that a large number of fine particles or powder was adsorbed onto the surface of the leached nickel-bearing pyrite (Figure 12a). Moreover, the cubic particles of the leached nickel-bearing pyrite became smaller and agglomerated, forming more spherical particles or clumps. EDS analysis showed that the adsorbed materials on the surface of the leached nickel-bearing pyrite mainly contained Cu, Fe, Ni, Au, S, and other elements, indicating that nickel-bearing pyrite adsorbed some gold complexes (Figure 12b,c).

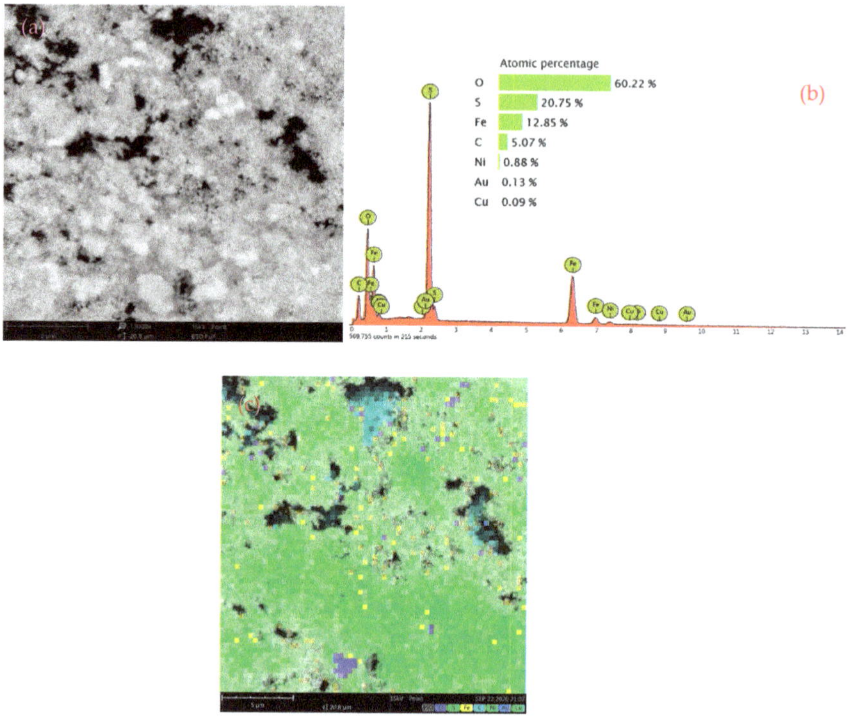

Figure 12. (**a**) SEM image of the leached nickel-doped pyrite surface in the Cu^{2+}-en-$S_2O_3{}^{2-}$ leaching system, (**b**) corresponding EDS spectrum, and (**c**) element map (pH 9.8–10, temperature 25 °C, time 35 h, solution: 0.1 mol/L $S_2O_3{}^{2-}$, 5 mmol/L $Cu(en)_2{}^{2+}$, 200 mL solution).

Owing to the high consumption of thiosulfate in the leaching system with nickel-bearing pyrite added, when the concentration of thiosulfate in the solution was very low or almost nonexistent, the stability of Au^+ and the number of adsorption sites for thiosulfate and gold thiosulfate complex ions decreased. Consequently, the adsorption of gold on the mineral surfaces increased, which led to a phenomenon known as gold robbing [20].

3.4.4. Mechanism of the Catalytic Decomposition of Thiosulfate by Nickel-Bearing Pyrite in the Cu^{2+}-en-$S_2O_3{}^{2-}$ Gold Leaching System

During gold leaching in the Cu^{2+}-NH_3-$S_2O_3{}^{2-}$ system, pyrite undergoes oxidation by dissolved oxygen and copper (II), which can be represented by the following simplified equation (based on the HSC 6.0 database):

$$\begin{gathered} 2FeS_2 + (13/2)O_2 + 4Cu(NH_3)_4^{2+} + 7H_2O + 12S_2O_3^{2-} \rightarrow \\ 2FeO{\cdot}OH + 4(NH_4)_2SO_4 + 4Cu(S_2O_3)_3^{5-} + 8NH_3 + 4H^+\ (\Delta G^\theta = -616.1\ \mathrm{kcal/mol}) \end{gathered} \quad (4)$$

Based on the effects of nickel-bearing pyrite on gold dissolution and thiosulfate and $Cu(en)_2{}^{2+}$ consumption in the Cu^{2+}-en-$S_2O_3{}^{2-}$ leaching system as well as the surface morphology and XPS analysis of the gold foil and nickel-bearing pyrite after leaching, the main adsorbed substances on the surface of the gold foil were determined to be NiS, $NiSO_4$, FeS_2, $FeSO_4$, and Cu_2S. This indicates that nickel dissolved in the system and the free nickel ions combined with $SO_4{}^{2-}$ and S^{2-}, resulting in the partial adsorption of NiS and $NiSO_4$ onto the gold foil, which adversely affected gold dissolution. Based on the changes in $Cu(en)_2{}^{2+}$ consumption in the system and the formation of Cu_2S on the surface of the leached gold foil, a portion of the nickel in the nickel-bearing pyrite dissolved in the system to form numerous free Ni^{2+}. These Ni ions accelerated the conversion of Cu^{2+} to Cu^+.

As the complexation of en with Ni^{2+} made chemical Equation (5) faster and Equation (6) more difficult, the concentration of $Cu(en)_2{}^{2+}$ continued to decrease over a short period of time. Cu(I) was reoxidized to Cu(II) by the dissolved oxygen in the leaching solution; however, the oxidation of the complex formed between Cu(I) and en by dissolved oxygen occurred at a slower rate than the complex reaction between nickel ions and free en (5); as shown in Table 2, the $Ni(en)_2{}^{2+}$ was not as stable as $Cu(en)_2{}^{2+}$. Therefore, after a certain period, the concentration of $Cu(en)_2{}^{2+}$ gradually increased.

$$Ni^{2+} + 2en + 2e^- \rightarrow Ni\,(en)_2^{2+} \tag{5}$$

$$Cu(en)_2^{2+} + e^- \rightarrow Cu\,(en)^+ + en \tag{6}$$

$$en + Cu(en)^+ + O_2 + 2H_2O + e^- \rightarrow Cu\,(en)_2^{2+} + 4OH^- \tag{7}$$

Table 2. The equilibrium constants (logK[a]) for the complexation of Ni^{2+}, Cu^+, Cu^{2+} with en: ionic strength (I = 0), 25 °C.

Metal Ion	Complexation	LogK
Ni^{2+}	$Ni(en)^{2+}$	7.52
	$Ni(en)_2{}^{2+}$	13.84
Cu^+	$Cu(en)_2{}^+$	10.8
Cu^{2+}	$Cu(en)^{2+}$	10.67
	$Cu(en)_2{}^{2+}$	20

a LogK taken from Lange's Handbook of Chemistry, 15th Ed.

The consumption of thiosulfate in the gold leaching system with the addition of nickel-containing pyrite increased sharply with increasing nickel content, indicating that the consumption of thiosulfate was related to nickel doping. First, the complexation of dissolved nickel ions with en accelerated the progress of Equation (8), thereby expediting the consumption of thiosulfate. Second, the nickel doping of pyrite increased the donor energy level of the pyrite semiconductor. The rate-controlling step of thiosulfate decomposition is the process in which oxygen accepts electrons provided by pyrite; therefore, a higher nickel doping content led to more donor impurities in the pyrite semiconductor and a narrower bandgap width, which was conducive to the rapid progress of the electron supply process in pyrite. Therefore, the process of electron acceptance by oxygen was promoted, and thiosulfate consumption increased with increasing Ni-doping content [22].

$$2Cu(en)_2^{2+} + 6S_2O_3^{2-} \rightarrow 2Cu\,(S_2O_3)_2^{5-} + 4en + S_4O_6^{2-} \tag{8}$$

4. Conclusions

The effects of the impurity content in nickel-containing pyrite on the light absorption and photocatalytic degradation properties as well as the catalytic degradation of thiosulfate and consumption of $Cu(en)_2{}^{2+}$ in the Cu^{2+}-en-$S_2O_3{}^{2-}$ system were studied. The main conclusions are as follows:

(1) In nickel-doped pyrite, the amount of nickel doping significantly affected the light absorption performance of pyrite. As the level of nickel doping increased, the photocatalytic performance of nickel-doped pyrite was enhanced between 2 and 4 hours.

(2) In the Cu^{2+}-en-$S_2O_3{}^{2-}$ gold leaching system, the passivation effect of nickel-bearing pyrite on gold dissolution was enhanced by increasing the nickel content.

(3) With an increase in the nickel content (from 2 to 10%) in pyrite, the catalytic degradation of thiosulfate was also enhanced. This is mainly because the nickel in nickeliferous pyrite forms complexes with en solution, resulting in the accelerated oxidation decomposition of thiosulfate to produce en and $SO_4{}^{2-}$. Additionally, a higher nickel content leads to an increased presence of donor impurities in pyrite semiconductors. The faster the electron

supply process of pyrite, the more it promotes the electron acceptance of oxygen, thereby accelerating the rate of oxidation and decomposition of thiosulfate.

Supplementary Materials: The following supporting information can be downloaded at: https://www.mdpi.com/article/10.3390/met14111256/s1.

Author Contributions: Conceptualization, X.Q. and W.Q.; Methodology, T.Z.; Investigation, T.Z.; Resources, T.Z. and W.Q.; Data curation, T.Z.; Writing—original draft, X.Q.; Writing—review & editing, X.Q.; Supervision, H.Z.; Project administration, H.Z.; Funding acquisition, T.Z. All authors have read and agreed to the published version of the manuscript.

Funding: This research was funded by Inner Mongolia University of Technology grant number [ZZ202115]. Basic Scientific Research Expenses Program of Universities directly under Inner Mongolia grant number [JY20220155]. Funded by Open Foundation of State Key Laboratory of Mineral Processing grant number [BGRIMM-KJSKL-2024-29].

Data Availability Statement: All relevant data are within the paper.

Conflicts of Interest: We declare that we have no known competing financial interests or personal relationships that could have appeared to influence the work reported in this paper.

References

1. Lampinen, M.; Laari, A.; Turunen, I. Ammoniacal thiosulfate leaching of pressure oxidized sulfide gold concentrate with low reagent consumption. *Hydrometallurgy* **2015**, *151*, 1–9. [CrossRef]
2. Dai, X.; Jeffrey, M.I. The effect of sulfide minerals on the leaching of gold in aerated cyanide solutions. *Hydrometallurgy* **2006**, *82*, 118–125. [CrossRef]
3. Hasab, M.G.; Rashchi, F.; Raygan, S. Chloride–hypochlorite leaching and hydrochloric acid washing in multi-stages for extraction of gold from a refractory concentrate. *Hydrometallurgy* **2014**, *142*, 56–59. [CrossRef]
4. Hashemzadehfini, M.; Ficeriová, J.; Abkhoshk, E.; Shahraki, B.K. Effect of mechanical activation on thiosulfate leaching of gold from complex sulfide concentrate. *Trans. Nonferrous Met. Soc. China* **2011**, *21*, 2744–2751. [CrossRef]
5. Hasab, M.G.; Raygan, S.; Rashchi, F. Chloride–hypochlorite leaching of gold from a mechanically activated refractory sulfide concentrate. *Hydrometallurgy* **2013**, *138*, 59–64. [CrossRef]
6. Mahandra, F.G.A. Novel Extraction Process for Gold Recovery from Thiosulfate Solution Using Phosphonium Ionic Liquids. *ACS Sustain. Chem. Eng.* **2021**, *9*, 8179–8185. [CrossRef]
7. Fleming, C.A.; McMullen, J.; Thomas, K.G.; Wells, J.A. Recent advances in the development of an alternative to the cyanidation process: Thiosulfate leaching and resin in pulp. *Min. Metall. Explor.* **2003**, *20*, 1–9. [CrossRef]
8. Feng, D.; Van Deventer, J.S. Ammoniacal thiosulphate leaching of gold in the presence of pyrite. *Hydrometallurgy* **2006**, *82*, 126–132. [CrossRef]
9. Yang, Y.B.; Zhang, X.; Bin, X.U.; Qian, L.I.; Jiang, T.; Wang, Y.X. Effect of arsenopyrite on thiosulfate leaching of gold. *Trans. Nonferrous Met. Soc. China* **2015**, *25*, 3454–3460. [CrossRef]
10. Xu, Y.; Schoonen, M.A. The stability of thiosulfate in the presence of pyrite in low-temperature aqueous solutions. *Geochim. Cosmochim. Acta* **1995**, *59*, 4605–4622. [CrossRef]
11. Nie, Y.; Wang, Q.; Zi, F.; Hu, X.; Chen, Y.; Guo, L.; Yu, H. The catalytic decomposition of thiosulfate by pyrite. *Appl. Surf. Sci.* **2018**, *436*, 217–223. [CrossRef]
12. Ma, W.; Chen, J.; Zhang, G.; Li, Y.; Feng, Q. The effect of As doping concentration on the electronic structure of FeS_2. *Chin. J. Nonferrous Met.* **2017**, *27*, 414–422. (In Chinese)
13. Chen, J.H.; Ye, C.; Li, Y.Q. Effect of vacancy defects on electronic properties and activation of sphalerite (110) surface by first-principles. *Trans. Nonferrous Met. Soc. China* **2010**, *20*, 502–506. [CrossRef]
14. Caban-Acevedo, M.; Faber, M.S.; Tan, Y.; Hamers, R.J.; Jin, S. Synthesis and properties of semiconducting iron pyrite (FeS_2) nanowires. *Nano Lett.* **2012**, *12*, 1977–1982. [CrossRef] [PubMed]
15. Chu, F.; Sun, G.; Hu, R.; Hu, D.; Wei, C. Electron Hole-Core Characteristics of Pyrites from the Main Types of Gold Deposits in China and Affecting Factors. *Acta Mineral. Sin.* **2004**, *24*, 211–217. (In Chinese)
16. Jefferson, M.; Yenial-Arslan, U.; Evans, C.; Curtis-Morar, C.; O'Donnell, R.; Parbhakar-Fox, A.; Forbes, E. Effect of pyrite textures and composition on flotation performance: A review. *Miner. Eng.* **2023**, *201*, 108234. [CrossRef]
17. Li, Y.Q.; Chen, J.H.; Guo, J. DFT study of influences of As, Co and Ni impurities on pyrite (100) surface oxidation by O_2 molecule. *Chem. Phys. Lett.* **2004**, *24*, 211–217. [CrossRef]
18. Wu, J.; Ma, W.; Zhang, G.-F.; Feng, Q.-M. First-Principle Study of Electronic Structure and Optical Property of Cu/Co Doped FeS_2. *Acta Opt. Sin.* **2016**, *36*, 1016001. (In Chinese)
19. Mingyue, Z. Ni-Doped FeS: Solvothermal Synthesis and the Visible-Light Photocatalytic Properties. *Chin. J. Inorg. Chem.* **2015**, *31*, 1119–1124. (In Chinese)

20. Wang, Q.; Hu, X.; Zi, F.; Yang, P.; Chen, Y.; Chen, S. Environmentally friendly extraction of gold from refractory concentrate using a copper–ethylenediamine–thiosulfate solution. *J. Clean. Prod.* **2019**, *214*, 860–872. [CrossRef]
21. Nie, Y.; Yu, Q.; Hu, X.; Zi, F.; Yu, H. The effect of ammonia on the anodic process of gold in copper-free thiosulfate solution. *J. Electrochem. Soc.* **2016**, *163*, E123–E129. [CrossRef]
22. Nie, Y.; Chi, H.; Zi, F.; Hu, X.; Yu, H.; He, S. The effect of cobalt and nickel ions on gold dissolution in a thiosulfate-ethylenediamine (en)-Cu^{2+} system. *Miner. Eng.* **2015**, *83*, 205–210. [CrossRef]
23. Cao, Y.; Sun, Y.; Gao, P.; Han, Y.; Li, Y. Mechanism for suspension magnetization roasting of iron ore using straw-type biomass reductant. *Int. J. Min. Sci. Technol.* **2021**, *31*, 9–14. [CrossRef]
24. Zhang, X.; Kou, J.; Sun, C. A comparative study of the thermal decomposition of pyrite under microwave and conventional heating with different temperatures. *J. Anal. Appl. Pyrolysis* **2019**, *138*, 41–53. [CrossRef]
25. Lin, X.; Xia, Y.; Wei, G.; Zhou, J.; Liang, X.; Xian, H.; Zhu, J.; He, H. Distinct effects of transition metal (cobalt, manganese and nickel) ion substitutions on the abiotic oxidation of pyrite: In view of hydroxyl radical production. *Geochim. Cosmochim. Acta* **2022**, *321*, 170–183. [CrossRef]
26. Xu, Y.; Schoonen, M.A.; Strongin, D.R. Thiosulfate oxidation: Catalysis of synthetic sphalerite doped with transition metals. *Geochim. Cosmochim. Acta* **1996**, *60*, 4701–4710. [CrossRef]
27. Ou, Y.; Yang, Y.; Li, K.; Gao, W.; Wang, L.; Li, Q.; Jiang, T. Eco-friendly and low-energy innovative scheme of self-generated thiosulfate by atmospheric oxidation for green gold extraction. *J. Clean. Prod.* **2023**, *387*, 135818. [CrossRef]
28. Feng, D.; Van Deventer, J.S. Thiosulphate leaching of gold in the presence of orthophosphate and polyphosphate. *Hydrometallurgy* **2011**, *106*, 38–45. [CrossRef]
29. Zhou, Z.; Chen, Z.; Weyer, S.; Horn, I.; Huo, H.; Zhang, W.; Li, N.; Zhang, Q.; Han, F.; Feng, H. Metal source and ore precipitation mechanism of the Ashawayi orogenic gold deposit, southwestern Tianshan Orogen, western China: Constraints from textures and trace elements in pyrite. *Ore Geol. Rev.* **2023**, *157*, 105452. [CrossRef]
30. Velásquez, P.; Leinen, D.; Pascual, J.; Ramos-Barrado, J.R.; Grez, P.; Gomez, H.; Schrebler, R.; Del Río, R.; Cordova, R. A chemical, morphological, and electrochemical (XPS, SEM/EDX, CV, and EIS) analysis of electrochemically modified electrode surfaces of natural chalcopyrite ($CuFeS_2$) and pyrite (FeS_2) in alkaline solutions. *J. Phys. Chem. B* **2005**, *109*, 4977–4988. [CrossRef]
31. Acres, R.G.; Harmer, S.L.; Beattie, D.A. Synchrotron XPS studies of solution exposed chalcopyrite, bornite, and heterogeneous chalcopyrite with bornite. *Int. J. Miner. Process.* **2010**, *94*, 43–51. [CrossRef]
32. Laajalehto, K.; Kartio, I.; Nowak, P. XPS study of clean metal sulfide surfaces. *Appl. Surf. Sci.* **1994**, *81*, 11–15. [CrossRef]
33. Deroubaix, G.; Marcus, P. X-ray photoelectron spectroscopy analysis of copper and zinc oxides and sulphides. *Surf. Interface Anal.* **1992**, *18*, 39–46. [CrossRef]

metals

MDPI

Article

Microscopic Simulation of RE^{3+} Migration in Ion-Type Rare Earth Ores Based on Navier–Stokes Equation—Exchange Reaction—Ion Migration Coupling

Dan Wang [1,2], Fuyu Wu [3,*], Yunzhang Rao [4,*], Zhilian Zhao [1], Wei Xu [4,5] and Min Han [4]

[1] Faculty of Resources and Civil Engineering, Gannan University of Science and Technology, Ganzhou 341000, China; 7120170006@mail.jxust.edu.cn (D.W.); 9320230084@gnust.edu.cn (Z.Z.)
[2] Ganzhou Key Laboratory of Mine Geological Disaster Prevention and Control and Ecological Restoration, Ganzhou 341000, China
[3] Jiangxi Burequ of Geology Non-Ferrous Geological Brigade, Ganzhou 341000, China
[4] School of Resources and Environmental Engineering, Jiangxi University of Science and Technology, Ganzhou 341000, China; 7120220067@mail.jxust.edu.cn (W.X.); han@jxust.edu.cn (M.H.)
[5] The Seventh Geological Brigade of Jiangxi Bureau of Geology, Ganzhou 341000, China
* Correspondence: 6720200478@mail.jxust.edu.cn (F.W.); raoyunzhang@jxust.edu.cn (Y.R.)

Abstract: In the in-situ leaching method of ionic rare earth, ion exchange reaction between rare earth ions and leaching agent ions is carried out, which allows the rare earth ions to be leached from the ore body as the leaching solution flows through the pores. This indicates that the leaching process of rare earth ions is closely related to the seepage field, ion exchange field, and ion migration process of the leaching solution. In this study, an ionic rare earth mine located in Longnan of Jiangxi Province was taken as the research object. By conducting nuclear magnetic resonance scanning on the ore samples of this mine and vectorizing the nuclear magnetic resonance images, a two-dimensional geometric model of pores was obtained. Then, COMSOL Multiphysics software was used to establish a coupled numerical model of seepage–exchange–migration of the ionic rare earth mine during the leaching process at the pore scale to study the seepage situation of leaching solution with different injection strengths and concentrations, as well as the exchange and migration process. The results show that increasing the concentration of magnesium ions can increase the difference of ion diffusion concentration, accelerate the forward exchange rate of ions, promote the forward exchange reaction, and improve the concentration gradient of rare earth ions in the leaching solution. The more significant the diffusion effect, the higher the ion migration rate, while at the same time inhibiting the reverse adsorption of rare earth ions, and accelerating the leaching efficiency of rare earth ions. In addition, increasing the strength of the injection solution allows rare earth ions to leach out of the ore body earlier, shortens the leaching cycle, and thus reduces the peak concentration of leached rare earth ions. By analyzing the effects of the strength of the injection solution and leaching concentration on ionic rare earth leaching, the influence of those two factors on engineering economy can be briefly evaluated, which can be provided as a reference for the optimization of ionic rare earth mining technology.

Keywords: ionic rare earth; nuclear magnetic resonance; pore mesoscopic structure; mesoscopic migration law; leaching efficiency

1. Introduction

Rare earth elements in ionic rare earth minerals are adsorbed on the surface of clay minerals in the ionic phase form, which has certain chemical stability, and which conventional methods cannot effectively enrich [1]. According to the physical and chemical characteristics of ionic rare earth minerals, after three generations of technological innovation, the in-situ leaching process has been widely promoted [2,3]. In-situ leaching technology mainly relies on a leaching solution to extract and collect rare earth ions in order to mine ionized rare earth ore. In the process of seepage in the pores of orebodies, the flow of the leaching solution is not only related to the type of liquid (liquid density, viscosity, and other factors), but also affected by pore friction, hydraulic shear, particle shape resistance, and natural gravity. The exchange reaction between active cations and rare earth ions in the leaching solution causes a change of solute content, which further affects the flow of the leaching solution.

The flow of leaching solution also affects the migration of various ions. In addition to convection, it can also affect the migration of ions through dispersion, diffusion, and other effects. Therefore, the leaching process of ionic rare earth is a multi-field coupling process of leaching fluid seepage, active cation and rare earth ion exchange, and ion migration.

Smooth seepage channels inside the ore body are conducive to the mutual contact between cations and rare earth ions, and also allow rare earth ions to be leached more smoothly with the leaching solution, thereby improving the leaching efficiency. Therefore, the hydraulic properties of clay minerals such as particle size (solid skeleton), pore structure, and permeability are important factors affecting the leaching process. At present, scholars mainly study the basic leaching theory of ionic rare earth based on laboratory tests, and focus on pore structure evolution, reduction of ammonia nitrogen pollution, and the mechanism of ore body strength weakening [4–6]. Most laboratory tests can only summarize the initial ore results to draw rules, and cannot intuitively show the hydrodynamic distribution of the leaching solution in the pores of the ore body, the ion exchange situation, the spatio-temporal distribution of ions during solute migration, etc. It is difficult to obtain instantaneous data in a non-equilibrium state. Therefore, it is difficult to describe the essential laws of the leaching process of ionic rare earth under multi field coupling relying solely on laboratory tests.

With the application and development of computer technology, numerical simulation test methods have gradually developed into an effective means of scientific research, as important as theoretical analysis and experimental methods [7,8]. Compared with laboratory tests, numerical simulation tests have the advantage of being able to consider multiple interactions and major control processes simultaneously. A model with complex initial parameters and boundary conditions can be simulated effectively [9]. Therefore, the use of a numerical simulation test can make up for the shortcomings of laboratory test research methods, directly display the transient data in the test process, and deepen the transient studies of the leaching process of ionic rare earth ore under the action of multi-field coupling. At present, in multi-field coupling simulation studies of the leaching process, the ore body is often regarded as a porous medium, and a uniform model is established with the help of the average concept to describe and simulate the leaching process [10–13]. However, without considering the non-uniformity of rare earth ore particle size and the disordered distribution of particles on seepage, it is impossible to accurately describe the spatiotemporal distribution of hydrodynamic parameters in various pores. With the development of modern testing technologies (CT, NMR, etc.), the internal pores of ionic rare earth ore bodies can be accurately displayed by these testing methods, laying a foundation for establishing the internal seepage channel model of ore bodies through pore images [14–16]. Sheikhzadeh [17] established an unsaturated seepage model of leach in two-dimensional uniform spherical ore particles considering the mass conservation relationship between the liquid phase in the particles and the ore layer. The model was solved by the implicit FDM method to study the influence of periodic infiltration of leach on vertical velocity distribution and saturation of ore body. Wu Chengyou [18] used the Boltzmann model of coupled ion exchange to study the seepage of leaching solution during the leaching process, and analyzed the influence of injection intensity and temperature on the rare earth ions transport. Liu Qingsheng [19] combined fluid dynamics, chemical reaction dynamics, and rock and soil mechanics to construct a coupled percolation–react–stress model of the leaching process, and studied the spatio-temporal evolution of seepage field, ion concentration field, and stress field during column leaching under different coaxial pressure, confining pressure and injection strength through COMSOL Multiphysics. Combining hydrodynamics and chemical reaction dynamics, Zeng Jia [20] explored the seepage law of leaching solution and the influence of ion exchange under different conditions by CFD simulation method, and put forward reasonable suggestions on the setting of injection strength. Wu [21] established a fully coupled model of flow–reaction–deformation–mass–transfer in the leaching process based on existing results with the consideration of the deformation factors of ore during the leaching process, and studied the distribution of porosity, saturation, leaching solution concentration, and leaching rare earth ion concentration in an ore heap under one-dimensional fixed spray and constant water head conditions. Liu Jinzhi [22] established the elastic deformation seepage control equation and mass transfer control equation of ore heap leaching, solved these two control equations through porosity coupling, and simulated the change law of leaching solution and leaching ion concentration during constant head leaching. Jiang Wang [23] constructed a real 3D model of an ionic rare earth mine using topographic data of the test stope, and constructed a coupling model of chemical react–dilute matter transfer based on Darcy's law through the reaction particle bed module in COMSOL Multiphysics to study the spatio-temporal distribution and evolution of process rare earth ions during the leaching process.

In this paper, NMRI (nuclear magnetic resonance imaging) technology was applied to construct a geometric model of the internal seepage channel of rare earth ore body at pore scale, combined with the momentum conservation equation of motion of viscous incompressible fluid (Navier–Stokes equation, hereinafter referred to as: N-S equation), considering the mechanism of ion exchange (adsorption and desorption) and solute convection and diffusion, and a coupled numerical model of ionic rare earth minerals at pore scale was established. With the help of COMSOL Multiphysics, the leaching process multi-field coupling under saturation leaching state was simulated, the seepage law, ion exchange law, and solute migration law of the leaching solution in the pores were summarized, and the process was directly explored. The research results obtained can provide reference for the optimization of ionic rare earth ore mining technology.

2. Theory and Experimental Process

2.1. Theory and Mathematical Model

1. Seepage control equation

In this study, we assume that the leaching solution is an incompressible fluid, and its flow process in the pores of the mineral soil can be described by the Navier–Stokes equation [24–26]. The continuity equation is as follow:

$$\nabla u = 0 \tag{1}$$

where u is the velocity vector of the fluid. The equation of motion is:

$$\rho\left[\frac{\partial u}{\partial t} + (u\nabla)u\right] = \nabla\left[-pI + \mu(\nabla u) + (\nabla u)^T\right] + F \tag{2}$$

where P is fluid pressure, Pa; ρ is the fluid density, kg/m^3; μ is hydrodynamic viscosity; I is the unit vector; t is time, s; and F is the volume force vector, N/m^3.

2. Control equation of reaction kinetics

The reaction relationship between $MgSO_4$ and ionic rare earth is expressed by the Arrhenius equation. At 25 °C, the activation energy of $MgSO_4$ and ionic rare earth is 7.77 kJ/mol, and the equilibrium constant of the reaction is 20.

$$R_{re} = 0.09528 \times C_0^{1.554} \times e^{-\frac{E}{RT}} \tag{3}$$

where R_{re} is the Arrhenius reaction rate of $MgSO_4$ leaching rare earth; E is the activation energy of chemical reaction, J/mol; R is the ideal gas constant, 8.314 J/(mol·K); and T is the temperature, 298.15 K.

When the leaching solution flows within the pores of the ore body, it also undergoes exchange reactions with rare earth ions adsorbed on the surface of the mineral soil particles [27]. Assuming that rare earth ions undergo tangential transfer along the surface, this process can be described by Fick's law:

$$N_{t,i} = -D_{s,i} \nabla_t c_{s,i} \tag{4}$$

where $N_{t,i}$ is the surface molar flux, mol/(m·s); $D_{s,i}$ is the surface diffusion coefficient of substance i, m^2/s; and $C_{s,i}$ is the concentration of surface substance i, mol/m^2. The controlling equation for the concentration of each substance on the surface of mineral soil particles is:

$$\frac{\partial c_{s,i}}{\partial t} = -\nabla_t N_{t,i} + R_{s,i} \tag{5}$$

where $R_{s,i}$ is the sum of source terms caused by surface reaction and adsorption analysis, $mol/(m^2 \cdot s)$.

3. Control equation of solute transport

After the ion exchange reaction, rare earth ions enter into the leaching solution and transfer in the pore of the ore during the seepage process [28]. This process is mainly controlled by convection and diffusion; therefore, the migration process of rare earth ions in the pores of the ore body during the leaching process can be described by the convection–diffusion equation:

$$\frac{\partial c_i}{\partial t} + u \nabla c_i = R_i + \nabla D_i \nabla c_i \tag{6}$$

where C_i is the concentration of substance i, mol/m^3; D_i is the diffusion coefficient of substance i, m^2/s; R_i is the reaction rate of substance i, $mol/(m^3 \cdot s)$; and u is the velocity vector of the fluid, m/s.

2.2. Column Leaching Experiment

2.2.1. Soil Sampling

The study area is an ionic rare earth mining area in Longnan, Ganzhou of Jiangxi Province, as shown in Figure 1. Longnan is situated in the southernmost part of southern Jiangxi Province, which has a typical subtropical monsoon climate with the same period of rain and heat. The landform of the mine is mainly hilly and is a weathering denudation landform. The distribution of ore bodies in the area is consistent with the granite weathering layer, and the ore types are mainly ion adsorption type. The mine profile is divided from top to bottom into topsoil, weathered layer containing rare earth elements (total regolith and semi-regolith) and bedrock. The vertical thickness of the topsoil is about 0.4~1.2 m, which is composed of humus, sand, clay and quartz gravel. Due to the fuzzy interface between weathered layers, it is difficult to classify them. Generally, rock layers other than topsoil and bedrock are considered as ore body layers. The vertical thickness of the ore layer is 6.0~10.0 m, and the thickest part can reach 13.0 m. The bedrock layer is mainly composed of biotite granite and muscovite granite, which has a compact structure and low water permeability and is a good natural water barrier floor. The main groundwater types in this area are loose rock pore water and bedrock fissure water, the former with a buried depth of 0.5~1.0 m, and the latter with a buried depth of 2.6~4.5 m, mainly diving type. The geological data of the mine show that the rare earth minerals in this area are mainly phosphorite, with an average grade of 0.0877%. At present, the mine adopts an in-situ leaching method, and the impurity elements in the mother liquor are mainly aluminum and iron.

Figure 1. Aerial view of the mining area.

To obtain the physical properties of the soil in the test stope more truly, a luoyang shovel with diameter of 160 mm and a shovel were used to remove the humus layer on the surface first, and the shovel was applied to shovel to 0.5 m of ore soil before sampling. The sampling process was as follows: Dig 1 m each time, take 3 soil samples at the same depth, totaling 9 soil samples, and wrap the extracted soil samples with plastic wrap to avoid excessive evaporation of water from soil samples during subsequent transportation, which may affect the determination of physical properties of undisturbed soil [29]. At the same time, the remaining loose soil after sampling of the same depth was packed and brought back with woven bags and marked on the outside. The sampling site of the mineral soil is shown in Figure 2.

Figure 2. Rare earth sampling site.

The ring tool method was used to drill a number of samples of different depths, and the density of the samples at each depth was tested [30]. Porosity is not only the main index to reflect the pore development of soil, but also a significant parameter to reflect the pore structure of mineral soil. Accurately measuring porosity and porosity ratio during this leaching process has significant practical significance. Based on the specific gravity and density ρ of the mineral soil, the porosity and porosity ratio were calculated as shown in Table 1. Using the vibrating screen method to screen the mineral soil at different depths, different mesh sieves were stacked on the vibrating screen machine. Samples of 500 g at different depths were placed in the top sieve, and the mineral soil was graded and vibrated for 10~15 min. The mineral soil mass on the sieve with different pore sizes was weighed and the proportion of particle size range was calculated. The particle size distribution results of samples at different depths are shown in Table 2.

Table 1. Soil sample density.

Soil Sample Number	Volume/cm^3	Weight/g	Natural Density/(g/cm^3)	Mean Density/(g/cm^3)	Porosity (n)	Void Ratio (e)
1	50.0	68.51	1.37			
2	50.0	67.85	1.36	1.38	0.473	0.898
3	50.0	70.38	1.41			

Table 2. Particle size distribution of soil samples at different depths.

Particle Size	Particle Diameter/mm					
	>5	2.5~5	1~2.5	0.5~1	0.075~0.5	<0.075
Percent	12.83%	33.70%	17.53%	16.87%	12.63%	6.43%

2.2.2. Column Leaching Test

The column leaching test device includes a liquid distribution/storage container, a disposable infusion tube, a loading stage, and a container for collecting leach liquid, as shown in Figure 3.

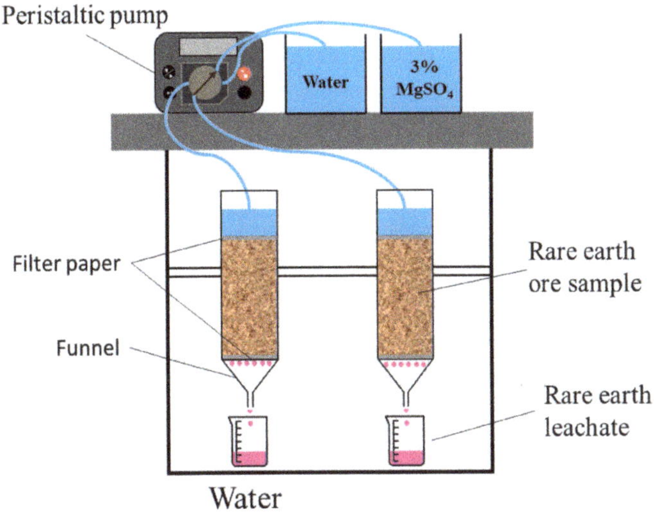

Figure 3. Indoor column immersion device.

After the above research on the leaching kinetics of $MgSO_4$ solution and combined with mine test stope experience, it can be concluded that using a 3% mass fraction $MgSO_4$ solution for leaching is more effective, and other ions in the product can also be significantly reduced. In this experiment, an $MgSO_4$ solution of this concentration was used for column leaching.

In this experiment, one terminal of the infusion tube was introduced into the liquid storage container, while the opposing terminal was positioned above the sample. The infusion rate was calibrated to 1 mL/min, and the infusion device was activated to replicate the leaching process. The leachate was subsequently collected at the outlet, and the concentration of rare earth ions in the leachate was quantified using the EDTA volumetric method. Measurements were conducted every 10 min during the initial phase and every 30 min during the subsequent phase, with data being recorded to elucidate the leaching process [31].

3. Model Establishment and Boundary Condition

The samples used in this study were taken from magnesium salt leaching test stope in Longnan. In order to obtain a more realistic internal pore microstructure of ionic rare earth ore body in the experimental leaching site, luoyang shovels were used to excavate to the bottom of the surface soil and a 2 m long, 42 mm diameter acrylic transparent tube was inserted into the ore body. In this way, an undisturbed sample was obtained when the tube body was extracted. Then, a cylindrical sample with a height of 100 mm with both ends was dug out, the two ends of the sample were sealed with permeable stone, and the pores of the cylindrical sample were filled with water after being treated with water for 48 h. In order to eliminate the disturbance caused by sample interception and closure, a nuclear magnetic resonance instrument was used to scan and image the samples with the height of 50 mm as the center and 30 mm at both ends along the axial direction. The higher the moisture content of the sample, the stronger the NMR signal, and the brighter the pixels displayed in the image, indicating the pore structure of the sample. In order to reflect the characteristics of the aggregate structure of the mineral soil, the gray image of the pore image generated at the position of the axial section was taken, and the image was processed by image processing software, such as light source correction, image segmentation, filtering, and denoising. Then, the solid skeleton and pore channel were digitized, and a two-dimensional geometric model of 42 mm × 60 mm was generated. The two-dimensional numerical model establishment process and the ion migration simulation calculation process are shown in Figure 4.

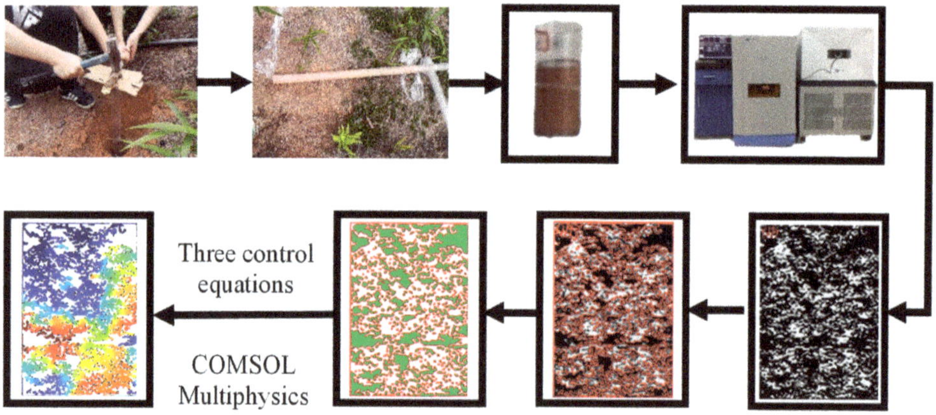

Figure 4. Simulation process of rare earth ion migration.

The NMR analysis and imaging system (MesomR23-060H-I V1, as shown in Figure 5) produced by Suzhou Niumai Analytical Instrument Co., Ltd. (Suzhou, China) was used to conduct NMR scanning and imaging of mineral samples. In this experiment, we set the magnetic field strength to 0.5 T, the probe coil diameter φ = 60 mm, and an effective testing length of 60 mm to conduct detection and analysis of the physical properties of samples, such as porosity, pore size distribution, oil/water saturation, and crack development, as well as the pore structure.

As the image analysis and processing technology has been widely used in aerospace, biomedicine, military and other fields, the development of this testing technology provides a possibility for the construction of ionic rare earth porous seepage channels [32–34]. In this study, Raster2Vector advanced raster image vectoring software system (hereinafter referred to as R2V) with good adaptability and high precision was adopted to vectorize the nuclear magnetic image of the pores of ionic rare earth ore, and complete the construction of its seepage channel, as shown in Figures 6 and 7. The detailed steps are as follows:

1. Image binarization and sharpening. NMR images were imported into R2V software and binarized. Image binarization was the process to control the grayscale value of each pixel in each image within the range of [0, 255], while using high-pass filtering to enhance the high-frequency semaphores of the image and filter the low frequency signals to achieve the purpose of sharpening the image, so as to make the outline of the ore body particle skeleton clearer and facilitate the edge detection and contour extraction in the later stage.
2. Threshold segmentation. The low threshold of the image obtained in the previous step was set to 0 and the high threshold to 56. Through the control of the high and low threshold, the area of the ore body particle skeleton located in the image was determined, which can effectively distinguish the area where the ore body particles and pores flow.
3. Edge detection. The Sobel operator was used to detect the edge of the image after threshold adjustment. The Sobel operator is a filter operator based on first derivative, which can detect the target edge contour effectively by fast convolution function. The Sobel operator usually includes a set of horizontal and vertical 3 × 3 matrices. The approximate brightness difference values of the horizontal and vertical images can be obtained by convolution of the image with the plane. The matrix formula is shown as follows:

$$S_x = \begin{bmatrix} -1 & 0 & +1 \\ -2 & 0 & +2 \\ -1 & 0 & +1 \end{bmatrix} S_y = \begin{bmatrix} +1 & +2 & +1 \\ 0 & 0 & 0 \\ -1 & -2 & -1 \end{bmatrix} \tag{7}$$

Horizontal and vertical edge detection images are expressed as: $G_x = S_x \times A$, $G_y = S_y \times A$, where A is the original image. The approximate values of horizontal and vertical gradients of each pixel in the edge detection image are calculated as follows: $G = \sqrt{G_x^2 + G_y^2}$. The direction of the gradient can be calculated by $\theta = tan^{-1}\left(\frac{G_y}{G_x}\right)$.

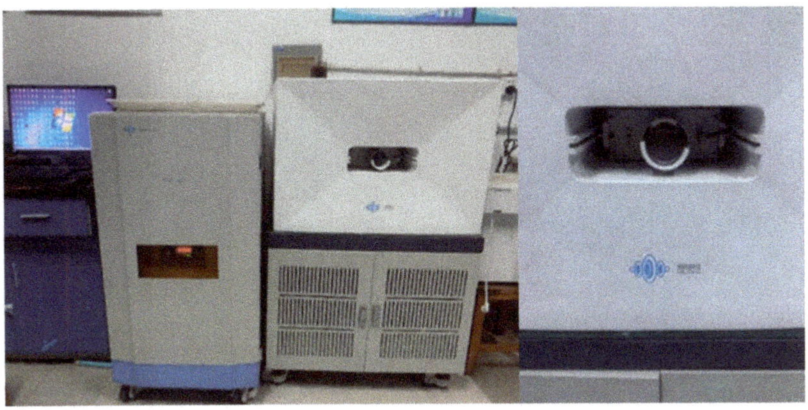

Figure 5. NMR scanning of rare earth pillars.

4. Contour extraction and edge processing. The frame edge of ore body particles detected by Sobel operator was vectorized by raster editing, and the edge contour vector of ore body particles was generated. The Sobel operator has a positive effect on the edge detection of the target, and the image noise after sharpening can be effectively suppressed, as shown in Figure 7.

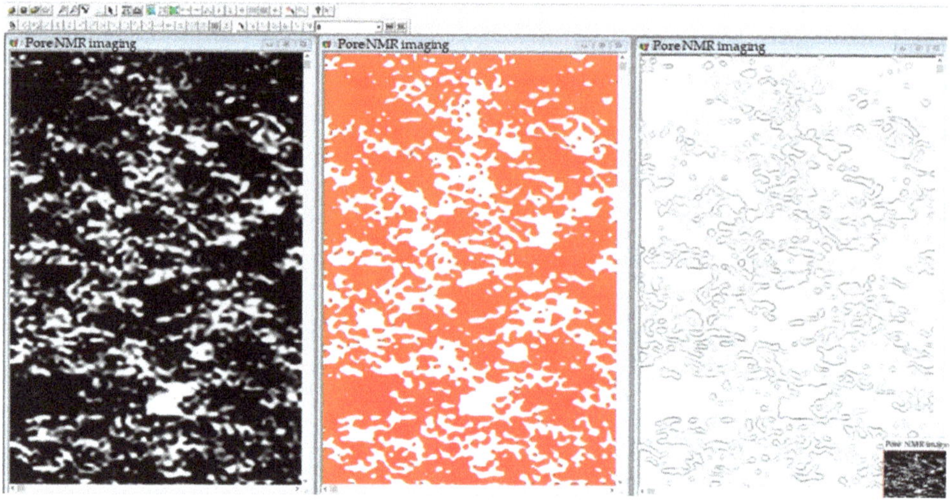

Figure 6. Nuclear magnetic image binarization, threshold segmentation and edge detection.

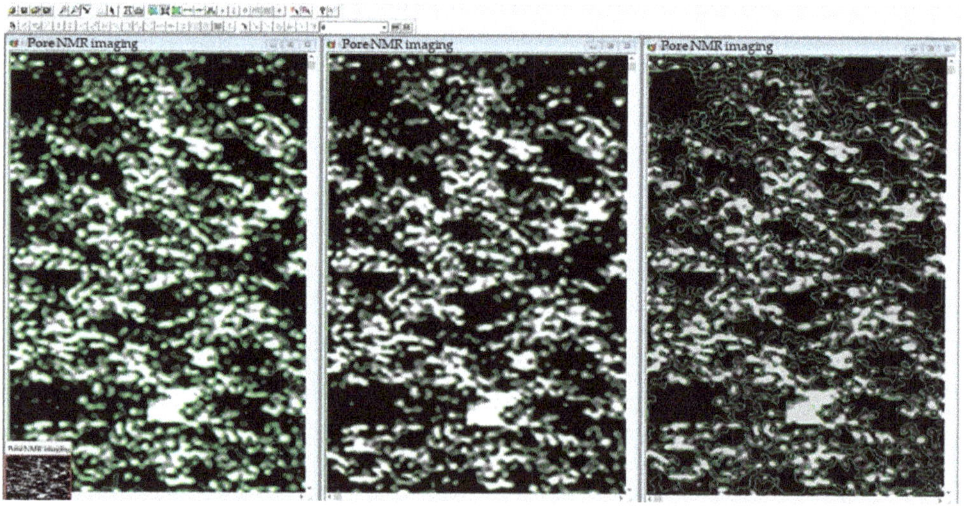

Figure 7. Contour extraction, edge processing and connectivity processing of nuclear magnetic images.

5. Pore channel connectivity processing. Since the NMR image is a two-dimensional image of the longitudinal section of the pillar, the vertical flow of the leaching solution can be reflected in the NMR image, while the seepage channel in the horizontal direction cannot be clearly reflected only through the image, and it is regarded as the particle part of the ore body.

The vector image obtained by the above steps was output into a DXF format file, the particle part of ore body was marked and distinguished, and the geometric model of the pore seepage channel of the ionic rare earth ore was obtained at pore scale. By comparison with the NMR image, the geometry model of the seepage channel can accurately restore the shape, size, and position distribution of the ore body particles and the distribution of pore channels, as shown in Figure 8.

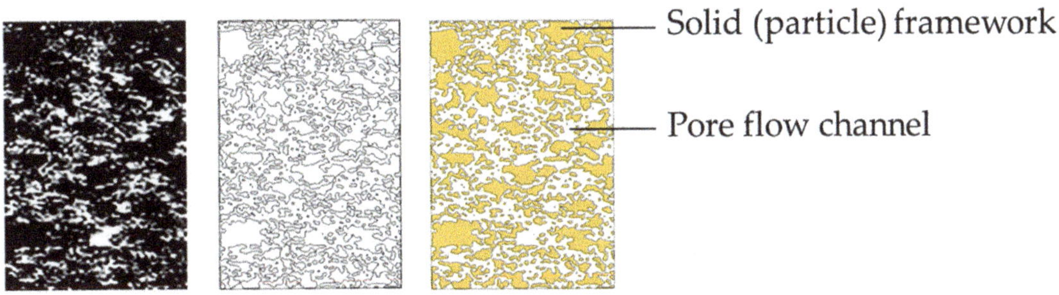

Figure 8. Nuclear magnetic resonance image, ore body particle profile, and pore model.

According to the results of rare earth ion concentration and the amount of liquid collected in indoor experiment, the total weight of rare earth ions obtained was calculated to be 140.438 mg with $MgSO_4$ at a mass concentration of 3%. At the same time, during the simulation, the concentration of rare earth ions on the surface of the mineral soil particles was calculated to be 0.12 mol/m^2 based on the particle surface area setting. Selecting the condition with 3% $MgSO_4$ and 1.0 mL/min injection intensity, the two-dimensional section model of undisturbed soil measured by nuclear magnetic resonance (hereinafter referred to as nuclear magnetic model) was compared with the results of column leaching experiment, as shown in Figure 9. The boundary conditions were set as follows: the top is the inflow boundary, and the type is the velocity boundary, with a velocity of 1.0 mL/min; the bottom boundary type is the pressure boundary, which is the free flow boundary, and the pressure is 0 Pa; both sides of the simulated tube wall are flux-free boundaries. According to the above three governing equations, the "Reaction engineering" interface in COMSOL Multiphysics software was used to calculate the relevant exchange reaction rate, diffusion coefficient, dispersion coefficient, and other parameters, and the chemical reaction process was coupled to the other two governing equations for solving.

Figure 9. Nuclear magnetic model.

The comparison of the changes in the average leached rare earth ion concentration (hereinafter referred to as leached concentration) over time between the experiment and simulation is shown in Figure 10. The leached concentration of the nuclear magnetic model and the column leaching experiment has a similar regular pattern with time, that is, in the initial stage, due to the fact that the leached mother liquor has not yet reached the outlet, the leached concentration maintains at a low level. As the mother liquor flows out from the outlet, the leached concentration gradually reaches its peak, and the peak leached concentration of the nuclear magnetic model and the indoor experiment are 9.31 g/L and 8.75 g/L, respectively. As the progress proceeds, the leached concentration continues to decrease and gradually approaches 0.0 g/L with the replacement of rare earth ions on the surface of the ore particles. Therefore, the nuclear magnetic model can reflect the micro-leaching process of in situ leaching well.

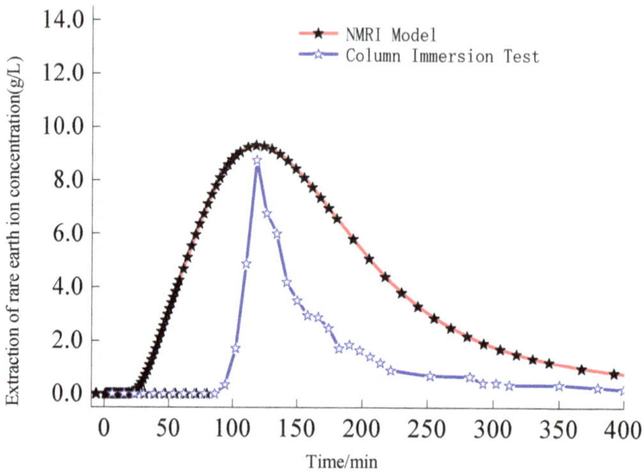

Figure 10. Comparison between experimental and simulated leaching concentration.

4. Results and Discussion

4.1. Rare Earth Ion Exchange and Migration Regular Patterns under Different Injection Intensity

In order to observe various ion exchange conditions on the particle surface of the ore body, 8887 probes were arranged in the model, as shown in Figure 11, to monitor the change of surface concentrations of rare earth ions, and magnesium ions over time on the particle surface of the ore body (i.e., the ion exchange reaction interface) during the leaching process. Nine points, A1, A2 (No. 4586), A3 (No. 7788), B1 (No. 1908), B2 (No. 4469), B3 (No. 7480), C1 (No. 1828), C2 (No. 5238), and C3 (No. 7880), were selected as representatives to explore the differences in ion exchange between dominant channels and non-dominant channels. In the vertical direction, the model was divided into three parts: upper, middle, and lower parts. In the horizontal direction, it was divided into left, middle, and right, and the nine monitoring points were selected from the dominant channel or non-dominant channel wall positions in these nine areas. Point A1 was taken from the non-dominant channel area on the left side of the upper part of the model, Point A2 from the middle dominant channel area in the upper part of the model, point A3 from the non-dominant channel area on the right side of the upper part of the model, point B1 from the non-dominant channel area on the left side of the upper part of the model, point B2 from the middle dominant channel area in the upper part of the model, point B3 from the non-dominant channel area on the right side of the upper part of the model, point C1 from the non-dominant channel area on the left side of the upper part of the model, point C2 from the middle dominant channel area in the upper part of the model, and point C3 was taken from the non-dominant channel area on the right side of the upper part of the model.

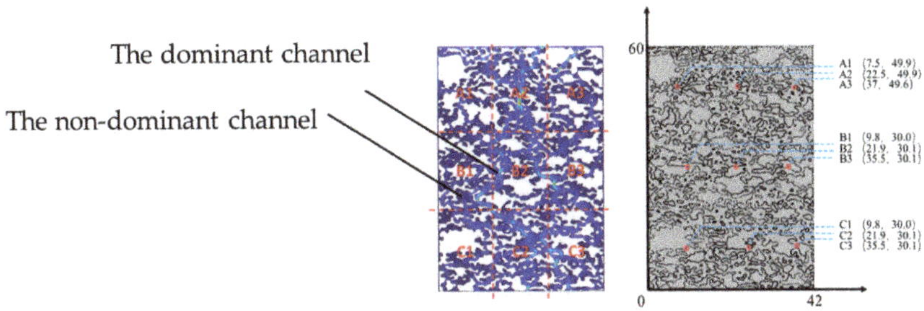

Figure 11. Probe point location and nine representative points in the region.

The variation curve of the surface rare earth ions concentration (hereinafter referred to as leached rare earth concentration) at point A1–C3 under the injection intensity of 0.25 mL/min, 0.50 mL/min, 1.00 mL/min, and 2.00 mL/min with time is shown in Figure 12. The initial surface rare earth ions concentration is 0.070 mol/m^2, and the same regular pattern can be obtained from the variation curve of point A1–C3 at each injection intensity. Due to limited space, the following only takes the injection intensity of 0.25 mL/min as an example for specific analysis.

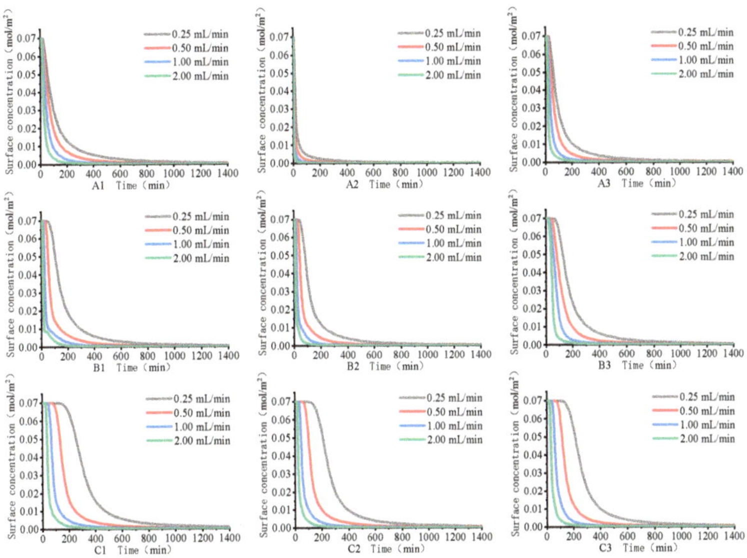

Figure 12. Surface concentrations of rare earth ions at points A1–C3 under different injection intensities.

From the variation curves in Figure 12, it can be seen that the surface concentrations of rare earth ions at point A1, A2, and A3 on the model begin to decrease at 10.7 min, 3.2 min, and 14.9 min, respectively. At this time, ion exchange at these three points is progressing forward, the desorption amount of rare earth ions is greater than the adsorption amount, and rare earth ions begin to leach out. The decline time of point A2 is much earlier than A1 and A3, indicating that rare earth ions are leached first in the upper middle dominant channel area where A2 is located. The final surface concentrations of rare earth ions at each monitoring point are not 0, indicating that there are still rare earth ions that cannot been exchanged. At the time of 1400 min, the surface concentrations of residual rare earth ions (hereinafter referred to as residual rare earth ion concentration) in point A1, A2, and A3 is 6.61×10^{-4} mol/m^2, 1.82×10^{-4} mol/m^2, and 4.83×10^{-4} mol/m^2, respectively, indicating that the leaching of rare earth ions in A2 region is more complete. The variation curves of B1, B2, and B3 in the middle point of the model began to decline at 33.6 min, 24.8 min, and 38.0 min respectively, and rare earth ions in the B2 region in the middle of the model are leached first. At the time of 1400 min, the residual rare earth ion concentrations of B1, B2, and B3 are 7.37×10^{-4} mol/m^2, 4.76×10^{-4} mol/m^2, and 6.79×10^{-4} mol/m^2, respectively, indicating that the leaching of rare earth ions in the dominant channel area is more complete. The curves of C1, C2, and C3 at the middle points of the model begin to decline at 92.9 min, 62.2 min, and 75.4 min, respectively, and rare earth ions in the C2 region in the middle of the model are leached first. At the time of 1400 min, the concentrations of residual rare earth ions in C1, C2, and C3 are 1.03×10^{-3} mol/m^2, 8.01×10^{-4} mol/m^2, and 8.22×10^{-4} mol/m^2, respectively; obviously, the leaching of rare earth ions in the dominant channel area is more complete. The flow rate of the fluid in the main flow channel is higher than that in other zones (hence its designation as the main flow channel). This increased flow rate allows more cations to pass through the zone, enabling a more thorough exchange between rare earth ions and cations. Additionally, the liquid entering the solution exits at a faster flow rate, thereby facilitating a more complete ion exchange. Comparing the curve decline time and residual rare earth ion concentration of the three groups of monitoring points (A1, B1, C1), (A2, B2, C2), and (A3, B3, C3), the flow velocity in the dominant channel is greater than that in the non-dominant channel, and more leaching solution passes through it per unit time. More active cations contact with rare earth ions, and the rare earth ions in the dominant channel always exchange preferentially. At the same time, the concentration of residual rare earth ions in the dominant channel is lower than that in the non-dominant channel at the same level and height. After the leaching is completed, the concentration of residual rare earth ions in the pores gradually increased from top to bottom, and the leaching of rare earth ions closer to the entrance end is more complete.

From the Figure 12, it can be seen that the point A1 curve at the injection intensities of 0.25 mL/min, 0.50 mL/min, 1.00 mL/min, and 2.00 mL/min start to decline at the time of 27.1 min, 6.5 min, 5.2 min, and 3.7 min, respectively, and the residual rare earth concentrations are 6.61×10^{-4} mol/m^2, 3.33×10^{-4} mol/m^2, 1.64×10^{-4} mol/m^2, and 8.10×10^{-5} mol/m^2, respectively. With the increase in liquid injection intensity, the point A1 curve decline time gradually advances, and the residual rare earth ion concentration decreases accordingly. This indicates that with the injection strength increases, the overall flow rate in the pores increases, and more active cations can exchange with rare earth ions per unit time, which can accelerate the desorption of rare earth ions, and is conducive to the complete leaching of rare earth ions, especially in non-dominant channels.

The variation curves of the leached rare earth concentration over time at 8887 monitoring points at injection intensities of 0.25 mL/min, 0.50 mL/min, 1.00 mL/min, and 2.00 mL/min are shown in Figure 13a–d. The surface concentrations curve at the initial moment of the four injection intensities begin to change, and the slope of curve decline is almost vertical. At the injection intensity of 0.25 mL/min, the descending time of the change curve at the latest is 197.0 min, and at 0.50 mL/min, 1.00mL/min, and 2.00 mL/min, the descending time is 154.0 min, 54.0 min, and 27.0 min, respectively. This shows that the increase of injection intensity can make ion exchange start earlier. The change rate of the curve reflects the rate of rare earth ion exchange. With the increase in liquid injection intensity, the decreasing slope of the 8887 curves becomes more vertical, and the area of the curve enclosed shape also decreases, indicating that the increase of liquid injection intensity is conducive to the rapid leaching of rare earth ions on the whole. At the end of the injection time ($t = 1400$ min), the lowest surface concentrations of rare earth ions is all 0 mol/m^2, and the highest surface concentrations of rare earth ions are 1.26×10^{-3} mol/m^2, 1.02×10^{-3} mol/m^2, 0.80×10^{-4} mol/m^2, and 0.67×10^{-4} mol/m^2, respectively; this means that there is still a small portion of rare earth ions in some places in the pores that has not been leached, and increasing the injection intensity helps this portion of rare earth ions to be more completely leached.

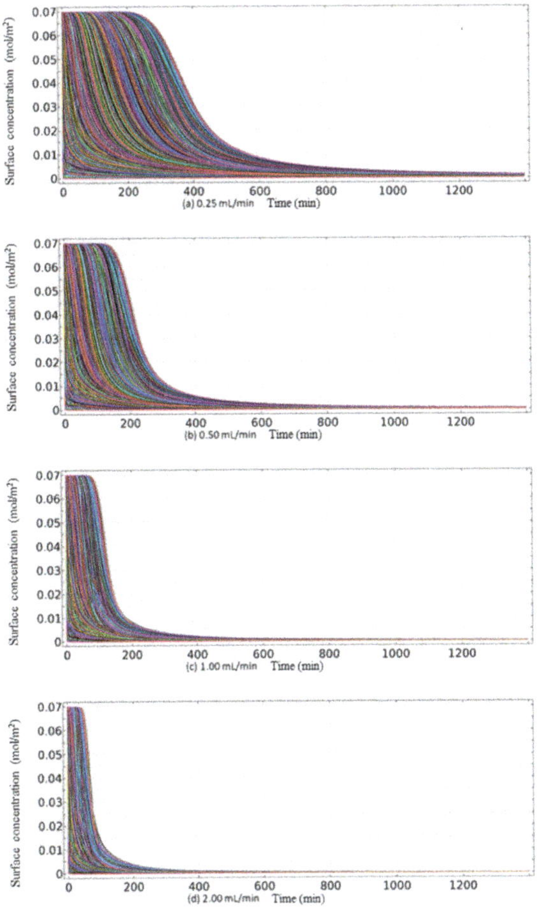

Figure 13. Surface concentration changes of rare earth ions at 8887 monitoring points under different injection intensities.

The variation curve of the leached rare earth concentration at the outlet under the injection intensities of 0.25 mL/min, 0.50 mL/min, 1.00 mL/min, and 2.00 mL/min changes over time as shown in Figure 14. All four curves are unsymmetrical unimodal curves with a tail, and there is an initial stage of gentle rise for a period of time. As the injection intensity increases, the initial stage gradually shrinks, the rare earth concentration rises sharply, and the tailing time shortens. When the injection strength is 0.25 mL/min, the peak concentration of leach rare earth ions is 9.22 g/L. When the injection strength increases to 0.50 mL/min, 1.00 mL/min, and 2.00 mL/min, the peak leaching concentration is 8.52 g/L, 7.89 g/L, and 7.22 g/L, respectively. At the initial stage of rare earth ion concentration, the exchange mainly occurs in the upper part of the model, and rare earth ions migrate toward the outlet end with the flow of the leaching solution. The rare earth ions are primarily distributed in the upper non-preferential channels and the middle preferential channels. In the non-preferential channels, the flow velocity is lower, and the ions are less affected by convection, leading to slower ion migration compared to the preferential channels, causing an accumulation of rare earth ions in this area. When the injection intensity increases, the distribution of rare earth ions remains relatively unchanged, but their concentration increases. During the rising phase of rare earth ion concentration, the migration speed of rare earth ions in the preferential channels is significantly higher than in the non-preferential channels. As the injection intensity increases, rare earth ions continue to accumulate in the middle non-preferential channels, and their concentration increases accordingly. At the peak concentration stage, rare earth ions are mainly distributed in the lower preferential channels and the middle/lower non-preferential channel regions. When the injection intensity increases, the area where rare earth ions are distributed in the lower part of the model decreases, and the ion concentration decreases. During the declining and tailing stages, ion exchange primarily occurs in the middle/lower non-preferential channels. The tailing phenomenon in the time-varying curve of the average rare earth ion concentration is caused by the late occurrence of ion exchange in the middle/lower non-preferential channels and the slower migration speed in this region. The increase of flow rate per unit time dilutes the rare earth ions, resulting in a decrease in rare earth concentration. With the increase of injection intensity, the corresponding peak concentration of rare earth leaching time is 268.6 min, 133.3 min, 61.2 min, and 29.7 min, respectively, indicating that by increasing the injection intensity can accelerate the whole leaching process to some extent.

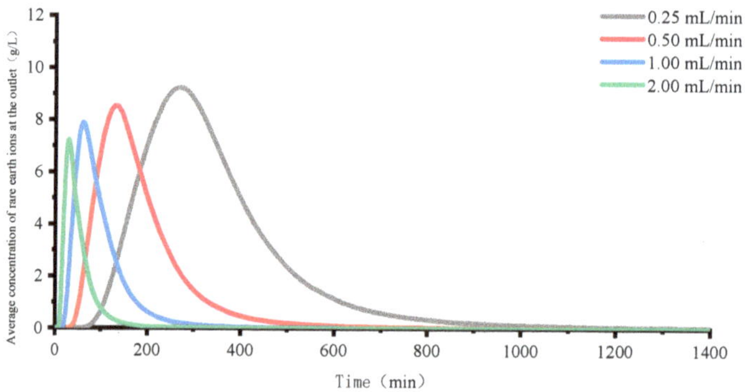

Figure 14. Average rare earth ion concentration at the outlet under different injection strength.

The comparison between the amount of magnesium ion and the total amount of rare earth ion leached under different injection intensity is shown in Figure 15. When the injection intensity increases from 0.25 mL/min to 0.50 mL/min, the amount of magnesium ion is doubled, and the total amount of rare earth ion leached is increased by 2.36%. When the injection intensity increases from 0.50 mL/min to 1.00 mL/min, the amount of magnesium ion is doubled, and the total amount of rare earth ion leached is increased by 1.18%. When the injection intensity increases from 1.00 mL/min to 2.00 mL/min, the amount of magnesium ion is doubled, and the total amount of rare earth ion leached is increased by 0.48%.

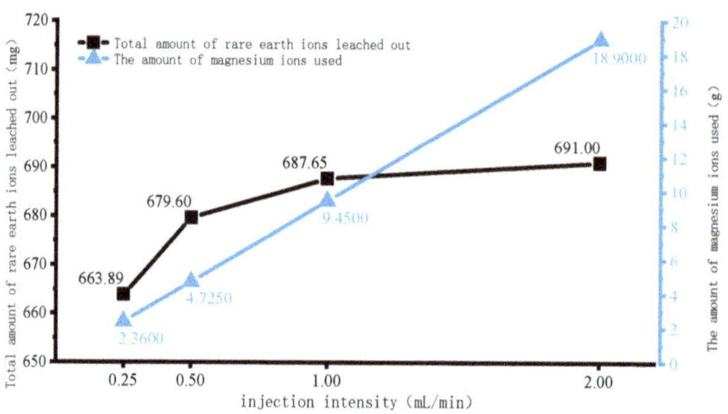

Figure 15. Magnesium ion dosage and total leach amount of rare earth ion under different injection strength.

The analysis results indicate that as the injection intensity increases, the amount of magnesium ions also increases. However, the doubling of the amount of magnesium ions does not exchange the doubling of rare earth ions, which undoubtedly increases production costs. Although the injection intensity can shorten the leaching cycle and improve production efficiency, the risk of a landslide is also increased, and the reduction of leached concentration makes the purification of rare earth elements difficult. Therefore, in practical production, considering the production cost on the premise of ensuring engineering safety and production efficiency, it is recommended to control the injection strength between 0.50 mL/min and 1.00 mL/min for this mining area.

4.2. Rare Earth Ion Exchange and Migration Regular Patterns under Different Injection Concentrations

The variation curves of surface concentration of rare earth ions of points A1–C3 over time when the concentrations of magnesium sulfate solution are 0.06 mol/L, 0.12 mol/L, 0.24 mol/L, and 0.48 mol/L are shown in Figure 16. From those curves, it can be seen that when the curve of points A1–C3 begins to decline, the concentration of the leaching solution does not change significantly. Moreover, the curves at the same point have significant differences in the solubility of the four leaching solutions, which are manifested in the curve decline rate and the surface concentrations of residual rare earth ions. Taking the four curves of point A1 as an example, with the increase of leaching solution concentration, the curve decline rate increases, and the surface concentrations of residual rare earth ions is 8.84×10^{-3} mol/m^2, 2.13×10^{-3} mol/m^2, 3.76×10^{-4} mol/m^2, and 6.08×10^{-5} mol/m^2, respectively. This indicates that the concentration of the leaching solution mainly affects the forward exchange rate of rare earth ions, and significantly promotes the more complete leaching of rare earth ions. The same regular pattern can be obtained by analyzing the changes in point A2–C3 curves at the $MgSO_4$ concentrations of 0.06 mol/L, 0.12 mol/L, 0.24 mol/L, and 0.48 mol/L. Therefore, the promotion effect of leaching solution on rare earth ion leaching is due to the increase in concentration, which increases the exchange rate of rare earth ions in all regions and leads to an increase in leaching efficiency accordingly.

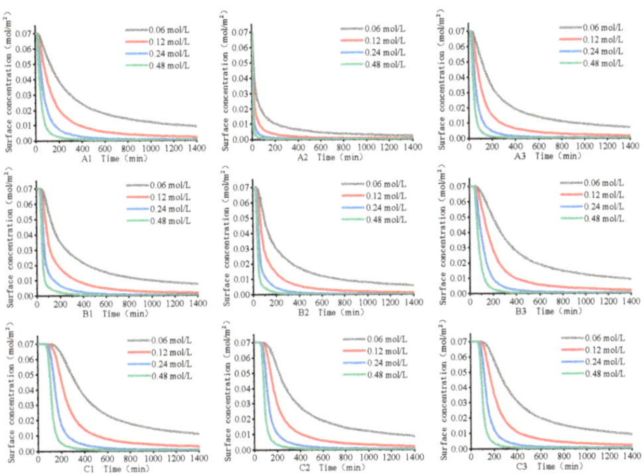

Figure 16. Surface concentrations of rare earth ions at points A1–C3 at different concentrations.

The variation curves of surface concentrations of rare earth ions at those 8887 monitoring points when the concentrations of magnesium sulfate solution are 0.06 mol/L, 0.12 mol/L, 0.24 mol/L, and 0.48 mol/L over time are shown in Figure 17a–d. It is not difficult to see from the graph that as the concentration of the solution increases, the decline rate of those 8889 curves increases. In the 0.06 mol/L concentration curve, the latest descent time is 201.0 min; the descending time of the curve at 0.12 mol/L, 0.24 mol/L, and

0.48 mol/L is 179.0 min, 171.0 min, and 164.0 min, respectively. This indicates that the increase of the concentration of the leaching solution is beneficial for the ion exchange to start earlier, but the effect is weaker than that of the injection intensity. Moreover, as the concentration of leaching solution increases, the overall slope of those 8887 curves decreases more vertically, indicating that the concentration change has an effect on rare earth ion leaching mainly because the concentration increase has an effect on the rate of rare earth ion exchange. In addition, the minimum surface concentrations of rare earth ions at the end time (t = 1400 min) of those four leaching solution concentrations is 0 mol/m^2, and the highest surface concentrations of rare earth ions are 1.41×10^{-2} mol/m^2, 3.78×10^{-3} mol/m^2, 7.86×10^{-4} mol/m^2, and 1.62×10^{-5} mol/m^2, respectively; this indicates that increasing the concentration of leaching solution can promote more complete leaching of rare earth ions, and its promoting effect is more significant than that of injection intensity.

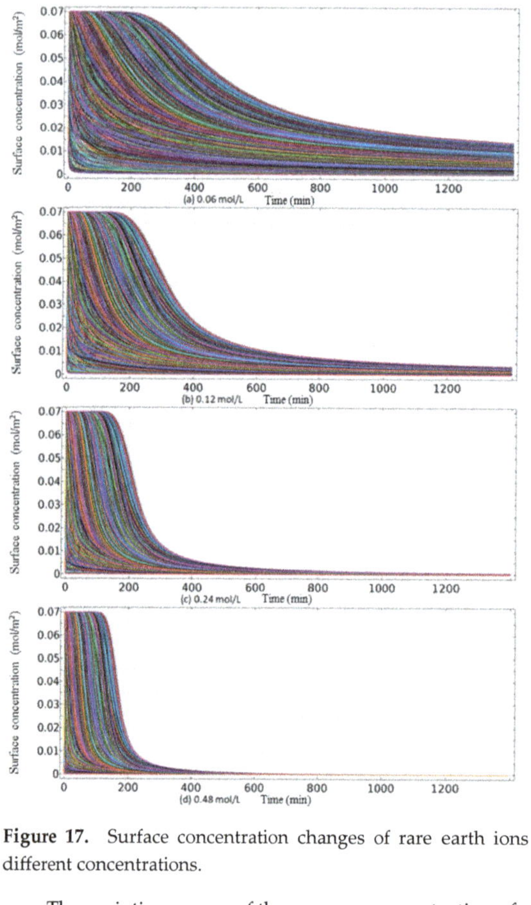

Figure 17. Surface concentration changes of rare earth ions at 8887 monitoring points under different concentrations.

The variation curves of the average concentration of rare earth ions at the outlet when the concentrations of magnesium sulfate solution are 0.06 mol/L, 0.12 mol/L, 0.24 mol/L, and 0.48 mol/L over time are shown in Figure 18. It is obvious that under those four different leaching concentrations, the starting time of the single peak leaching curve is the same, and the time of the peak appearing is basically within the same position range. In addition, as the concentration of the leaching solution increases, the curve rises faster and faster, and the peak concentration increases as well. When the concentration of leaching solution is 0.06 mol/L, the peak concentration of leached rare earth ions is 3.06 g/L. When the concentration increases to 0.12 mol/L, 0.24 mol/L, and 0.48 mol/L, the peak leaching concentrations is 5.30 g/L, 8.33 g/L, and 11.63 g/L, with an increase of 73.20%, 57.16%, and 39.62%, respectively. During the continuous leaching process, rare earth elements continuously accumulate in the lower part of the ore column, causing the exchange zone to gradually move downward until the entire area completes the exchange reaction. As the concentration of magnesium sulfate increases, the leaching agent solute Mg^{2+} enters an excess state, significantly raising the Mg^{2+} content in the solution. The concentration of Mg^{2+} in the exchange zone also increases, and the concentration gradient grows accordingly, further enhancing the diffusion capacity of Mg^{2+}. This accelerates its internal diffusion rate and speeds up the exchange reaction. As a result, the total amount of rare earth ions leached per unit time increases, and the peak concentration of leached rare earth ions rises with the concentration of the injected solution. As the concentration of the leaching solution increases, the corresponding moments of peak concentration are 196.6 min, 161.3 min, 131.4 min, and 108.2 min, respectively; this indicates that effect of increasing the concentration of the leaching solution on the leaching process is limited, and the leaching period remains constant.

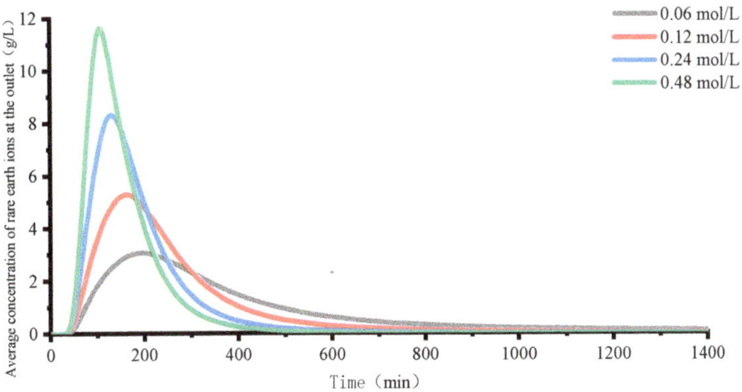

Figure 18. Average rare earth ion concentration at the outlet under different injection concentrations.

The amount of magnesium ion and the total amount of rare earth ion leached under different leaching solution concentration are shown in Figure 19. When the concentration of leaching solution increases from 0.06 mol/L to 0.12 mol/L, the total amount of rare earth ions leached increases by 11.52%. When the concentration of leaching solution increases from 0.12 mol/L to 0.24 mol/L, the amount of magnesium ion is doubled, and the rare earth ions leached increases by 4.84%. When the concentration of leaching solution increases from 0.24 mol/L to 0.48 mol/L, the amount of magnesium ion is doubled, and the rare earth ions leached increases by 4.65%.

The amount of magnesium ions increases with the increase of the concentration of the leaching solution. Although the growth rate of the total leached rare earth ions is very impressive in the range of 0.06 mol/L to 0.12 mol/L, the total leached rare earth ions does not increase linearly with the increase of the amount of magnesium ions. Increasing the concentration of magnesium sulfate has a significant effect on increasing the concentration of rare earth leaching, but it is very limited in accelerating the leaching cycle. Therefore, it is not advisable to continuously increase the concentration of $MgSO_4$ to improve production efficiency and leaching concentration, as it will not only raise the cost of mineral leaching but also increase the concentration of residual magnesium ions in the soil, thereby raising the cost of leaching magnesium ions from the soil. For this mining area, it is recommended to control the concentration of leaching solution in the production process within the range of 0.12~0.24 mol/L (molar concentration of 1.4~2.8%).

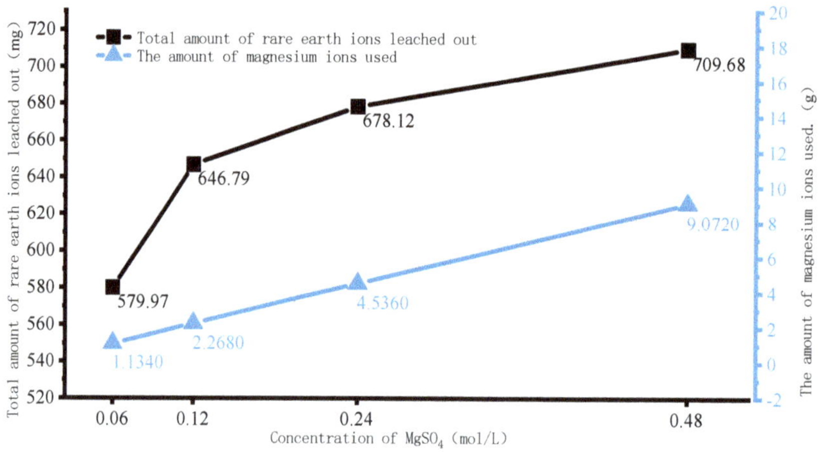

Figure 19. The amount of magnesium ions and the total amount of rare earth leached at different concentrations.

5. Conclusions

This paper took an ionic rare earth mine in Longnan, southern Jiangxi Province as the research object, and constructed a porous scale ionic rare earth percolation–exchange–migration model based on nuclear magnetic images. Through numerical simulation test, the influence of injection intensity and leaching concentration on rare earth leaching process, leaching fluid seepage, exchange, and migration of rare earth were studied, and the regular law of leaching fluid seepage, ion exchange, and migration were summarized. Based on the percolation–exchange–migration model of ionic rare earth ore at pore scale, reasonable suggestions for the setting of injection intensity and leaching concentration were put forward. The conclusions are as follows:

1. The samples were scanned by NMR, the scanned NMR images were vectorized by image processing technology, and a two-dimensional pore percolation channel geometric model of ionic rare earth ore was established, and combined with control equations of ionic rare earth pore percolation–exchange–migration to simulate the meso-leaching-process of rare earth ions.
2. Increasing the injection intensity can facilitate rare earth ions to leach out earlier and outflow from the ore body, thereby shortening the leaching period and expediting the leaching process. However, it can also lead to a decrease in the peak concentration of rare earth ions in the leaching process. For this mining area, it is recommended to control the injection intensity between 0.50 mL/min and 1.00 mL/min.
3. Increasing the concentration of leaching solution accelerates the forward exchange rate of leaching agent cations and rare earth ions, thereby enhancing the leaching efficiency of rare earth ions. The greater the concentration gradient of rare earth ions in the leaching solution, the more significant the molecular diffusion effect and the faster the ion migration rate. Although the high concentration leaching solution is conducive to the leaching of rare earth ions, once the rare earth ions are essentially leached from the ore body, on the one hand, the exchange competition between rare earth and impurity ions is weakened, and impurity ions are leached out under better exchange conditions. On the other hand, the continuous injection of leaching solution causes a waste of resources. For this mining area, it is recommended to control the concentration of leaching solution between 0.12 mol/L and 0.24 mol/L.

Author Contributions: Conceptualization, D.W. and Y.R.; methodology, D.W.; software, F.W. and W.X.; validation, Z.Z., D.W., and W.X.; formal analysis, M.H., Z.Z., and D.W.; investigation, W.X. and M.H.; resources, Y.R.; data curation, F.W. and W.X.; writing—original draft preparation, D.W.; writing—review and editing, Y.R. and Z.Z.; visualization, D.W. and M.H.; supervision, D.W.; project administration, Y.R.; funding acquisition, Y.R. All authors have read and agreed to the published version of the manuscript.

Funding: This research was funded by the National Natural Science Foundation of China, grant number 51964014.

Data Availability Statement: The data that support the findings of this study are available from the corresponding author (V.S.) upon reasonable request.

Acknowledgments: The authors appreciate the editors and anonymous reviewers for their great efforts on the manuscript.

Conflicts of Interest: The authors declare no conflicts of interest.

References

1. Liu, L.; Rao, Y.Z.; Tian, C.S.; Huang, T.; Lu, J.C.; Zhang, M.D.; Han, M. Adsorption Performance of La(III) and Y(III) on Orange Peel: Impact of Experimental Variables, Isotherms, and Kinetics. *Adsorpt. Sci. Technol.* **2021**, *2021*, 7189639. [CrossRef]
2. Zhou, F.; Zhang, Y.X.; Liu, Q.; Huang, S.H.; Wu, X.Y.; Wang, Z.W.; Zhang, L.S.; Chi, R.A. Modified tailings of weathered crust elution-deposited rare earth ores as adsorbents for recovery of rare earth ions from solutions: Kinetics and thermodynamics studies. *Miner. Eng.* **2023**, *191*, 107937. [CrossRef]
3. He, Z.Y.; Zhang, R.; Sha, A.Y.; Zuo, Q.; Xu, Z.G.; Wu, M.; Chi, R.A. Anti-swelling mechanism of DMDACC on weathered crust elution-deposited rare earth ore. *J. Rare Earths* **2022**, *40*, 1803–1811. [CrossRef]
4. Han, M.; Wang, D.; Rao, Y.; Xu, W.; Nie, W. An Experimental Study on the Kinetics of Leaching Ion-Adsorbed REE Deposits with Different Concentrations of Magnesium Sulfate. *Metals* **2023**, *13*, 1906. [CrossRef]
5. Luo, X.P.; Yuan, X.Q.; He, K.Z.; Zhang, Y.B.; Luo, C.G.; Liu, Z.S.; Zhou, H.P. Precipitation process for combined impurity removal from a magnesium sulphate-based leachate of ionic rare earth ore. *Miner. Eng.* **2022**, *189*, 107911. [CrossRef]
6. Li, L.Q.; Liu, C.H.; Zhang, H.P.; Huang, B.; Luo, B.; Bie, C.; Sun, X.Q. The enrichment of rare earth from magnesium salt leaching solution of ion-adsorbed type deposit: A waste-free process for removing impurities. *J. Environ. Manag.* **2022**, *310*, 114743. [CrossRef]
7. Wu, X.Y.; Zhou, F.; Liu, C.F. Effect of polyacrylamide on the process of removing impurities in the rare earth leachate. *Physicochem. Probl. Miner. Process.* **2021**, *57*, 182–191. [CrossRef]
8. Bezaatpour, J.; Fatehifar, E.; Rasoulzadeh, A. Coarse-grained geological porous media structure modeling using heuristic algorithm and evaluation of porosity, hydraulic conductivity, and pressure drop with experimental results. *Environ. Earth Sci.* **2021**, *80*, 1–14. [CrossRef]
9. Zhao, K.; Zhuo, Y.L.; Wang, X.J.; Zhong, W. Aggregate Evolution Mechanism during Ion-Adsorption Rare Earth Ore Leaching. *Adv. Mater. Sci. Eng.* **2018**, *2018*, 4206836. [CrossRef]
10. Wang, S.F.; Lv, L.; Xue, K.X.; Zhang, D.J.; Li, M.D.; Li, D.W.; Yuan, C. Micropore Structure and Fractal Characteristics of Clays Due to Freeze-Thaw and Compression Based on Mercury Intrusion Porosimetry. *Front. Earth Sci.* **2022**, *10*, 851673. [CrossRef]
11. Gong, L.X.; Nie, L.; Xu, Y.; Ji, X.K.; Liu, B.M. Characterization of Micro-Scale Pore Structure and Permeability Simulation of Peat Soil Based on 2D/3D X-ray Computed Tomography Images. *Eurasian Soil Sci.* **2022**, *55*, 790–801. [CrossRef]
12. Yin, S.H.; Chen, X.; Yan, R.F.; Wang, L.M. Pore structure characterization of undisturbed weathered crust elution-deposited rare earth ore based on X-ray micro-CT scanning. *Minerals* **2021**, *11*, 236. [CrossRef]
13. Kihm, J.H.; Hwang, G. Numerical Simulation of Water Table Drawdown due to Groundwater Pumping in a Contaminated Aquifer System at a Shooting Test Site, Pocheon, Korea. *Econ. Environ. Geol.* **2021**, *54*, 247–257. [CrossRef]
14. Lu, D.C.; Li, X.Q.; Du, X.L.; Lin, Q.T.; Gong, Q.M. Numerical simulation and analysis on the mechanical responses of the urban existing subway tunnel during the rising groundwater. *Tunn. Undergr. Space Technol. Inc. Trenchless Technol. Res.* **2020**, *98*, 103297. [CrossRef]
15. Xu, J.; Lan, W.; Ren, C.; Zhou, X.G.; Wang, S.H.; Yuan, J. Modeling of coupled transfer of water, heat and solute in saline loess considering sodium sulfate crystallization. *Cold Reg. Sci. Technol.* **2021**, *189*, 103335. [CrossRef]
16. Wu, Z.J.; Lu, H.; Weng, L.; Liu, Q.S.; Shen, J.Q. Study on the seepage characteristics of fractured sandstone based on real-time imaging technology of nuclear magnetic resonance. *Chin. J. Rock Mech. Eng.* **2021**, *40*, 263–275.
17. Sheikhzadeh, G.A.; Mehrabian, M.A.; Mansouri, S.H.; Sarrafi, A. Computational modelling of unsaturated flow of liquid in heap leaching—Using the results of column tests to calibrate the model. *Int. J. Heat Mass Transf.* **2005**, *48*, 279–292. [CrossRef]
18. Wu, C.Y. Simulation and Analysis for Percolation Leaching Process of Ionic Rare Earth Ore. Master's Thesis, Jiangxi University of Science and Technology, Ganzhou, China, 2016.
19. Liu, Q.S.; Xiao, H.; Tan, C.L.; Yu, X.Y.; Wang, D.L.; Qiu, T.S. Coupled simulation of percolation-reaction-stress in leaching process of ion-type rare earth ore. *J. Chin. Soc. Rare Earths* **2022**, *40*, 880–892.

20. Zeng, J. Numerical simulation of seepage characteristics and ion exchange reaction in leaching process of ionic rare earth ore. Master's Thesis, Jiangxi University of Science and Technology, Ganzhou, China, 2022.
21. Wu, A.X.; Yin, S.H.; Wang, H.J.; Su, D.Y. Solutes transport mechanism and model of dump leaching. *J. Cent. South Univ.* **2006**, *2*, 385–389.
22. Wu, A.X.; Liu, J.Z.; Tang, L.Y. Simulation of Coupled Flowing-Reaction-Deformation with Mass Transfer in Heap Leaching Processes. *Appl. Math. Mech.* **2007**, *28*, 327–335. [CrossRef]
23. Wang, D.; Wu, F.Y.; Rao, Y.Z.; Xu, W.; Han, M.; Shi, L. Simulation of an Ionic Rare Earth Leaching Process Based on the Darcy Law-Chemical Reaction Engineering-Transfer of Dilute Substance Coupling. *Minerals* **2022**, *12*, 1500. [CrossRef]
24. Hu, S.L.; Cao, X.J.; Wang, G.S.; Long, P.; Zhou, X.Y. An Ion Exchange Model for Leaching Process of Weathered Crust Elution-deposited Rare Earth. *Min. Metall. Eng.* **2018**, *38*, 1–5.
25. Long, P.; Wang, G.S.; Tian, J.; Hu, S.L.; Luo, S.H. Simulation of one-dimensional column leaching of weathered crust elution-deposited rare earth ore. *Trans. Nonferrous Met. Soc. China* **2019**, *29*, 625–633. [CrossRef]
26. Li, Q.; Qin, L.; Wang, G.S.; Long, P.; Peng, C.L. Leaching Mechanism of Ion-Adsorption Rare Earth. *J. Chin. Soc. Rare Earths* **2021**, *39*, 543–554.
27. Guo, Z.Q.; Zhao, K.; Jin, J.F.; Zhu, Z.C.; Li, G. Solute Transport Mechanism of Ion-Adsorption Type R are Earth in-situ Leaching Mining. *J. Chin. Soc. Rare Earths* **2019**, *37*, 121–128.
28. Liu, J.; Brady, B.H. Evaluation of velocity-dependent in situ leaching processes: Single-porosity model. *Metall. Mater. Trans. B. Process Metall. Mater. Process. Sci.* **1998**, *29*, 1227–1234. [CrossRef]
29. Phong, P.H.; Anh, P.B.V.; Ha, V.T.T.; Hung, L.Q.; Thanh, L. Simulating and Monitoring the Temporal and Spatial Transfer of NPK Fertilizer in Agricultural Soils Using a Mathematical Model and Multi-Channel Electrical Conductivity Measurement. *J. Soil Sci. Plant Nutr.* **2020**, *21*, 374–388. [CrossRef]
30. Yang, Z.B.; He, Q.Q.; Wang, K.; Liu, N. Application of Portable X-ray Fluorescence Spectrometer in the Investigation of Contaminated Site. *Environ. Monit. Forewarning* **2023**, *15*, 23–26+51.
31. Gao, L.H.; Xing, R.R.; Zhu, Y.R.; Zhao, W.Y. The total amount of rare earth in rare earth ferroalloy was determined by EDTA titration. *Met. Funct. Mater.* **2022**, *29*, 98–102.
32. Yang, B.H.; Wu, A.X.; Miu, X.X.; Liu, J.Z. 3D characterization and analysis of pore structure of packed ore particle beds based on computed tomography images. *Trans. Nonferrous Met. Soc. China* **2014**, *24*, 833–838. [CrossRef]
33. Wu, A.X.; Yang, B.H.; Liu, J.Z.; Zhang, J. Analysis of Pore Structure Evolution of Ore Granular Media during Leaching Based on X-ray Computed Tomography. *Chin. J. Process Eng.* **2007**, *5*, 960–966.
34. Luo, X.P.; Zhang, Y.B.; Zhou, H.P.; He, K.Z.; Zhang, B.Y.; Zhang, D.M.; Xiao, W.J. Pore structure characterization and seepage analysis of ionic rare earth orebodies based on computed tomography images. *Int. J. Min. Sci. Technol.* **2022**, *32*, 411–421. [CrossRef]

metals MDPI

Review

Mineralogy of Deep-Sea Manganese Nodules and Advances in Extraction Technology of Valuable Elements from Manganese Nodules

Xu Wang [1], Wenqing Qin [1], Maolin Li [2,3], Xueduan Liu [1], Yangrui Cheng [2,3], Shiping Chen [2,3] and Congren Yang [1,*]

1 School of Minerals Processing and Bioengineering, Central South University, Changsha 410083, China; 175601022@csu.edu.cn (X.W.); qinwenqing369@126.com (W.Q.); liuxueduan@csu.edu.cn (X.L.)
2 State Key Laboratory of Exploitation and Utilization of Deep-Sea Mineral Resources, Changsha 410083, China; 13974851980@vip.163.com (M.L.); chengyr@minmetals.com (Y.C.); chshiping@minmetals.com (S.C.)
3 Changsha Research Institute of Mining and Metallurgy Co., Ltd., Changsha 410083, China
* Correspondence: yangcongren@csu.edu.cn

Abstract: The vast seabed holds tremendous resource potential that can provide necessary materials for future human societal development. This study focuses on the mineralogy of seafloor manganese nodules off the coast of China in the Western Pacific and the primary techniques for extracting valuable metal elements from manganese nodules. The research indicates that the main valuable metal elements in the manganese nodules from this region include Cu, Co, Ni, Mn, Fe, etc. The key to extracting these valuable metals lies in reducing Mn(IV) to Mn(II) to disrupt the structure of the nodules, thereby releasing the valuable elements. The extraction processes for the main valuable metal elements of manganese nodules are mainly divided into two categories: pyrometallurgical–hydrometallurgical and solely hydrometallurgical. In order to cope with the challenges of environmental change and improve utilization efficiency, bioleaching, hydrogen metallurgy, and co-extraction are gaining increasing attention. For promoting commercialization, the future development of manganese nodule resources can refer to the technical route of efficient short-process extraction technology, the comprehensive recovery of associated resources, and tail-free utilization.

Keywords: manganese nodule; mineralogy; extraction; pyrometallurgical–hydrometallurgical; hydrometallurgical

1. Introduction

The vast depths of the ocean harbor rich reserves of energy resources, mineral resources, biological resources, and genetic resources. These resources offer potential solutions to humanity's energy and food crises, providing opportunities for future survival and development. As terrestrial resources diminish day by day, humanity's reliance on marine resources continues to grow. The utilization and development of deep-sea resources have become a crucial stage for countries worldwide to demonstrate their comprehensive national strength and international influence [1–6].

Manganese nodules, also known as polymetallic nodules, contain over 30 types of metal elements and represent a major form of deep-sea mineral resources. These nodules are extensively distributed across the seafloors of the Pacific, Atlantic, and Indian Oceans at depths ranging from 3500 to 6000 m [7]. The reserves of some metals in manganese nodules are dozens or even hundreds of times that of terrestrial resources. The most commercially valuable ones include copper, cobalt, nickel, and manganese. It happens that these four metals are energy metals that are currently in short supply [8]. Relevant data shows that the abundance of manganese nodules at the bottom of the ocean is mostly between 5~30 kg/m^2. The reserves of manganese nodules in the Pacific Ocean alone can reach 1.7 trillion tons, of which the metal content of manganese (Mn) is 400 billion tons, the metal content of nickel (Ni) is 16.4 billion tons, the metal amount of copper (Cu) is 8.8 billion tons, and the metal amount of cobalt (Co) is 5.8 billion tons. These reserves are equivalent to more than 400 times the current terrestrial Mn reserves, more than 1000 times the Ni reserves, 88 times the Cu reserves, and more than 5000 times the Co reserves. Therefore, manganese nodules can provide sufficient raw materials for new energy industries such as batteries, which is a beneficial supplement to the shortage of land-based energy metals. It can help humans cope with climate change and achieve the goal of "carbon peaking and carbon neutrality" [9–15]. Figure 1 presents the distribution of major mineral resources in the deep sea and their spatial overlay with the exclusive economic zones.

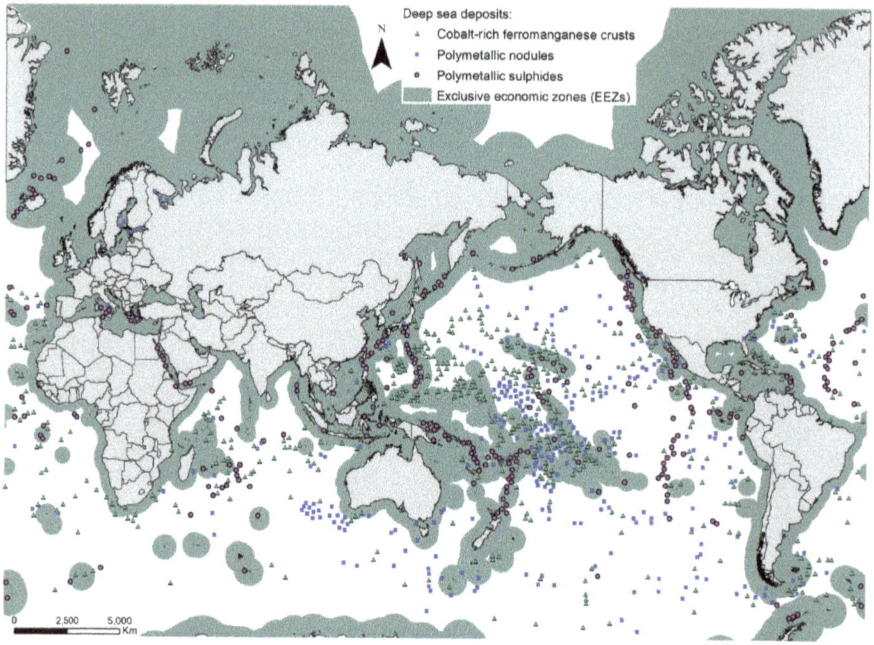

Figure 1. Location of deep-sea deposits and their spatial overlay with the exclusive economic zones. Reprinted with permission from ref. [16]. 2021 Elsevier.

The mining of seabed mineral resources from water depths below thousands of meters faces huge technical challenges [17]. It involves multi-disciplinary fields such as exploration, mining, mineral processing, metallurgy, biology, environment, machinery, materials, and automation, which is a collection of cutting-edge technologies. Due to mining costs and marine environmental protection, manganese nodules have not yet been recovered and utilized on a large scale [18–21]. However, due to earlier awareness of the importance of seabed mineral resources, some countries and institutions have mastered key technologies and core equipment manufacturing capabilities for deep-sea mineral mining. For example, The Metals Company (TMC) of Canada has conducted an in-situ mining test at a sea depth of 4300 m in the Clarion–Clipperton zone (CCZ) in October 2022, and collected about 3600 t of deep-sea manganese nodules. At the same time, the corresponding environmental assessment is underway [22–27]. TMC plans to achieve commercial mining of manganese nodules on a scale of tens of millions by 2025, which means that the development and utilization of deep-sea mineral resources has entered a new stage.

With the advancement of mining technology, the extraction technology of valuable metal elements from manganese nodules has also attracted more and more attention. Unlike the occurrence state of terrestrial mineral resources, even though deep-sea manganese nodules contain higher concentrations of valuable metal elements, they are highly dispersed within the nodules in the form of fine particles, making it challenging to achieve effective enrichment through traditional ore beneficiation methods [28]. Detailed mineralogical studies play an important role in guiding the extraction process of valuable metal elements from manganese nodules. Building upon previous research, this paper conducts a detailed investigation of the mineralogy of manganese nodules along the southeastern coast of China, and reviews the existing extraction processes of valuable elements from manganese nodules. The objective is to serve as a reference for the subsequent improvement of low-cost, high-efficiency, and environmentally friendly extraction technologies for manganese nodules, thus promoting the commercial development and utilization of manganese nodules.

2. Mineralogy of Deep-Sea Manganese Nodules

The manganese nodules in this study were obtained from the southeastern coast of China in the Western Pacific Ocean during sea trials by the Changsha Institute of Mining and Metallurgy. As shown in Figure 2, the color of manganese nodules is often black or brown-black, and the shapes are mostly spherical, elliptical, potato-shaped, flat, slag-shaped, etc., with different sizes. It can be observed that the manganese nodules are composed of a loose core and a dense shell after cutting the manganese nodules in half, which is consistent with relevant reports. The main minerals of manganese nodules are cryptocrystalline iron-hydroxide manganese ore (δ-$MnO_2 \cdot H_2O$) and amorphous ferrihydrite (FeOOH), which are usually epitaxially intertwined [29–31]. The rest of the minerals are fine-grained clastic minerals, including quartz, feldspar, pyroxene, clay, zeolite, and authigenic mineral carbon fluoroapatite (CFA). The elements in manganese nodules in the Pacific can be roughly classified according to their mass fractions, as shown in Table 1. The element contents indicate that the metal elements with greater economic value in deep-sea manganese nodules are mainly Mn, Fe, Ni, and Co, followed by Cu, etc. [32–34].

Figure 2. Physical picture of manganese nodule.

Table 1. Classification of elements in manganese nodule. Adapted from Ref. [35].

Classification	Element
Main element (>1%)	Mn (13~27%), Fe (6~18%), SiO_2 (5%), Al_2O_3 (1.2%), MgO (1.5%), Na_2O (1.8%)
Minor element (600×10^{-6}~1%)	Co (0.3~1.2%), Fe (1670×10^{-6}~7250×10^{-6}%), Cu (573×10^{-6}%), Ba, Sr, Pb
Trace element (<600×10^{-6}%)	Mo, W, Pt, Pd, Nb, Ga, Te, Sc, Y

2.1. Composition and Content of Manganese Nodules

The representative manganese nodules were crushed and homogenized for X-ray fluorescence (XRF) analysis, multi-element chemical analysis, X-ray diffraction (XRD) analysis and chemical phase analysis. As present in results of XRF semi-quantitative analysis (Table 2), the element contents of Mn, Fe, Ni, Co, and Cu that are valuable for recovery in this manganese nodule sample are 29.92%, 9.92%, 1.44%, 0.27%, and 1.18%, respectively. The gangue elements with high content include Si, Al, Ca, Na, Mg, K, etc. The elements with lower content such as Ti, Ba, Pb, Zn, Mo, W, Ce, etc. can be considered for comprehensive recovery.

Table 2. Results of XRF analysis (mass fraction, %).

O	Mn	Si	Fe	Al	Ca
33	29.91	10.28	9.19	3.34	2.45
Na	Mg	Ni	K	Cu	Cl
2.45	2.17	1.44	1.28	1.18	0.94
Ti	Ba	Co	P	S	Zn
0.68	0.342	0.27	0.27	0.22	0.144
Sr	V	Mo	Pb	Ce	Zr
0.0765	0.0707	0.0573	0.559	0.051	0.0387
Tl	W	La	Y	Nb	Rb
0.0335	0.032	0.023	0.0129	0.003	0.002

Further chemical analysis results (Table 3) indicate that the contents of Mn, Fe, Ni, Co, and Cu are 24.68%, 7.85%, 1.05%, 0.23%, and 0.92%, respectively. Coupled with the phase analysis results (Tables 4–6), it can be concluded that Mn mainly exists in the aggregate of manganese iron oxides, accounting for 65.23%, followed by 27.81% of manganese dioxide, 6.19% of manganese oxide and 0.76% of manganese silicate minerals. More than 99% of Co and Ni are distributed in manganese oxides, which is related to the formation mechanism of manganese nodules. During the formation of manganese nodules, Co^{2+} in seawater is adsorbed by the negatively charged hydrated MnO_2 colloid under the action of external coordination adsorption and oxidized to Co^{3+}. Ni^{2+} is also fixed on MnO_2 through adsorption. Therefore, the distribution of Co and Ni is positively correlated with Mn. However, iron oxyhydroxide that is positively charged is easily adsorbed by certain hydrated metal ions that are negatively charged. The distribution of these metal elements is positively related to Fe and mainly includes Pb, V, As, Mo, W, Ti, Th, Zr, rare earth elements (REE), etc. [36–40]. The formation mechanism of manganese nodule is shown in Figure 3, which illustrates the co-occurrence relationship between various elements.

Table 3. Results of multi-element chemical analysis (mass fraction, %).

Mn	Cu	Co	Ni	Na	Mg
24.68	0.92	0.23	1.05	1.80	2.19
K	Ca	Si	Fe	Al	W
1.01	1.86	7.76	7.85	2.65	0.0072
V	Mo	Pb	Zn	P	S
0.04	0.047	0.052	0.13	0.30	0.15

Table 4. Phase analysis results of manganese.

Phase of Mn	Manganous Silicate	Manganese Dioxide	Ferromanganese Oxide	Manganous Oxide	Total
Content/%	0.19	6.92	16.23	1.54	24.88
Distribution/%	0.76	27.81	65.23	6.19	100.00

Table 5. Phase analysis results of cobalt and nickel.

Phase	In Manganese Oxides	In Iron Oxides	Total
Content of Co/%	0.23	0.0005	0.2305
Distribution of Co/%	99.78	0.22	100.00
Content of Ni/%	1.04	0.0082	1.0482
Distribution of Ni/%	99.22	0.78	100.00

Table 6. Phase analysis results of copper.

Phase of Cu	Free Copper Oxide	Bound Copper Oxide	Primary Copper Sulfide	Secondary Copper Sulfide	Total
Content/%	0.89	0.02	0.005	0.005	0.92
Distribution/%	96.74	2.17	0.54	0.54	100.00

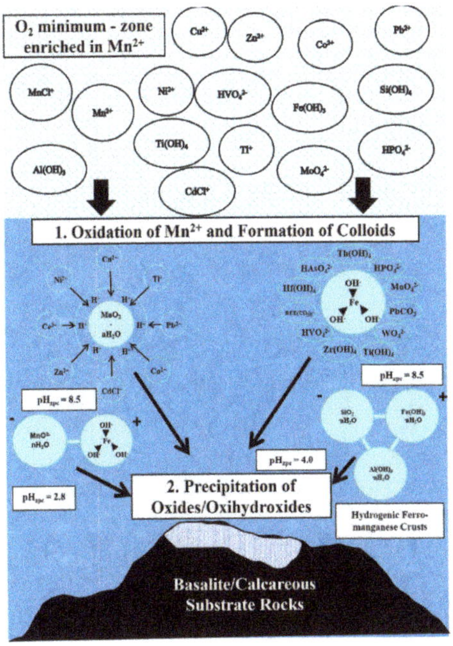

Figure 3. Schematic diagram of deep-sea manganese nodule formation. Adapted from Ref. [38].

The Cu content in this nodule is 0.92%, which is far higher than the average Cu content of the Pacific manganese nodule (573×10^{-6}%), and has great recovery value. Cu in seawater exists primarily in the form of hydrated divalent cations or carbonate complexes. Its concentration and distribution are influenced by geographical location, water depth, sediment types, and other environmental factors. It does not exhibit a strong correlation with the distribution patterns of Mn or Fe. A total of 96.02% of Cu in the manganese nodule in this area is in the form of free copper oxide.

As can be seen from Figure 4, the intensity of the XRD peak of the manganese nodule is very weak, and amorphous minerals such as vernadite are difficult to detect, which is consistent with the relevant research findings. However, hydromanganese mineral groups, such as todorokite and asbolane, and zeolite mineral groups, such as laumontite and leonhardite, can still be observed [41,42].

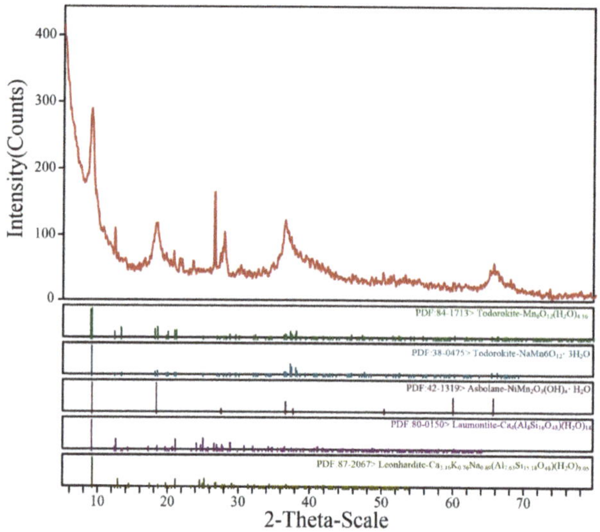

Figure 4. XRD pattern of manganese nodule.

2.2. Occurrence of Main Elements of Manganese Nodules

The distribution status of the main elements in manganese nodules were investigated in detail using Multimode V scanning microscope. As shown in Figure 5, the dense shell and loose core of the cross-sectioned manganese nodule under the microscope show obvious boundaries. An area in the core and shell is randomly selected for surface scanning to analyze the content and distribution of the elements, respectively. The results show that the main element contents in the core from high to low are: O (34.28%), Si (27.36%), Fe (13.95%), Al (9.35%), Mn (5.60%), K (2.81%), Mg (2.53%), Ca (1.10%), Na (1.09%), Cu (0.72%), Ni (0.68%), Co (0.54%), the main elements in the shell are Mn (34.59%), O (27.81%), Fe (15.10%), Si (5.13%), Mo (4.71%), Ca (2.54%), Al (2.42%), Mg (2.19%), Ni (1.36%), Na (0.88%), Cu (0.86%), K (0.68%), Zn (0.61%), W (0.57%), Co (0.49%), and Pb (0.05%). The contents of Si and Al in the core significantly exceed those in the shell, and the distributions of Al and Si in the core are closely related (Figure 6), indicating that the main minerals in the core are aluminosilicate minerals such as feldspar, pyroxene, clay, zeolite, etc. The contents of Mn and Fe in the shell are much higher than those in the core, and the Mn and Fe in the shell are highly dispersed (Figure 7), indicating that the main mineral in the shell is the iron–manganese mineral phase. The main valuable metal elements Ni, Co, and Cu are scattered in the core and shell of nodules. Comparing the content of each element in the core and the shell, when the content of Mn and Fe is high, the content of elements related to them is also high, which is consistent with the research results on the formation mechanism of manganese nodules.

Line scans were performed on different areas of manganese nodules to further study the changes in the content of each element. As presented in Figure 8, the change pattern of element characteristic peak intensity also proves that Al and Si are closely associated, and Fe and Mn are closely associated. From the core to the shell, the contents of Al and Si gradually decrease, while the contents of Fe and Mn gradually increase.

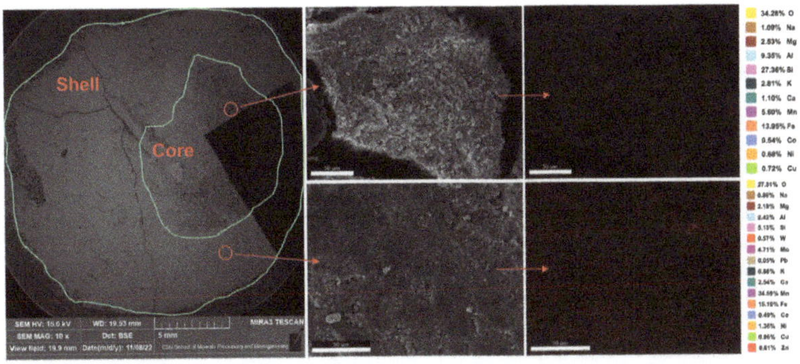

Figure 5. Occurrence of elements in manganese nodule.

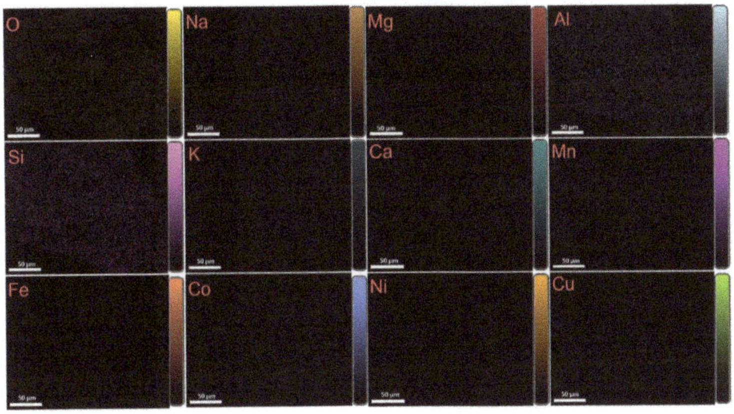

Figure 6. Occurrence of elements in the core of manganese nodule.

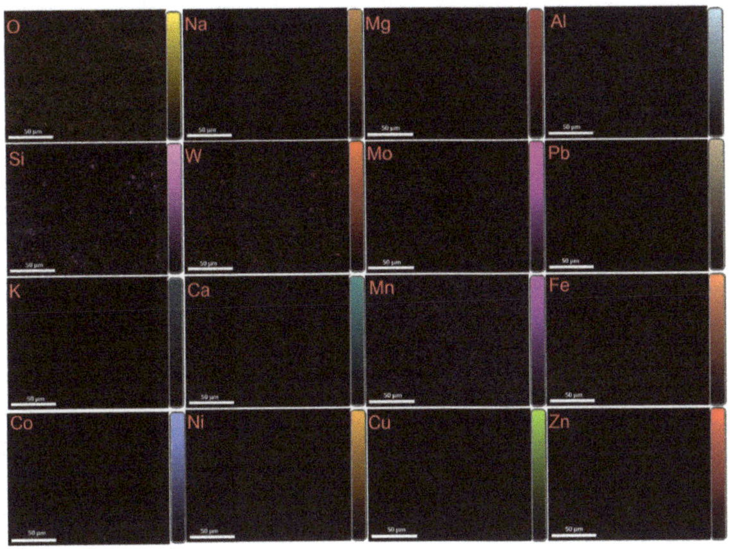

Figure 7. Occurrence of elements in the shell of manganese nodule.

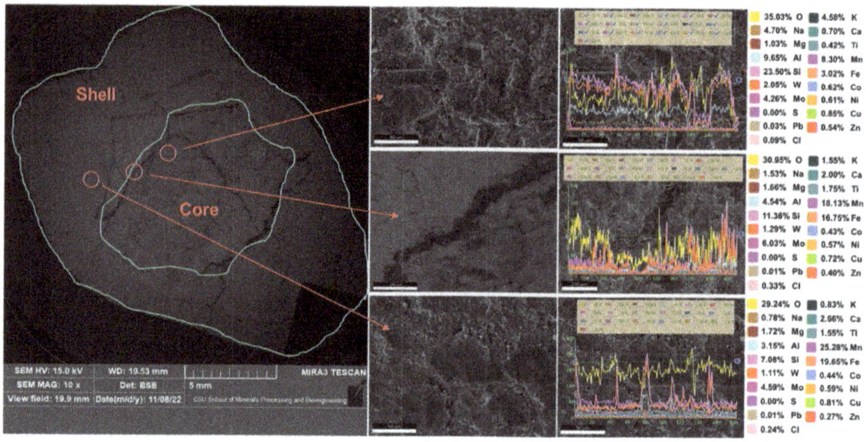

Figure 8. Occurrence of elements in different parts of manganese nodule.

3. Extraction of Main Valuable Elements from Manganese Nodules

Compared with the same metal minerals on land, the main metal content in manganese nodules is more abundant, which provides the possibility of direct metallurgical recovery. In the 1970s, four organizations, namely International Nickel Company (INCO), Kennecott Copper Corporation (KCC), Metallurgie Hoboken Overpelt Company (MHO), and Deep Sea Ventures (DSV), were among the pioneers to commence the metallurgical recovery of Cu, Co, Ni, and Mn from manganese nodules. Their efforts laid the foundation for the subsequent development of extraction technologies. At present, the mainstream extraction processes are mainly divided into two categories: pyrometallurgical–hydrometallurgical and hydrometallurgical [43–45]. In order to reduce costs or improve the comprehensive utilization value of manganese nodules, researchers have proposed new technologies such as microbial extraction, short-process for directly producing metal materials from manganese nodule alloys, and the preparation of environmentally friendly materials from metallurgical slag. The critical step in these extraction processes is to reduce Mn(IV) to Mn(II), destroying the structure of manganese nodules, so that the valuable elements such as Co, Ni, and Cu are released into the solution [46]. Figure 9 illustrates the chemical species distribution of Mn and Fe under varying redox potential (Eh) and pH conditions. At high redox potentials, manganese primarily exists in high oxidation states, such as MnO_2(s). In contrast, under low redox potentials or acidic conditions, manganese is reduced to its dissolved form, Mn^{2+}. To convert MnO_2(s) or Mn_3O_4(s) into Mn^{2+}, it is necessary to lower the redox potential to a reductive environment (e.g., negative Eh values) while maintaining acidic conditions (pH < 5). Therefore, using reductants to decrease the Eh value is a key strategy for achieving efficient manganese leaching. Iron primarily exists in dissolved forms (Fe^{2+} or Fe^{3+}) under acidic conditions. In neutral or alkaline environments, iron tends to form hydroxide or oxide precipitates, such as $Fe(OH)_3$ and Fe_2O_3. At low Eh and acidic conditions, iron readily dissolves as Fe^{2+}, which is critical for facilitating the reductive leaching of MnO_2. However, as pH increases or Eh rises, iron is more likely to transition into Fe^{3+}, forming precipitates that hinder the leaching process. After leaching is completed, impurities are removed from the leachate and elements are separated using treatment processes similar to those used for terrestrial resources, such as solvent extraction, electrolytic deposition, and so on [47–50].

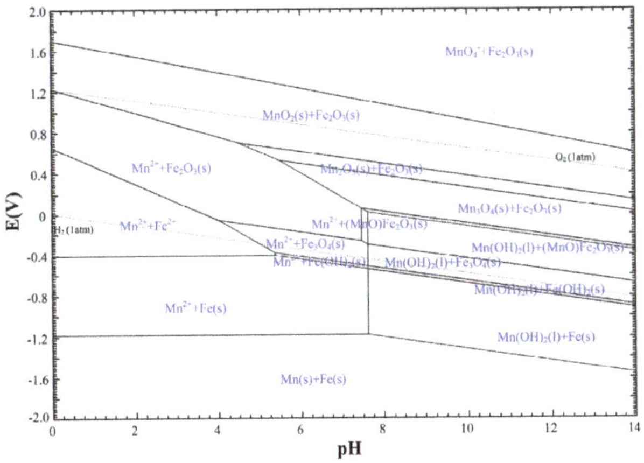

Figure 9. Potential–pH diagram for Mn–Fe–H_2O system at 25 °C and unit activity (based on HSC Chemistry 6.1).

3.1. *Pyrometallurgical–Hydrometallurgical Process*

3.1.1. Reduction Roasting–Acid Leaching Process

(1) INCO process. This pyrometallurgical–hydrometallurgical process is developed by INCO. As shown in Figure 10, the dried and ground manganese nodules are roasted reductively, and then smelted in an electric furnace to produce slag rich in manganese and iron, as well as alloys containing copper, nickel, cobalt, and small amounts of iron and manganese. The slag is further utilized to produce ferromanganese. The alloy is sulfurized to produce matte, and then leached with sulfuric acid. Subsequently, the leach solution undergoes extraction, electrowinning, hydrogen reduction, and other processes to separate and produce copper, nickel, and cobalt [51,52].

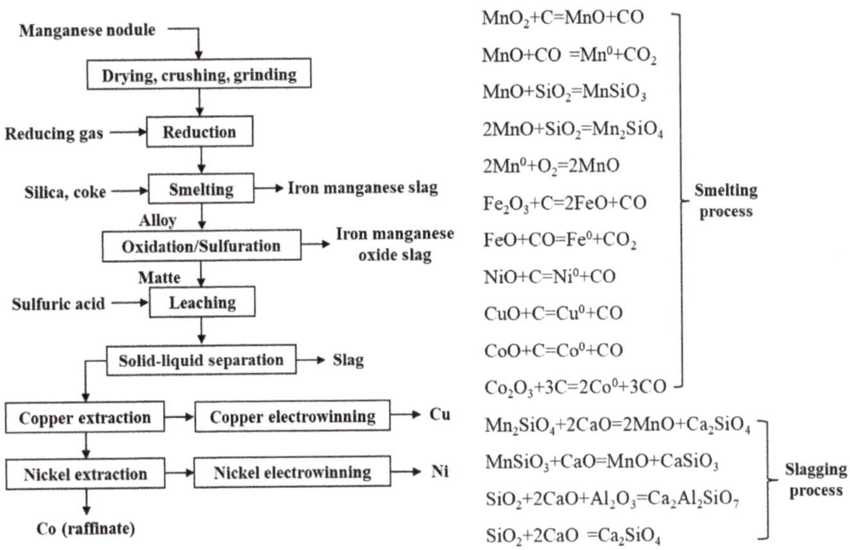

Figure 10. INCO smelting–leaching process and reactions involved. Adapted from Refs. [51,52].

(2) Improved INCO process. Based on the INCO process, China Ocean Mineral Resources Research & Development Association (COMRA) has made improvements to

the subsequent hydrometallurgical leaching of polymetallic alloys. The hydrochloric acid corrosion method was employed to precipitate most of the iron, while copper, cobalt, nickel, and manganese enter the solution. The metals were then extracted and separated. In small-scale trials, the metal recovery rates were as follows: 94.21% of Cu, 95.23% of Co, 96.49% of Ni, 97.68% of Mn, and 96.16% of Fe [53]. Similarly, Interoceanmetal Joint Organization (IOM) employed H_2SO_4-H_2SO_3 instead of H_2SO_4 to dissolve the alloy of copper, nickel, and cobalt produced by electrosmelting. Nickel, cobalt, and a small amount of manganese and iron were selectively dissolved, while copper precipitated as CuS. The manganese-rich slag generated during electrosmelting was recovered in the form of Si-Mn alloy. Ultimately, the recovery rates of Cu, Ni, and Co can reach 92%, 93%, and 89%, respectively [54].

The related researchers has improved the INCO process by adding a second smelting step on the basis of optimizing the original smelting process. This enhancement further processed the manganese-rich slag to produce high-carbon ferromanganese, which is used in the production of manganese steel. In addition, this process reduced the heavy metal content in the waste slag, making it suitable for use in the construction industry. Simultaneously, the Fe-Ni-Cu-Co alloy was directly subjected to pressurized sulfuric acid leaching without requiring conversion to sulfide, followed by solvent extraction of copper, iron precipitation, and cobalt solvent extraction. Copper sulfate was added during leaching to prevent the formation of hydrogen gas. The waste generated in the hydrometallurgical process was recycled back into the pyrometallurgical process, achieving zero waste discharge, as shown in Figure 11 [55].

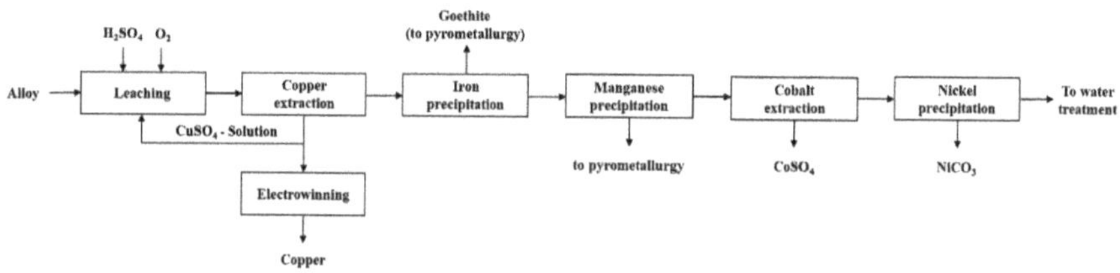

Figure 11. Improved hydrometallurgical process of INCO. Adapted from Ref. [55].

Xue et al. utilized residual carbon from coal gasification slag for the reduction roasting–sulfuric acid leaching treatment of manganese nodules, successfully extracting valuable metals such as manganese, copper, cobalt, and nickel. The leaching efficiency exceeded 98% for manganese and 96% for copper, cobalt, and nickel. The reduction and leaching processes were dominated by chemical interfacial reactions, with SO_2 generated from the residual carbon significantly enhancing manganese reduction. In addition to its high efficiency, this process offers substantial environmental and economic benefits by repurposing industrial by-products like coal gasification slag, minimizing waste, and reducing energy costs [56]. The selective carbothermal reduction and high-pressure sulfuric acid leaching has also been employed to recover cobalt, nickel, manganese, and copper from cobalt-rich deep-sea nodules. At 600 °C, 60 min and 3 wt% carbon, the method achieved leaching efficiencies of 98.47% cobalt, 95.72% nickel, 97.24% manganese, and 94.99% copper, with minimal extraction of iron and aluminum. This provides an efficient and sustainable method for recovering valuable metals [57].

3.1.2. Reduction Roasting–Ammonia Leaching Process

Inspired by the Caron process using ammonia solution to treat laterite nickel ores, the reduction roasting method has been introduced for processing manganese nodules. The products of reduction roasting of manganese nodules were subjected to ammonia leaching. Subsequently, copper, nickel, and cobalt formed soluble complexes with ammonia in the liquid phase. Finally, extraction and precipitation were used to achieve separation between different metals. Han et al. [56] performed reduction roasting of different manganese nodules at 600 °C in an atmosphere with a CO/CO_2 ratio of 60/40 for 2 h, followed by leaching with aqueous NH_3-$(NH_4)_2CO_3$ and NH_3-$(NH_4)_2SO_4$ at room temperature and atmospheric pressure. The extraction rate of nickel can reach more than 80%, and the extraction rate of cobalt was close to 50%. The mechanism study also confirmed that the leaching rate of cuprous oxide is faster than that of copper peroxide and metallic copper. Therefore, the excessive reduction at high temperature was not conducive to the ammonia leaching of copper. In the NH_3-$(NH_4)_2SO_4$ system, the extraction efficiency of metals is higher. However, since manganese, iron, and carbonate can form precipitates, manganese and iron are basically not extracted in the NH_3-$(NH_4)_2CO_3$ system, which is beneficial to the subsequent separation and purification of different metals. The dissolution reactions that may be involved are shown in Table 7.

Table 7. Dissolution reactions of copper, nickel, cobalt, and their oxides in ammonia solution. Adapted from Ref. [58].

Reaction	ΔG/kcal·mol^{-1}
$Cu + \frac{1}{2}O_2 + 2NH_3 + 2NH_4^+ \Leftrightarrow Cu(NH_3)_4^{2+} + H_2O$	−34.2
$CuO + 2NH_3 + 2NH_4^+ \Leftrightarrow Cu(NH_3)_4^{2+} + H_2O$	−3.8
$Cu_2O + \frac{1}{2}O_2 + 4NH_4^+ \Leftrightarrow 2Cu(NH_3)_2^{2+} + 2H_2O$	−33.44
$Cu_2O + 2NH_3 + 2NH_4^+ \Leftrightarrow 2Cu(NH_3)_2^+ + H_2O$	−2.05
$Ni + \frac{1}{2}O_2 + 4NH_3 + 2NH_4^+ \Leftrightarrow Ni(NH_3)_6^{2+} + H_2O$	−54.53
$NiO + 4NH_3 + 2NH_4^+ \Leftrightarrow Ni(NH_3)_6^{2+} + H_2O$	−3.23
$Ni_3O_4 + 10NH_3 + 8NH_4^+ \Leftrightarrow 3Ni(NH_3)_6^{2+} + H_2O + 2H^+ + 3H_2O + \frac{1}{2}O_2$	+31.38
$Co + \frac{3}{4}O_2 + 3NH_3 + 3NH_4^+ \Leftrightarrow Co(NH_3)_6^{3+} + \frac{3}{2}H_2O$	−62.72
$CoO + \frac{1}{2}O_2 + 3NH_3 + 3NH_4^+ \Leftrightarrow Co(NH_3)_6^{3+} + \frac{3}{2}H_2O$	−13.76
$Co_3O_4 + \frac{1}{4}O_2 + 9NH_3 + 9NH_4^+ \Leftrightarrow 3Co(NH_3)_6^{3+} + \frac{9}{2}H_2O$	−20.47
$Co_2O_3 + 6NH_3 + 6NH_4^+ \Leftrightarrow 2Co(NH_3)_6^{3+} + 3H_2O$	−10.4

Kmetova [59] et al. separately investigated the performance of wood charcoal and natural gas in the reduction roasting–ammonia leaching process. The results indicated that the extraction efficiencies of nickel, copper, and molybdenum exceeded 90%, 70%, and 60%, respectively, when the roasting was at 1073 K of temperature, with a 6% addition ratio of wood charcoal for 120 min, and the leaching at 318 K of temperature, with a 1 M solution of $(NH_4)_2CO_3$ in 10% NH_3, 20 of L/S for 210 min. To achieve an extraction efficiency of over 90% for cobalt, the roasting temperature needed to be increased to 1123 K. However, the performance of natural gas was not as effective as wood charcoal. At a roasting temperature of 873~973 K, the extraction efficiencies of Cu and Mo were the highest, 60% and 85%, respectively. At a roasting temperature of 1173 K, the extraction efficiency of Ni and Co reached the maximum, more than 40%.

India's National Metallurgical Laboratory (NML) has also been dedicated to improving the metal recovery rates of the reduction roasting–ammonia leaching process. The study revealed that excessive reduction roasting can lead to the formation of alloys between iron and nickel or cobalt. These alloys hindered the reaction between the leaching reagent and nickel or cobalt, resulting in a decrease in extraction efficiency. Therefore, it was essential to select appropriate roasting temperatures and durations. Another factor contributing to low leaching rates was that as the leaching time extends, the small amount of Cu $(NH_3)_2^+$ that was released rapidly elevated the system's redox potential, causing the precipitation of iron and manganese. The recovery rates of nickel and cobalt decreased due to co-precipitation/adsorption, especially cobalt [60,61]. Improvements in extraction efficiency have been achieved by employing a two-stage ammonium leaching process using high-concentration ammonium solution followed by low-concentration ammonium solution, while strictly controlling the redox potential in the first-stage leaching system. However, the high iron content in the leaching solution was not favorable for the subsequent extraction and electrowinning for purifying the various metals. Therefore, the recovery rate of Co needed to be sacrificed for the rationality of the entire process. After 16 leaching cycles, the recovery rates of Cu, Ni, and Co reached 92%, 90%, and 56% [62]. In order to improve the recovery rate of Co, Mishra et al. introduced anionic surfactants in the pretreatment process of roasted manganese nodules, which reduced the adsorption of Co on the ferromanganese precipitate and achieved an impressive average metal recovery rates of 92.5% of Cu, 91.5% of Ni, 71.35% of Co [63]. The optimized process is shown in Figure 12. In order to study the interactive influences of the leaching behavior of various metals in NH_3-CO_2 solution after the reduction roasting of manganese nodules, Jana et al. used metal powders of copper, nickel, cobalt, manganese, and iron to simulate ammonia leaching experiments. The results showed that, in the absence of iron and manganese, the combined leaching behavior of copper, nickel, and cobalt was the same as their individual leaching behaviors. However, when manganese was added, the leaching of each metal decreased. The reduced recovery rate was due to the reducing atmosphere caused by the hydrogen gas produced when metallic manganese reacted with the ammonia solution to form an aminomethanoate compound. The presence of iron alone had little effect on the leaching behavior of copper, nickel, and cobalt. In the actual leaching of manganese nodules after reduction roasting, the recovery of copper was still satisfactory (97%), while the recovery of nickel and cobalt were lower. The high recovery of copper may be due to the fact that the metallic manganese in the nodules after reduction roasting was still less than the oxidized manganese, which cannot create a reducing atmosphere by generating hydrogen. Additionally, when preconditioned in the presence of air, a passivation film formed on the surface of the iron–cobalt and iron–nickel alloys, inhibiting the dissolution of cobalt and nickel into the solution. By conducting the pretreatment in an air-free environment and then leaching in the presence of air, the formation of the passivation film can be avoided, and iron can quickly precipitate in the initial stage of leaching, improving the recovery of nickel and cobalt [64].

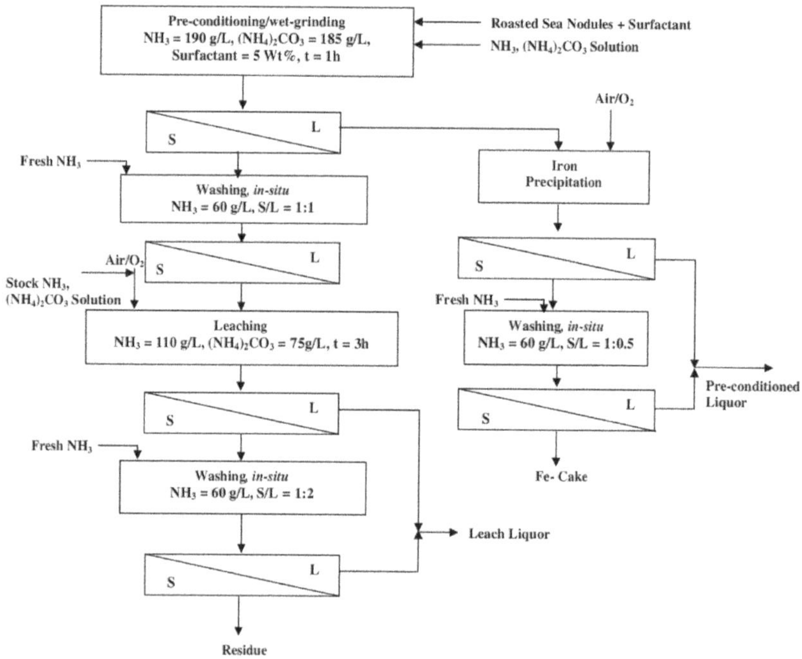

Figure 12. Optimized leaching process under lock-cycle mode. Reprinted with permission from ref. [63]. 2011 Elsevier.

3.2. *Hydrometallurgical Process*

3.2.1. Ammonia Leaching

KCC initially employed the ammonia leaching process to extract valuable metals from manganese nodules, known as the Cuprion process. As presented in Figure 13, Cu(II) in the solution was reduced to Cu(I) by CO, and then Mn(IV) was reduced to Mn(II) by Cu(I). Copper, nickel, and cobalt formed soluble complexes with ammonia, while manganese and iron formed carbonates and precipitated into the leaching residue. Subsequently, a solvent extraction–electrowinning method was employed to produce copper and nickel, followed by the recovery of cobalt and molybdenum from the raffinate. The process had mild reaction conditions and good selectivity, but it also had disadvantages such as low cobalt recovery (50%), low pulp concentration, and the potential hazards associated with CO [65,66]. Other reductants have been introduced to improve the Cuprion process include glucose [67], manganese ions [68], ferrous sulfate [69], elemental sulfur [70], sulfur dioxide [71], ammonium thiosulfate, and ammonium sulfite [72]. The subsequent separation and purification of valuable metals in the leachate and leaching residue also adopted conventional metallurgical methods such as solvent extraction and precipitation [73,74].

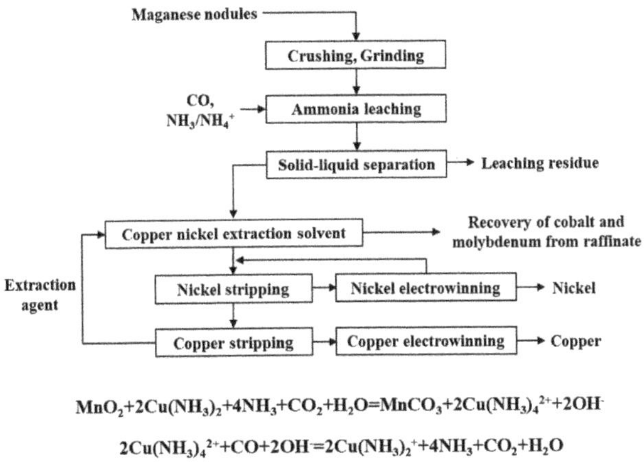

Figure 13. The Cuprion process and main reactions involved. Adapted from Refs. [63,64].

The National Institute for Resources and Environmental Research (NIRE) in Japan combined the advantages of NH_3-CO and NH_3-SO_2 processes to leach manganese nodules and cobalt-rich crusts in solutions of $(NH_4)_2CO_3$ and $(NH_4)_2SO_3$. During the leaching process, the dissolved Mn, Fe, and CO_3^{2-} formed precipitates into the residue, and then Cu and Ni were simultaneously extracted from the leachate. Finally, methanol was added to the leachate containing Co, $(NH_4)_2CO_3$, and $(NH_4)_2SO_4$ to precipitate cobalt, and the remaining leachate was recycled [50,75].

India has perfected the NH_3-SO_2 process and built a 500kg/d industrial demonstration line. The average recovery of copper, nickel, and cobalt reached 85%, 90%, and 80%, respectively. As shown in Figure 14, the process mainly included: ① leaching manganese nodules in the presence of NH_3 and SO_2; ② solvent extraction and electrowinning of Cu; ③ sulfide co-precipitation of Ni and Co, and high-pressure acid leaching of precipitation; ④ extraction separation of Ni and Co; ⑤ electrowinning of Ni and Co. This process can selectively recover manganese from silicomanganese residue [76,77]. The leaching process involves reactions as Equations (1)–(6):

$$CuO + (NH_4)_2SO_3 + 2NH_3 = Cu(NH_3)_4SO_4 + H_2O \quad (1)$$

$$NiO + (NH_4)_2SO_3 + 2NH_3 = Ni(NH_3)_4SO_4 + H_2O \quad (2)$$

$$CoO + (NH_4)_2SO_3 + 2NH_3 = Co(NH_3)_4SO_4 + H_2O \quad (3)$$

$$MnO_2 + (NH_4)_2SO_3 = MnO + (NH_4)_2SO_4 \quad (4)$$

$$MnO_2 + (NH_4)_2SO_3 + 2NH_3 = Mn(NH_3)_4SO_4 + H_2O \quad (5)$$

$$2FeOOH + (NH_4)_2SO_3 + H_2O = 2Fe(OH)_2 + (NH_4)_2SO_4 \quad (6)$$

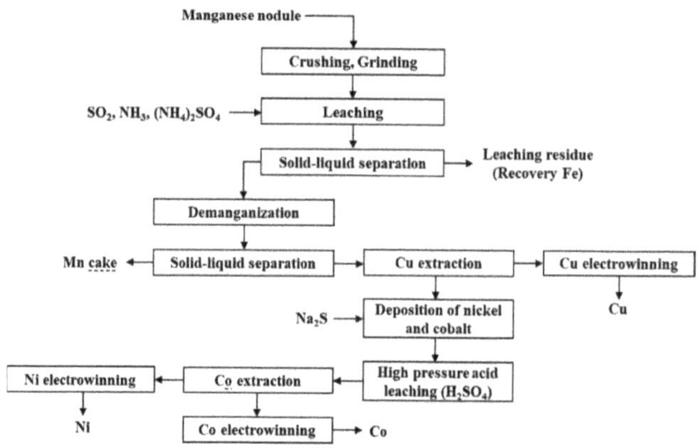

Figure 14. NH_3-SO_2 process developed in India. Adapted from Ref. [77].

3.2.2. Hydrochloric Acid Leaching

DSV and MHO earlier developed extraction processes that used concentrated hydrochloric acid to dissolve all metals in manganese nodules [35,74]. The DSV process leached the ground nodules with concentrated hydrochloric acid, where Mn(IV) was reduced to Mn(II), as shown in Equation (7). Then, a series of solvent extraction and electrowinning operations were used to separate the various metals in the leachate: (1) extract iron and copper separately; (2) co-extract cobalt and nickel, with $MnCl_2$ entering the raffinate; (3) selectively strip nickel from the loaded organic phase, followed by cobalt; (4) electro-deposit copper, nickel, and cobalt from their respective chloride solutions. Meanwhile, HCl and Cl_2 were regenerated through high-temperature hydrolysis for recycling. This process achieved high metal recovery, but the chlorides were highly corrosive, requiring equipment that can withstand these conditions.

$$MnO_2 + 4HCl = MnCl_2 + Cl_2 + H_2O \quad (7)$$

The MHO process is similar to the DSV process but uses the Cl_2 produced in the reaction to oxidize the dissolved Mn(II) back to Mn(IV), which then precipitates as MnO_2. In addition to copper, cobalt, nickel, and manganese, the MHO process also recovers lower concentrations of vanadium, molybdenum, and zinc from the chloride leachate.

To address the high energy consumption of the reduction roasting process, and the drawbacks of direct leaching with concentrated hydrochloric acid, the leaching of manganese nodules in dilute hydrochloric acid using iron minerals, sodium sulfite, and carbon as reducing agents has been investigated [78]. The results showed that using pyrite as a reducing agent, more than 80% of Mn and Co can be dissolved in 1.5 M HCl solution at temperatures of 80~90 °C over 3 h. With sodium sulfite (Na_2SO_3) as the reducing agent, similar extraction efficiencies can be achieved at lower leaching temperatures (50~60 °C) and a shorter leaching time (60 min). Furthermore, a combination of 10% carbon and 15% pyrite was more effective than using either carbon or pyrite alone.

The main advantage of the reductive leaching method using dilute hydrochloric acid was its ability to efficiently dissolve manganese while maintaining a low iron dissolution rate, which facilitated the subsequent separation process. In this process, the primary role of the reducing agents was to reduce dissolved $FeCl_3$ back to $FeCl_2$, which then promoted the reduction of MnO_2. For sodium sulfite, the SO_2 generated in the acidic solution also participated in the reduction of MnO_2. Therefore, lower temperature was sufficient for maximum metal extraction. The optimal leaching times for manganese nodules in 1.5 M HCl with different reducing agents were: sodium sulfite (60 min) < combination of carbon and pyrite (100 min) < pyrite (150 min) < carbon (180 min). The main reactions are shown in Equations (8)–(14).

Stepwise reactions with pyrite:

$$FeS_2 + 14FeCl_3 + 8H_2O \rightarrow 15FeCl_2 + 12HCl + 2H_2SO_4 \tag{8}$$

$$7MnO_2 + 14FeCl_2 + 28HCl \rightarrow 7MnCl_2 + 14FeCl_3 + 14H_2O \tag{9}$$

Overall reaction with pyrite:

$$FeS_2 + 7MnO_2 + 16HCl \rightarrow FeCl_2 + 7MnCl_2 + 2H_2SO_4 + 6H_2O \tag{10}$$

Stepwise reactions with sodium sulfite:

$$Na_2SO_3 + 2HCl \rightarrow 2NaCl + H_2O + SO_2 \tag{11}$$

$$2FeCl_3 + SO_2 + 2H_2O \rightarrow 2FeCl_2 + 2HCl + H_2SO_4 \tag{12}$$

$$MnO_2 + 2FeCl_2 + 4HCl \rightarrow MnCl_2 + 2FeCl_3 + 2H_2O \tag{13}$$

Overall reaction with sodium sulfite:

$$MnO_2 + Na_2SO_3 + 4HCl \rightarrow MnCl_2 + 2NaCl + H_2SO_4 + H_2O \tag{14}$$

Alcohols can also serve as reductants to enhance the leaching rate of Cu, Fe, Mn, Ni, and Co. In dilute hydrochloric acid (2.75 M), only 80% of Cu and 60% of Fe can be extracted from manganese nodules. However, by simply adding 9% propanol or 4.5% butanol to dilute the hydrochloric acid (2.75 M), the recovery rate of each metal can be increased to over 90%. Moreover, as the chain length of the alcohol increased, the required dosage of alcohol decreased [79]. The reactions involved are as follows:

Without alcohol:

$$MnO_2 + 4HCl \rightarrow MnCl_2 + Cl_2 \tag{15}$$

With alcohol:

$$MnO_2 + 4(CH_3)_2CHOH_2^+ + 4Cl^- \rightarrow MnCl_2 + 4(CH_3)_2CHOH + 2H_2O + Cl_2 \tag{16}$$

The commercial alkylphosphines, Cyanex, can be utilized for subsequently extracting pure Co(II), Ni(II), and Cu(II) from the concentrated hydrochloric acid leachate of manganese nodules. Co(II) and Cu(II) were extracted using Cyanex 923, and Ni(II) using Cyanex 301. These extractants exhibited excellent stability even after extended contact with hydrochloric acid and maintain their superior extraction capabilities after 20 cycles of reuse. Consequently, they were widely employed for the separation of Co(II), Ni(II), and Cu(II) from other metal ions, such as Ti(IV), Al(III), Fe(III), Mn(II), and Zn(II). In the organic phase, Co(II) and Cu(II) were present as $H_2CoCl_4{\cdot}2$ Cyanex 923 and $CuCl_2{\cdot}2$ Cyanex 923 complexes, respectively, while Ni(II) was extracted as NiR_2 (HR = Cyanex 301). The Stripping of Co(II) and Cu(II) was accomplished using a 0.001 M H_2SO_4, and Ni(II) was stripped with a 5% NH_4Cl in 75% NH_3. This method effectively recovered about 90% of Co(II), Ni(II), and Cu(II) from sea nodules, achieving an approximate purity of 99% for the metal ions [80].

3.2.3. Sulfuric Acid Leaching

Sulfuric acid is also widely used as an acidic medium for reduction leaching of manganese nodules. IOM has conducted extensive research on sulfuric acid leaching of manganese nodules from CCZ. As shown in Figure 15, the main operations were as follows: (I) preliminary grinding of wet raw materials in a ball mill to reach 86% particles of −0.2 mm; (II) selective leaching of Cu, Ni, Co, and Mn under normal pressure containing SO_2, thickening of the leached pulp with flocculant, and filtration of the underflow of the thickener; (III) selective precipitation of Cu from the leachate by introducing active sulfur powder and feeding of sulfuric anhydride to the reactor under normal pressure; (IV) precipitation of Ni and Co concentrate by introducing powdered sulfur and metallic manganese; (V) precipitation of MnO_2 during the neutralization reaction with ammonia water, thickening, filtration, washing of filter with manganese hydroxide concentrate, drying in a rotating furnace, and briquetting [81,82].

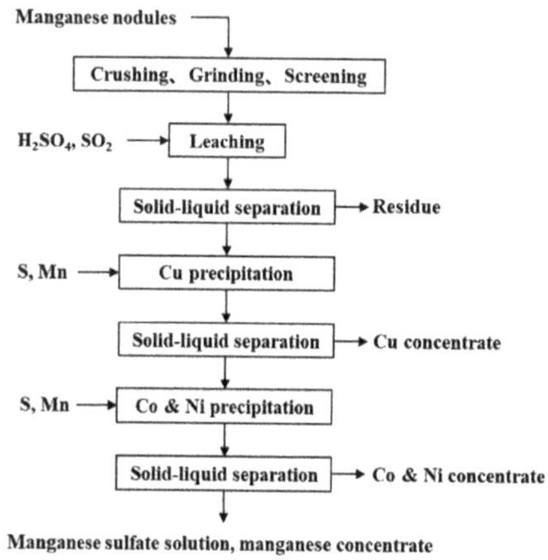

Figure 15. Simplified scheme for sulfuric acid leaching using SO_2 as a reductant. Adapted from Refs. [81,82].

The chemical reactions involved in the process of leaching manganese nodules with sulfuric acid using SO_2 as a reductant are as follows:

$$SO_2 + H_2O = H_2SO_3 \tag{17}$$

$$MnO_2 + 2H_2SO_3 = MnS_2O_6 + 2H_2O \tag{18}$$

$$MeO + H_2SO_3 = MeSO_3 + H_2O \tag{19}$$

$$Me_2O + H_2SO_3 = Me_2SO_3 + H_2O \tag{20}$$

$$MeO + H_2SO_4 = MeSO_4 + H_2O \tag{21}$$

$$Me_2O + H_2SO_4 = Me_2SO_4 + H_2O \tag{22}$$

where Me is Cu, Ni, Co, Fe, Ca, and Mg.

$$Cu_2SO_3 + CuSO_3 + 2H_2O = Cu_2SO_3 \cdot CuSO_3 \cdot 2H_2O \tag{23}$$

$$MnS_2O_6 + 2H_2SO_4 = MnSO_4 + H_2S_2O_6 \tag{24}$$

$$H_2S_2O_6 = H_2SO_4 + SO_2 \uparrow \tag{25}$$

Other reactions may occur in the presence of dissolved oxygen:

$$H_2O + \frac{1}{2}O_2 + 2SO_2 = H_2S_2O_6 \tag{26}$$

$$H_2SO_3 + \frac{1}{2}O_2 = H_2SO_4 \tag{27}$$

The chemical reactions during the selective precipitation of Cu with activated sulfur powder and SO_2 are as follows:

$$2CuSO_4 + 2SO_2 + S + 4H_2O = Cu_2S \downarrow + 4H_2SO_4 \tag{28}$$

$$CuSO_4 + SO_2 + S + 2H_2O = Cu_2S \downarrow + 2H_2SO_4 \tag{29}$$

The chemical reactions during Ni-Co concentrate precipitation with activated sulfur powder and metallic manganese are followed:

$$MeSO_4 + Mn_{met.} + S^0 = MeS \downarrow + MnSO_4 \tag{30}$$

where Me is Ni and Co.

The chemical reactions during the precipitation of MnO_2 are:

$$MnSO_4 + 2NH_4OH + \frac{1}{2}O_2 = MnO_2 + (NH_4)_2SO_4 + H_2O \tag{31}$$

Finally, a copper concentrate with a Cu content of 40%, a nickel–cobalt concentrate with a Ni and Co content of 20.8% and 2.7% C, respectively, and a manganese concentrate with a Mn content of 62% can be obtained through this process. The recovery rates of each metal were 92% of Cu, 96% of Ni, 92% of Co, and 96% of Mn. Using a similar process for leaching manganese nodules from the Indian Ocean, over 85% of Mn, Ni, and Co, as well as more than 75% of Cu were successfully extracted. The target metal extraction process was generally completed within 5~10 min. However, extending the leaching duration to 25~30 min proved to be more beneficial for the precipitation and separation of Fe [83].

IOM has also developed a high-pressure sulfuric acid leaching process, using molasses and pyrite as reducing agents [81]. Ion exchange resin was used to extract and separate metal ions in the high-pressure sulfuric acid leachate, followed by sulfide precipitation to extract copper, solvent extraction to extract zinc, and finally sulfide precipitation of nickel and cobalt, and calcining to obtain manganese oxide. The optimal operating parameters and indicators for the high-pressure sulfuric acid leaching process with molasses and pyrite can be seen in Table 7.

The principal reactions that occur in the leaching process with molasses are:

$$24MnO_2 + C_{12}H_{22}O_{11} + 24H_2SO_4 = 24MnSO_4 + 12CO_2 + 35H_2O \tag{32}$$

$$24Me_2O_3 + C_{12}H_{22}O_{11} + 48H_2SO_4 = 48MeSO_4 + 12CO_2 + 59H_2O \tag{33}$$

where Me are Co and Fe.

$$MnO_2 + FeSO_4 + H_2SO_4 = MnSO_4 + Fe_2(SO_4)_3 + H_2O \tag{34}$$

$$MeO + H_2SO_4 = MeSO_4 + H_2O \tag{35}$$

where Me are Cu, Ni, Co, Zn, and Al.

The principal reactions that occur in the leaching process with pyrite are:

$$FeS_2 + 3.5O_2 + H_2O = FeSO_4 + H_2SO_4 \tag{36}$$

$$2FeSO_4 + H_2SO_4 + 0.5O_2 = Fe_2(SO_4)_3 + H_2O \tag{37}$$

$$MnO_2 + 2Fe^{2+} + 4H^+ = Mn^{2+} + 2Fe^{3+} + 2H_2O \tag{38}$$

The research conducted by Anand et al. also explored the leaching of manganese nodules using dilute sulfuric acid under conditions of high temperature, high pressure, and oxygen. At the specific conditions of 423 K, a sulfuric acid concentration of 0.46 g/g of manganese nodules, an oxygen partial pressure of 0.55 MPa, and a leaching duration of 4 h, the process can extract nearly all the copper and nickel, as well as 88% of the cobalt. However, the leaching rates for manganese and iron were only 28% and 5.7%, respectively. To effectively extract all four metals—copper, nickel, cobalt, and manganese—charcoal can be utilized as a reducing agent to convert manganese dioxide into soluble manganese sulfate. When the leaching was conducted for 4 h at a temperature of 423 K, with a sulfuric acid concentration of 0.66 g/g of manganese nodules and an oxygen partial pressure of 0.55 MPa, the extraction for copper, nickel, cobalt, and manganese reached 77.8%, 99.8%, 88%, and 99.8%, respectively [84].

Barik et al. investigated the potential of using a micellar high-temperature acid dissolution process to selectively leach manganese from manganese nodules. It was found that cetyltrimethylammonium Bromide (CTAB) significantly enhanced the recovery of Cu, Ni, Co, Zn, and Mn. Moreover, the surfactant-assisted high-temperature sulfuric acid leaching method was employed for the effective removal of iron, aluminum, and silicon dioxide. Increasing the temperature of the medium had a notable impact on the selective leaching of Mn, Cu, Ni, and Co. The ideal conditions for maximum metal extraction were determined to be a pulp density of 10%, a leaching duration of 2 h, a leaching temperature of 160 °C, and a sulfuric acid concentration of 5.0% (v/v), with CTAB reaching its critical micelle concentration (CMC). Under these conditions, the recovery for Mn was 99%, and for Cu, Co, and Ni, it was also 99%. As presented in Figure 16, the addition of CTAB increased the rate of mass transfer during leaching, greatly enhancing the efficiency of the process. However, forming surfactant micelles in the leaching system beyond the CMC slightly reduced the mass transfer rate and leaching efficiency. Additionally, CTAB likely adsorbed effectively on iron oxide phases, inhibiting their dissolution, and thereby increasing the selectivity of the leaching system [85].

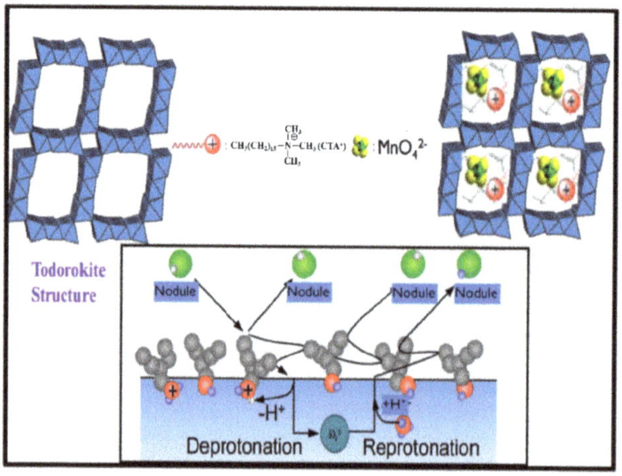

Figure 16. Mechanism of the sulfuric acid leaching of manganese nodule in the presence of CTAB. Reprinted with permission from ref. [85]. 2016 Elsevier.

Torres et al. conducted a comparative study on the effectiveness of various iron-reducing agents in the sulfuric acid leaching process of manganese nodules. The research revealed that Fe^0 (FeC) was the most efficient agent for this process. Its efficacy was attributed to the direct contact of Fe in the solution, which promoted the regeneration of ferrous ions due to the high concentrations of both ferrous and ferric ions. Remarkably, 97% of Mn was extracted using a MnO_2/FeC ratio of 1/2 and 0.1 mol/L H_2SO_4 in just 20 min. When employing Fe^{2+}, FeC, and Fe_2O_3 as high-concentration reducing agents (with MnO_2 to reducing agent ratios of 1/2 or lower), it was possible to maintain low potential values, enabling effective leaching at a low acid concentration of 0.1 mol/L. However, better results were achieved with pyrite (FeS_2) at higher ratios of MnO_2/FeS_2 (1/3) and acid concentrations of 1 mol/L, likely due to the refractoriness of pyrite. Additionally, maintaining the potential and pH within the ranges of −0.4~1.4 V and −2~0.1, respectively, when using various iron-reducing agents, proved beneficial for dissolving manganese nodules and simultaneously prevented the formation of iron oxides [86]. Additionally, the studies confirmed that the leaching process of nodules in $FeSO_4$-H_2SO_4-H_2O solutions was remarkably swift [87]. The optimal conditions identified were using the stoichiometric amount of $FeSO_4$, 1.6 times the stoichiometric amount of H_2SO_4, a temperature of 90 °C, a liquid-to-solid ratio of 7:1, and a manganese nodule particle size of −1000 μm. Under these conditions, it was possible to extract more than 85% of cobalt and over 90% of nickel, copper, and manganese within 30 min. The concentration of cobalt in the leachate was about 146 mg/L, nickel was 1.63 g/L, copper was 1.69 g/L, and manganese was 30 g/L. The primary reaction involved in this process is represented by Equation (39):

$$MnO_2 + 2FeSO_4 + 2H_2SO_4 = MnSO_4 + Fe_2(SO_4)_3 + 2H_2O \quad (39)$$

In subsequent research, $(NH_4)_2S$ was also used for sulfide precipitation to precisely separate Cu, Ni, and Co from the leachate of manganese nodules in the $FeSO_4$-H_2SO_4-H_2O solution after iron removal. At a concentration of 5.5% $(NH_4)_2S$, room temperature, pH = 1.0 (for copper precipitation), and pH = 3.0 (for nickel-cobalt precipitation), and with the optimized flow rate of 0.15 g S^{2-}/min per L of processed solution, ideal results were achieved with about 98% of Cu, Ni, and Co being effectively separated. This provided high-quality raw materials to produce the three metals [88].

Some commercial extractants can also be used directly for subsequent separation of metals. Shen et al. confirmed the effectiveness of sulfuric acid leaching of manganese nodules using pyrrhotite as the reducing agent and studied the subsequent separation and purification of manganese, nickel, and cobalt. The process involves: ① leaching manganese nodules at 90 °C for 4 h with a 1.25 mol/L sulfuric acid concentration, a 3:2 nodule-to-pyrrhotite ratio, a 2:1 liquid-to-solid ratio, stirring at 300 rpm; ② removing impurities such as iron and aluminum through oxidation precipitation; ③ adding sodium sulfide and stirring at 50 °C for 2 h to precipitate cobalt and nickel from the leachate, and the resulting solution is concentrated to obtain manganese sulfate; ④ redissolving cobalt and nickel precipitates and using an eight-stage process with di(2-ethylhexyl) phosphoric acid (D2EHPA) at an organic-to-aqueous ratio of 3:5 to eliminate impurities; ⑤ extracting cobalt using 2-ethylhexylphosphonic acid mono-2-ethylhexyl ester (PC88A or P507) in a three-stage counter-current operation with an O/A volume ratio of 2:3, followed by the washing and stripping of cobalt and nickel. The maximum recovery of manganese, cobalt, and nickel were 85%, 75%, and 78%, respectively [89]. A similar metallurgical process was used to separate valuable metals from a leachate produced by sulfuric acid leaching of manganese nodules, with starch as the reducing agent. The leachate contains 22.85 g/L of manganese, 6.38 g/L of iron, 1.01 g/L of copper, 0.023 g/L of zinc, 0.09 g/L of cobalt, and 1.44 g/L of nickel. Initially, iron was precipitated and removed from the solution using $Ca(OH)_2$ at a pH of 3.8. Subsequently, copper was extracted using the LIX 84I extractant. After the removal of iron and copper, zinc was removed using kerosene-based di-(2-ethylhexyl) phosphoric acid (D2EHPA). Following this, manganese was extracted using saponified D2EHPA (NaD2EHPA). Employing a two-stage extraction at A:O ratio of 3:4, highly efficient manganese extraction is achieved with an efficiency of 99.93%. During this process, cobalt was co-extracted with manganese at a concentration of 43 mg/L. To remove this cobalt, the solution was washed twice using a $MnSO_4$ solution. Furthermore, manganese is further extracted using 4% H_2SO_4 at an A:O ratio of 1:1 through a two-stage process. Nickel was extracted from the residual liquid after manganese extraction using NaD2EHPA at an A:O ratio of 3:2. This two-stage extraction results in a combined extraction quantity of cobalt and nickel of 47 mg/L. The nickel load was then washed with a $NiSO_4$ solution. Finally, nickel was efficiently leached at a rate of 99.93% using 0.5% H_2SO_4 at an A:O ratio of 1:1 in a two-stage process [90].

Other reductants like phenols [91], aromatic amines [92], and saw dust [93] employed for sulfuric acid leaching of manganese nodules have also been intensively studied. A summary of the sulfuric acid leaching process of manganese nodules under different reductants is shown in Table 8.

Table 8. Summary of the sulfuric acid leaching process of manganese nodules under different reductants/auxiliaries.

Reductant/Auxiliary	Leaching Condition	Leaching Efficiency/%	Reference
SO_2	Manganese nodules from CCZ, −0.2 mm particles accounted for 86%, temperature 343~353 k	In total, 92% of Cu, 96% of Ni, 92% of Co, 96% of Mn	[82]
SO_2	Manganese nodules from Indian Ocean, liquid/solid 10:1, 10 mL 1 M H_2SO_4 added per 100 mL, SO_2 concentration 5% (*wt*/*v*), leaching time 15 min, temperature 31 °C	Over 85% of Mn, Ni, and Co, over 75% of Cu	[83]
SO_2	Manganese nodules from Cook Islands Exclusive Economic Zone, 30 °C, L/S ratio of 6:1 mL/g, H_2SO_4 dosage of 37.5 wt% and SO_2 partial pressure of 200 kPa	In total, 98.7% of Mn, 91.4% of Fe, 95.3% of La, 99.2% of Ce, 99.3% of Ni, 95.9% of Co	[94]
Molasses	Manganese nodules from CCZ, concentration 31%, H_2SO_4 0.8 kg/kg nodule, molasses 0.12 kg/kg nodule, leaching time 60 min, temperature 140 °C	In total, 97.70% of Ni, 98.31% of Co, 91.77% of Cu, 97.99% of Zn, 97.07% of Mn, 24.54% of Fe	[81]
Pyrite	Manganese nodules from CCZ, concentration 35%, H_2SO_4 0.81 kg/kg nodule, pyrite 0.12 kg/kg nodule, leaching time 90 min, pressure 10 atm, temperature 160 °C	In total, 98.42% of Ni, 91.67% of Co, 95.98% of Cu, 96.75% of Zn, 27.52% of Mn, 27.22% of Fe	[81]
Water	Manganese nodules from Indian Ocean, −250 μm particles accounted for 100%, concentration 15% (*wt*/*v*), H_2SO_4 0.46 g/g nodule, leaching time 4 h, oxygen partial pressure 0.55 MPa, temperature 423 K	Nearly 100% of Cu and Ni, 88% of Co, 28% of Mn, 5.7% of Fe	[84]
CTAB	Manganese nodules from Indian Ocean, −100 μm particles accounted for 100%, concentration 10% (*wt*/*v*), H_2SO_4 5% (*v*/*v*), leaching time 2 h, CTAB at critical micelle concentration, temperature 160 °C	In total, 99% of Mn, Cu, Co, and Ni	[85]
FeC	Manganese nodules from the Blake Plateau in the Atlantic Ocean, particle size of −140 + 100 μm, liquid/solid 100:1, FeC/MnO_2 2:1, H_2SO_4 0.1mol/L, leaching time 20 min, room temperature	In total, 97% of Mn	[86]

Table 8. *Cont.*

Reductant/Auxiliary	Leaching Condition	Leaching Efficiency/%	Reference
$FeSO_4$	Manganese nodules from CCZ, particle size of −1000 μm, liquid/solid 7:1, $FeSO_4$ at stoichiometric amount, H_2SO_4 at 1.6 times stoichiometric amount, leaching time 30 min, temperature 90 °C	More than 85% of Co, over 90% of Ni, Co, Mn	[87]
Phenols	Manganese nodules from the central Pacific Basin, −74 μm particles accounted for 77%, phenol 0.25~0.4 g/g nodules, H_2SO_4 0.925g/g nodules, liquid/solid 4:1, leaching time 10~20 min, room temperature	Over or nearly 95% of Mn, Co, Ni, Cu	[91]
Aromatic amines	Manganese nodules from the central Pacific Basin, −74 μm particles accounted for 77%, aromatic amines 0.3 g/g nodules, H_2SO_4 0.925 g/g nodules, liquid/solid 4:1, leaching time 10~20 min, room temperature	Over 97% of Mn, Co, Ni, Cu	[92]
Sawdust	Manganese nodules from Indian Ocean, −100 μm particles accounted for 100%, pulp concentration 10% (*wt*/*v*), sawdust 0.5 g/g nodules, H_2SO_4 5% (*v*/*v*), leaching time 2 h, temperature 105 °C	In total, 99.5% of Mn, 99.1% of Cu, 99.6% of Ni, 93% of Co, 64.6% of Fe	[93]
Paper	Manganese nodules from Indian Ocean, −150 μm particles accounted for 100%, pulp concentration 20% (*wt*/*v*), paper 0.3 g/g nodules, H_2SO_4 7.56% (*v*/*v*), leaching time 2 h, temperature 90 °C	In total, 97.28% of Cu, 98.66% of Ni, 97.90% of Co and 99.00% of Mn	[95]
Glycerol	Manganese nodules from Indian Ocean, −150 μm particles accounted for 100%, pulp concentration 10% (*wt*/*v*), glycerol 1% (*v*/*v*), H_2SO_4 10% (*v*/*v*), leaching time 1 h, temperature 80 °C	Over 95% of Ni, over 98% of Cu, Co, Mn	[96]

3.3. Other Extraction Process

In order to enhance resource recovery efficiency and address the increasingly severe challenges of environmental protection, various new extraction technologies are continuously being developed.

(1) Co-extraction with other mineral resources. Tomas et al. investigated the leaching effect of chalcopyrite in hydrochloric acid when manganese nodules were used as oxidants. The results indicate that chalcopyrite is not dissolved independently, but is oxidatively leached by Cl_2 released by the reaction of MnO_2 with HCl, see DSV and MHO. Drawing inspiration from this research, it was possible to utilize terrestrial resources such as low-grade sulfide ore or tailings that are challenging to recover through conventional beneficiation methods as reducing agents to achieve the reduction leaching of valuable elements in manganese nodules. This can not only reduce the recovery cost of manganese nodules, but also maximize the resource recovery of low-grade sulfide ores on land [97]. Using manganese nodules as an oxidant and high chloride ion concentration wastewater for

the sulfuric acid leaching of chalcopyrite, the high chloride ion content in the wastewater successfully addressed the passivation issue typically encountered in sulfuric acid leaching. This approach not only utilized the wastewater efficiently but also achieved synergistic leaching of manganese nodules and chalcopyrite. As a result, the leaching rate of copper from chalcopyrite can reach up to 77% [98].

The latest research involved co-leaching of manganese nodules with flotation tailings in sulfuric acid solution. The tailings primarily consist of iron minerals, with 58.52% magnetite and 4.47% limonite. The $FeSO_4$ produced during the leaching process acts as an effective reducing agent, shortening the dissolution time of MnO_2. High concentrations of $FeSO_4$ maintained the leaching system's potential and pH within the ranges of −0.2~1.2V and −1.8~0.1, respectively, which was conducive to the formation of Fe^{2+} and Fe^{3+}. Under these conditions, with a MnO_2/Fe_2O_3 ratio of 1:3 and an H_2SO_4 concentration of 0.1 M, the manganese nodules achieved rapid leaching within a short period (5~20 min), with a manganese leaching rate of 68% to 73%. The addition of tailings for the reductive leaching of marine nodules in an acidic medium presented an attractive and cost-efficient alternative [99,100]. The main reaction and its standard Gibbs free energy change (25 °C) are presented in Table 9.

Table 9. Main reaction and thermodynamic information.

Reaction	ΔG° (kJ)
$Fe_2O_{3(s)} + 3H_2SO_{4(aq)} = Fe_2(SO_4)_{3(s)} + 3H_2O_{(l)}$	−163.37
$Fe_3O_{4(s)} + 4H_2SO_{4(l)} = FeSO_{4(aq)} + Fe_2(SO_4)_{3(s)} + 4H_2O_{(l)}$	−261.30
$2FeSO_{4(aq)} + 2H_2SO_{4(aq)} + MnO_{2(s)} =$ $Fe_2(SO_4)_{3(s)} + MnSO_{4(aq)} + 2H_2O_{(l)}$	−199.52

(2) Solid-state metalized reduction technology. Zhao et al. proposed a process that combined solid-state metallized reduction treatment and magnetic separation to recover valuable metals in marine polymetallic nodules. Under the conditions of CaF_2 dosage of 4%, anthracite coal dosage of 7%, SiO_2 dosage of 5%, FeS dosage of 6%, temperature of 1100 °C, and time of 2.5 h, most of the valuable metals were selectively reduced to the metallic state. Magnetic separation was then used to recover the reduced metals as concentrate. The optimal recovery rates of Ni, Co, Cu, Mn, and Fe in the concentrate were 86.48%, 86.74%, 83.91%, 5.63%, and 91.46%, respectively [101]. The introduction of magnetic separation made the process of extracting valuable metals from manganese nodules more economical and environmentally friendly.

(3) Segregation roasting. In 1974, the University of California pioneered the application of segregation roasting for extracting metals from manganese nodules. The results indicated that using $CaCl_2$ as the chlorinating agent and petroleum coke as the reductant, the optimal conditions for metal separation were achieved with a roasting time of about 2 h. When the roasting temperature was set to 850 °C, the recovery of copper reached a maximum of 75%, while the recovery of nickel and cobalt only reached around 25%. At 1050 °C, the recovery of nickel and cobalt increased to about 60%, while copper dropped to 35%. The metal oxides in the nodules mainly went through chlorination and reduction. Based on reaction energies, the metal response order was Cu > Ni > Co > Fe > Mn. Table 10 depicts the energy for this reaction in the metal oxide–hydrogen–chloride system. Despite current limited studies on this method, it is effective for producing high-purity metals from refractory materials, making it a top choice for deriving ultra-pure metals from manganese nodules.

Table 10. Standard free energy for the chloritization reduction cycle in metal oxide–hydrogen–chloride systems.

Reaction	ΔG^{θ} (kcal)	Temp (°K)
$Cu_2O + 2HCl = 2/3Cu_3Cl_3 + H_2O$	−14.9	
$2/3Cu_3Cl_3 + H_2 = 2Cu + 2HCl$	−8.3	1100
$Cu_2O + H_2 = 2Cu + H_2O$	−23.2	
$NiO + 2HCl = NiCl_2 + H_2O$	+8.2	
$NiCl_2 + H_2 = Ni + 2HCl$	−20.0	1200
$NiO + H_2 = Ni + H_2O$	−11.8	
$CoO + 2HCl = CoCl_2 + H_2O$	+3.0	
$CoCl_2 + H_2 = Co + 2HCl$	−10.7	1200
$CoO + H_2 = Co + H_2O$	+7.7	
$FeO + 2HCl = FeCl_2 + H_2O$	+4.1	
$FeCl_2 + H_2 = Fe + 2HCl$	−2.9	1200
$FeO + H_2 = Fe + H_2O$	+1.2	
$MnO + 2HCl = MnCl_2 + H_2O$	+3.16	
$MnCl_2(g) + H_2(g) = Mn + 2HCl$	+50.62	1200
$MnO + H_2 = Mn + H_2O$	+53.78	

(4) Hydrogen metallurgy. Hydrogen metallurgy is an attractive method for reducing carbon emissions and is equally applicable to the reduction roasting of deep-sea manganese nodules. The result has revealed that the presence of H_2 can disrupt the lattice structure of manganese minerals, thereby releasing NiO, CoO, and CuO, and facilitating the formation of Fe-Ni-Co-Cu alloys. However, the reducing capacity of H_2 was limited, leading to the proposal of a gas–solid synergistic reduction process. Here, H_2 and C acted together as reducing agents, effectively separating Ni, Co, and Cu from the manganese minerals. The maximum contents of Ni, Co, and Cu in the alloy can reach 29.72%, 4.35%, and 15.44%, respectively. Moreover, the reduction temperature played a vital role in the reduction process of pellets. As the temperature increased, the mass loss rate of the pellets raised, SiO_2 in the nodules gradually disappeared, and a large amount of $FeMn(SiO_4)$ and MnO formed. Particularly at 1100 °C, nearly all Ni, Co, and Cu in the manganese minerals were reduced and entered the alloy. Below 1000 °C, the reduction of pellets was still in its initial stage, primarily driven by H_2, with excess C not participating in the reaction. When the temperature exceeded 1100 °C, the carbon gasification reaction intensified, weakening the effect of H_2 on the reduction of carbon-containing multi-metallic nodule pellets. Further increasing the temperature only promoted the sintering and aggregation of the pellets, but had little effect on enhancing the reduction reaction itself. Since the reduction at low temperatures was mainly accomplished by H_2 with minimal participation of C, increasing the carbon-to-oxygen ratio had a limited impact on the efficiency of pellet reduction, thus helping to reduce carbon consumption. At higher temperatures, the enhanced carbon gasification reaction released a significant amount of heat, aiding the separation of Fe and Mn. At this stage, manganese primarily existed in the form of $FeMn(SiO_4)$ and MnO [102].

(5) Bioleaching. Bioleaching is well-suited for processing low-grade, complex ores and has achieved commercial success in extracting base metals from low-grade sulfide ores and in the pre-treatment of refractory sulfide gold ores. Additionally, various bacteria play a crucial role in the formation of manganese nodules [103]. These nodules host manganese-oxidizing and manganese-reducing bacteria, as well as bacteria that do not interact with manganese. The bacteria are key in oxidizing Mn(II) to Mn(IV) oxides, a critical component of the nodules. Furthermore, under specific conditions, some bacteria can directly or indirectly reduce Mn(IV) to Mn(II), disrupting the structure of the nodules and releasing elements like Cu, Co, and Ni. Developing biohydrometallurgical methods to extract metals from manganese nodules holds great potential. Compared to traditional methods, this

approach is more sustainable, consumes less energy, helps reduce carbon emissions, and supports the development of a circular economy [104,105].

The bioleaching of manganese nodules from CCZ was conducted using anaerobic Mn-reducing bacteria. The experimental results demonstrated that the leaching efficiencies of Mn, Co, and Ni were significantly increased from 18%, 7%, and 10% to 77%, 70%, and 75%, respectively, following the inoculation of the Mn-reducing bacterial enrichment culture. The research identified that the optimal temperature range for effective leaching was between 30 °C to 45 °C, with the most suitable pH being slightly acidic. Moreover, the particle size of the manganese nodules significantly impacted the leaching efficiency, with smaller sizes yielding higher efficiencies. Notably, the experiment did not require the addition of mineral salts, indicating that the bioleaching process only necessitated the addition of a carbon source. After 48 h of anaerobic treatment, the leaching efficiencies of Mn, Co, and Ni were greatly enhanced. This suggests that anaerobic Mn-reducing microorganisms, using metal ions as electron acceptors and organic compounds as electron donors, can effectively leach these metals from manganese nodules [106].

Mehta et al. [107] studied the biodissolution kinetics of precious metals such as copper, nickel, and cobalt from Indian Ocean nodules using Thiobacillus ferrooxidans and Thiobacillus thiooxidans in the presence of pyrite and sulfur. Ferrous sulfate and sulfuric acid produced, respectively, when pyrite and sulfur were oxidized by bacteria reduce Mn(IV) in nodules to Mn(II), destroying the structure of manganese nodules and indirectly dissolving metals. The bacterial oxidation and Mn(IV) reduction reactions involved are shown in Equations (40)–(45).

$$FeS_2 \rightarrow Fe^{2+} + 2S^0 + 2e^- \tag{40}$$

$$S^0 + O_2 + H_2O \rightarrow H_2SO_3 \tag{41}$$

$$S^0 + 3/2\,O_2 + H_2O \rightarrow H_2SO_4 \tag{42}$$

$$MnO_2 + 2Fe^{2+} + 4H^+ \rightarrow Mn^{2+} + 2Fe^{3+} + 2H_2O \tag{43}$$

$$MnO_2 + 2e^- + 4H^+ \rightarrow Mn^{2+} + 2H_2O \tag{44}$$

$$MnO_2 + H_2SO_3 + 2H^+ \rightarrow Mn^{2+} + H_2SO_4 + 2H_2O \tag{45}$$

The dissolution of metals followed the shrinking core kinetic model, with leaching reactions controlled by internal diffusion. Pyrite was more effective than sulfur for the biorecovery of the metals. At a pH of 2, with a stirring speed of 100 rpm, a slurry concentration of 5%, a feed particle size ranging from 75 to 300 µm, a 10% addition of pyrite, and a temperature of 308K, the recovery of various metals were as follows: 96% for Cu, 79% for Ni, 49.5% for Mn, 48% for Co, and 13% for Fe. In a sulfuric acid solution, pyrite/sulfur undergoes redox reactions via a galvanic action [FeS_2/S (anode): MnO_2 (cathode)] to reduce MnO_2, as seen in Equations (16) and (17). Additionally, pyrite can also reduce Fe^{3+} to Fe^{2+}, forming a Mn^{4+}—Fe^{2+} redox system, as seen in Equation (48). In the pyrite system, bacteria will also slowly oxidize Fe^{2+}, as seen in Equation (49). Nakazawa and Sato also confirmed the effectiveness of sulfur and pyrite as reducing agents in their experiments on the bioleaching of cobalt-rich crusts [108].

$$15MnO_2 + O_2 + 2FeS_2 + 14H_2SO_4 \rightarrow 15MnSO_4 + Fe_2(SO_4)_3 + 14H_2O \tag{46}$$

$$2MnO_2 + S^0 + 1/2O_2 + H_2SO_4 \rightarrow 2MnSO_4 + H_2O \tag{47}$$

$$Fe_2(SO_4)_3 + FeS_2 \rightarrow 3FeSO_4 + 2S^0 \tag{48}$$

$$Fe^{2+} \rightarrow Fe^{3+} + e^- \tag{49}$$

Beolchini et al. conducted a study on the fungal leaching of metals from manganese nodules located approximately 5000 m deep in CCZ. The research thoroughly investigated the bioleaching efficiency of Aspergillus and a mixed cultures of *A. niger* and *Trichoderma sp.* under various growth conditions. These fungi primarily dissolve metals from the manganese nodules by forming complexes with organic acids produced during metabolism, such as oxalic acid and citric acid. Under optimal culture conditions, after 11 days of growth, Penicillium achieved extraction exceeding 80% for Mn, Cu, and Ni, while the extraction for Co and Fe were approximately 70% and 30%, respectively. Compared to chemical strategies, the biotechnological process offers a more sustainable method for reducing carbon footprint [10].

The National Metallurgical Laboratory in India conducted a study on leaching copper, nickel, and cobalt from polymetallic manganese nodules using the fungus Aspergillus niger. The findings revealed that under conditions of an initial pH of 4.5, a temperature of 35 °C, and a pulp density of 5% (*w*/*v*), Aspergillus niger can extract 97% of copper, 98% of nickel, 86% of cobalt, 91% of manganese, and 36% of iron within 30 days. This extraction rate was significantly higher compared to control experiments that directly used oxalic and citric acids. The dissolution process of metals from the nodules by Aspergillus niger was indirect, predominantly driven by the secretion of oxalic and citric acids. These organic acids disrupted the main structure of manganese (IV) and iron (III) oxides, effectively liberating valuable metals, particularly copper, nickel, and cobalt [109]. The involved reactions are as follows:

$$MnO_2 + HOOC\text{-}COOH + 2H^+ \rightarrow Mn^{2+} + 2CO_2 + 2H_2O \quad (50)$$

$$2Fe\text{-}O\text{-}OH + HOOC\text{-}COOH + 4H^+ \rightarrow 2Fe^{2+} + 2CO_2 + 4H_2O \quad (51)$$

The mechano-chemical activation's effect on the bioleaching of metals such as copper, cobalt, and nickel has been further studied. After 10 min of treatment with a high-energy mill, the surface potential of manganese nodules decreased from −18 mV to −34 mV, and the duration to leach out over 95% of copper, nickel, and cobalt from the nodules was reduced to 15 days [110].

Kang et al. investigated the role of A. ferrooxidans in the reductive dissolution of polymetallic nodules. Using Fe^{2+} ions as reductants, the extraction rates of manganese, nickel, copper, and cobalt exceeded 95%, with an apparent activation energy of 13.9 $kJ \cdot mol^{-1}$ for manganese oxide dissolution. The role of A. ferrooxidans in these processes can be summarized in three main aspects. I. Electron-proton carrier or buffer: facilitates the transfer of interfacial species, enhancing reaction efficiency. II. Catalyst: accelerates the transformation of Fe^{3+}/Fe^{2+}, increasing the reduction rate of manganese oxides. III. Product manager: optimizes the migration and transformation of intermediates and final products. This study provides valuable insights into the mechanisms underlying the bio-assisted reductive dissolution of polymetallic nodules [111].

(6) Electrochemical leaching and electrobioleaching. Kumari and Natarajan achieved almost complete dissolution of copper, nickel, cobalt, and other metals in sulfuric acid medium or the action of microorganisms by applying an external electric field and a negative direct current of −600 mV or lower. Under the condition of an external electric field, the reduction of iron and manganese oxides was the main mechanism for the release of other non-ferrous metals. Such electrochemical or electrobioleaching processes developed under laboratory conditions may be cost-effective, energy-saving, and environmentally friendly [112–115]. The electroleaching system is shown in Figure 17, the unmarked A, B, and C in the figure are three pyrex glass compartments.

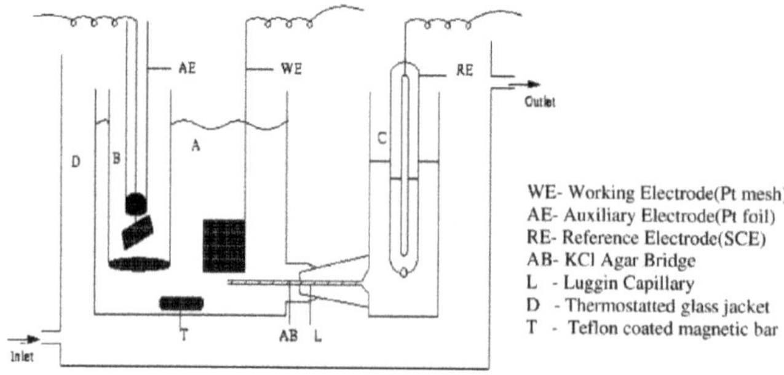

Figure 17. Schematic diagram of an electrochemical cell. Reprinted with permission from ref. [114]. 2001 Elsevier.

4. Other Utilization Potentials of Manganese Nodules

In order to maximize the utilization value of manganese nodules, other possible utilization potentials of manganese nodules have also been widely explored.

(1) Used as catalyst or adsorbent. The fact that manganese nodules have a porous structure and are composed of a variety of transition metal oxides has long attracted attention for their potential as catalysts or adsorbent. Relevant studies indicate that manganese nodules can be used to catalytically decompose isopropyl alcohol at a roasting temperature of 200~300 °C. At this time, the specific surface area of manganese nodules can reach 300 m^2/g, which is comparable to commercial aluminum-based or silicon-aluminum base catalyst. The large amount of surface excess oxygen contained in manganese oxide played a major role in the catalytic process, and trace amounts of metal oxides in manganese nodules can enhance the reactivity [116].

The adsorption and oxidation capabilities of manganese nodules towards Cr(III) at varying concentrations were also investigated. When the initial concentration of Cr(III) was set at 200 mg/L, the adsorption concentration and capacity of Cr(III) reached their maximums at 154 mg/L and 30.7 mg/g, respectively. In different gaseous environments, the catalytic oxidation capability of manganese followed the sequence of oxygen < air < nitrogen. As the pH increased, the oxidation capability of manganese nodules for Cr(III) diminished. The total concentration of Cr(VI) decreased from 8.2 mg/L at pH 2.0 to 4.0 mg/L at pH 6.0. The presence of Ca^{2+} and Mg^{2+} slightly reduced the oxidation capacity to Cr(III), while the presence of silicates significantly lowered it. As presented in Figure 18, manganese oxides primarily facilitate the oxidation of Cr(III), while iron oxides predominantly aid in the adsorption and stabilization of Cr(VI). Most of the Cr(III) was adsorbed by manganese nodules, with only a minor proportion being oxidized during the reaction, after which the newly formed Cr(VI) was released into the solution. Additionally, the released Mn(II) was re-oxidized to form Mn(III) intermediates or Mn(IV) oxides under the catalysis of manganese nodules [117].

The washed manganese nodule leached residue (WMNLR) had a larger specific surface area than manganese nodules, exhibited good affinity for phosphate, and can be used as an adsorbent to remove phosphate contaminants in aqueous phases [118]. It can also be used to adsorb aqueous selenite ($SeO_3{}^{2-}$), with a removal efficiency of over 90% [119,120]. The molecular sieve (OMN-C), obtained by heating 5 g of manganese nodules with 0.02 mol of potassium permanganate at 1000 °C for 6 h, exhibited excellent selective adsorption capabilities for K^+ in seawater. The adsorption capacity of OMN-C for K^+ reached 22.1 mg/g, while its adsorption capacity for Na^+ was only 0.4 mg/g. The primary composition of OMN-C was birnessite-type potassium manganese oxides, characterized by a high micropore volume, which was the key factor contributing to its efficiency and high selectivity [121]. In addition, the manganese nodule extraction residue had good adsorption capacity for Cr(VI) [122], Cd(II) [123].

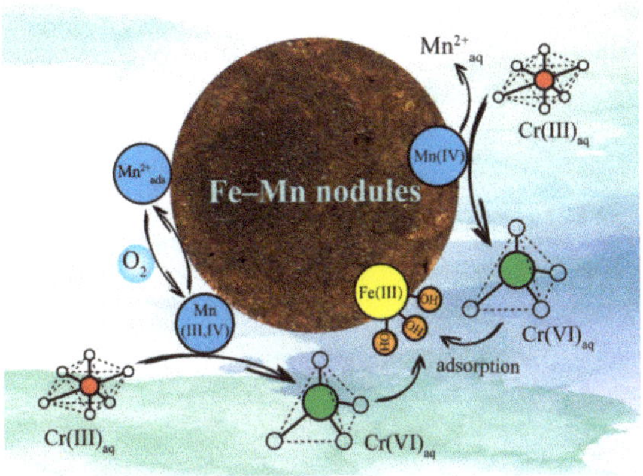

Figure 18. Mechanism model of Cr(III) adsorption and oxidation by manganese nodules. Reprinted with permission from ref. [117]. 2020 Elsevier.

A summary and generalization of the adsorption mechanisms of manganese nodules can be divided into two main types. As illustrated by Equations (52) and (53), one type involved the exchange between ligands in the aqueous solution and surface hydroxyl groups, forming inner-sphere complexes in a process driven by not only electrostatic forces but also surface coordination interactions. This mechanism primarily occurred during the adsorption of strongly binding anions, such as phosphates on manganese nodules. The other type involved the formation of outer-sphere complexes, mainly occurring during the adsorption of weakly binding anions like chlorides and nitrate ions on manganese nodules. Water molecules remained retained between surface sites and the adsorbed ligands, and the adsorption forces are predominantly electrostatic [120].

$$SOH + L^{2-} + H^{+} \leftrightarrow SL^{-} + H_2O \tag{52}$$

$$SOH + L^{2-} + H^{+} \leftrightarrow SOH_2{}^{+}{-}L^{2-} \tag{53}$$

where S-OH is a surface hydroxyl group and L is the ligand-adsorbed species.

Transforming leach residues into environmentally friendly materials not only addresses the disposal challenges of manganese nodule metallurgical residues but also further increases the utilization value of manganese nodules.

(2) Preparing material precursors. To minimize the potential reagent consumption and wastewater generation during the separation process of valuable metals from manganese nodule leachate, a mixture of pyridine carboxylate and naphthalene sulfonic acid was employed as the extracting agent to selectively co-extract Ni, Co, and Mn and directly prepare battery-grade Ni/Co/Mn sulfate solutions (Ca < 0.02 mg/L and Mg < 0.05 mg/L). Through counter-current extraction and washing with varied sulfuric acid concentrations, the final recovery rates in the battery-grade solution were 96.84% for Ni, 95.11% for Co, and 5.30% for Mn. This innovative method not only maximized metal recovery but also significantly reduced operational costs, holding vast potential for industrial applications, which offered a new approach for the short-process, high-value utilization of manganese nodules [124]. Alena et al. explored the development of natural alloys from deep-sea manganese nodules using aluminothermic reduction and rapid solidification. The resulting alloy was processed into thin ribbons via melt spinning, followed by rapid cooling, which significantly altered its phase composition. The alloy, primarily composed of β-Mn, Mn_2FeSi, and $(Cu,Mn)_3(Al,Si)$, demonstrated excellent thermal stability during annealing at 500–750 °C for up to 100 h. Rapid cooling effectively prevented the formation of brittle α-Mn, and the alloy retained outstanding hardness exceeding 800 HV0.1 even after prolonged annealing.

This study highlights the potential of deep-sea nodules as a resource for high-performance structural materials while supporting sustainable resource utilization [125].

(3) Recovery of other co-associated elements. Beyond manganese, iron, copper, cobalt, and nickel, the comprehensive recovery of other minor non-ferrous and rare earth metal elements from manganese nodules can further enhance their economic value. Parhi et al. separated and recovered molybdenum from a leachate of sea nodules containing Mo(VI) at 0.505 kg/m^3, Fe(III) at 0.212 kg/m^3, Cu(II) at 12.08 kg/m^3, Co(II) at 2.012 kg/m^3, and Ni(II) at 15.16 kg/m^3. Alamine 304-1, dissolved in kerosene, was used to extract Mo(VI) from the leachate through solvent extraction, followed by the preparation of ammonium molybdate through crystallization. The ammonium molybdate crystals were then thermally decomposed to successfully produce MoO_3 with a purity of 99.9% [126]. The specific process flow and related parameters can be seen in Figure 19.

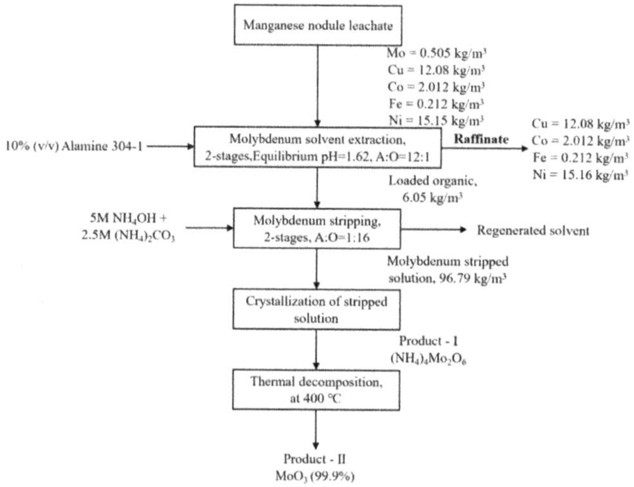

Figure 19. Flowsheet of the process for the recovery of molybdenum from manganese nodule leaching liquor. Adapted from Ref. [126].

In the experiment of leaching rare earth metals from deep-sea nodules using sulfuric acid solution, it was found that the extraction efficiency of rare earth metals from sea nodules using a combination of H_2SO_4 and H_2SO_3 was almost similar to that using only H_2SO_4. This suggested that most of the rare earth (RE) metals were likely to be present in the iron oxide phase rather than the manganese oxide phase. The concentration of H_2SO_4 and the leaching temperature significantly affected the leaching efficiency. The maximum leaching efficiency (total of 68.0%) was achieved in 3 M H_2SO_4, with most RE leaching rates around 90%, and the remaining metals being Ce 44.6%, La 55.3%, Nd 84.7%, Eu 85.8%, Pr 76.4%, and Th 78.7%, determining an average RE leaching efficiency of 83.03%. Subsequently, dilute H_2SO_4 (0.2 M) was chosen as the optimal lixiviant concentration for selective leaching of REE, thereby minimizing the co-extraction of base metals Mn, Cu, Co, and Ni. Under the optimal conditions of 0.2 M H_2SO_4 and a temperature of 45 °C, a total REE extraction of approximately 58% was achieved, along with the co-extraction of Mn 0.3%, Fe 4.63%, Cu 23.7%, Co 0.2%, and Ni 31.8% [127].

5. Conclusions

With the depletion of terrestrial mineral resources, the development of seabed mineral resources will inevitably become a critical support for human society's advancement. This study focuses on the mineralogy of seabed manganese nodules off the coast of China in the Western Pacific and the main technical methods for extracting valuable metal elements from manganese nodules (as shown in Figure 20). It reviews the primary techniques for extracting valuable metal elements from manganese nodules and, based on current research, summarizes new directions for the extraction and comprehensive utilization of valuable metal elements from manganese nodules in the future. The main conclusions are as follows:

(1) The mineralogy study results indicate that the manganese nodules off the coast of China in the Western Pacific are composed of a shell of dense ferromanganese phase and a core of loose silicate phase. The primary valuable metal elements mainly include Cu, Co, Ni, Mn, Fe, etc. Among them, Co and Ni are mainly distributed in the manganese oxide phase, and Cu mainly exists in the form of free copper oxide. To extract valuable metals, the key is to reduce Mn(IV) to Mn(II), which disrupts the crystal structure of the nodules and releases the valuable metal elements. During the recovery process of the primary valuable metals, the recovery of a small amount of associated non-ferrous metals and rare earth elements can also be considered to enhance the recovery and utilization value of manganese nodules.

(2) The extraction processes for the main valuable metal elements of manganese nodules are mainly divided into two categories: pyrometallurgical–hydrometallurgical and sole hydrometallurgical. The pyrometallurgical–hydrometallurgical process combines the high reduction efficiency of pyrometallurgy with the high recovery rates of hydrometallurgy, but it is associated with high energy consumption and carbon emissions. By introducing hydrogen metallurgy, partially replacing carbon with hydrogen can reduce carbon emissions and improve reduction efficiency. Additionally, utilizing residual heat from the pyrometallurgical process to support the hydrometallurgical stage can further reduce energy consumption. The hydrometallurgical process avoids high-temperature operations but faces challenges such as high acid consumption, complex wastewater treatment, difficulties in handling leaching residues, and complicated downstream separation processes. These issues can be addressed by developing highly selective leaching reagents, optimizing wastewater treatment technologies, and achieving resource utilization of leaching residues to enhance economic and environmental performance.

(3) Bioleaching, co-extraction technology, short-process extraction technology, etc. have attracted widespread attention due to their environmentally friendly and efficient characteristics. In the future, manganese nodule extraction technology should focus on integrated development, promote the application of technologies such as the comprehensive recovery of associated resources and tailings-free utilization, improve process stability and efficiency through intelligent control systems, increase the utilization value of manganese nodules, reduce development costs, and accelerate its commercialization process, thereby achieving more efficient and sustainable resource development.

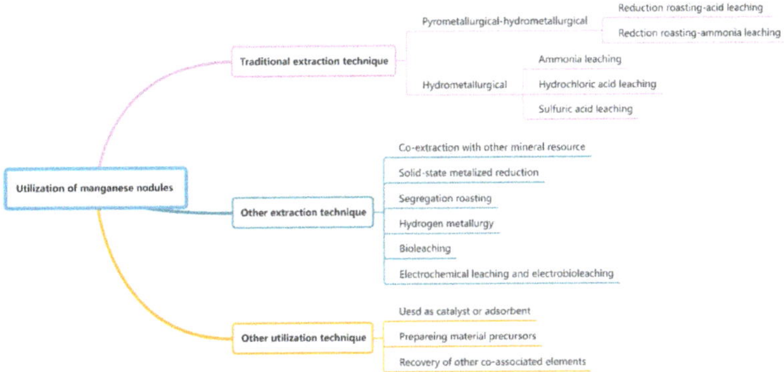

Figure 20. Summary of utilization techniques of manganese nodule.

Author Contributions: X.W.: Conceptualization, Investigation, Data curation, Writing—original draft preparation, Funding acquisition; W.Q.: Writing—review and editing, Supervision, Project administration; M.L.: Resources, Supervision, Project administration; X.L.: Conceptualization, Supervision, Project administration; Y.C.: Investigation, Resources; S.C.: Investigation, Resources; C.Y.: Conceptualization, Investigation, Data curation, Writing—review and editing. All authors have read and agreed to the published version of the manuscript.

Funding: This research was funded by the National key research and development plan (2022YFC280 6800), Central South University postdoctoral research fund.

Conflicts of Interest: Authors Maolin Li, Yangrui Cheng, Shiping Chen were employed by the company Changsha Research Institute of Mining and Metallurgy Co., Ltd. The remaining authors declare that the research was conducted in the absence of any commercial or financial relationships that could be construed as a potential conflict of interest.

References

1. Sparenberg, O. A historical perspective on deep-sea mining for manganese nodules, 1965–2019. *Extr. Ind. Soc.* **2019**, *6*, 842–854. [CrossRef]
2. Savinova, E.; Evans, C.; Lèbre, É.; Stringer, M.; Azadi, M.; Valenta, R.K. Will global cobalt supply meet demand? The geological, mineral processing, production and geographic risk profile of cobalt. *Resour. Conserv. Recycl.* **2023**, *190*, 106855. [CrossRef]
3. Petterson, M.G.; Tawake, A. The Cook Islands (South Pacific) experience in governance of seabed manganese nodule mining. *Ocean. Coast. Manag.* **2019**, *167*, 271–287. [CrossRef]
4. Toro, N.; Gálvez, E.; Saldaña, M.; Jeldres, R.I. Submarine mineral resources: A potential solution to political conflicts and global warming. *Miner. Eng.* **2022**, *179*, 107441. [CrossRef]
5. Hein, J.R.; Koschinsky, A.; Kuhn, T. Deep-ocean polymetallic nodules as a resource for critical materials. *Nat. Rev. Earth Environ.* **2020**, *1*, 158–169. [CrossRef]
6. Budnik, V.; Chernyi, S. Future Development of the World Ocean Mining for the Industry. *Procedia Eng.* **2016**, *150*, 2150–2156. [CrossRef]
7. Heinrich, L.; Koschinsky, A.; Markus, T.; Singh, P. Quantifying the fuel consumption, greenhouse gas emissions and air pollution of a potential commercial manganese nodule mining operation. *Mar. Policy* **2020**, *114*, 103678. [CrossRef]
8. Paulikas, D.; Katona, S.; Ilves, E.; Ali, S.H. Life cycle climate change impacts of producing battery metals from land ores versus deep-sea polymetallic nodules. *J. Clean. Prod.* **2020**, *275*, 123822. [CrossRef]
9. Alvarenga, R.A.F.; Préat, N.; Duhayon, C.; Dewulf, J. Prospective life cycle assessment of metal commodities obtained from deep-sea polymetallic nodules. *J. Clean. Prod.* **2022**, *330*, 129884. [CrossRef]
10. Beolchini, F.; Becci, A.; Barone, G.; Amato, A.; Hekeu, M.; Danovaro, R.; Dell'Anno, A. High fungal-mediated leaching efficiency of valuable metals from deep-sea polymetallic nodules. *Environ. Technol. Innov.* **2020**, *20*, 101037. [CrossRef]
11. Verlaan, P.A.; Cronan, D.S. Origin and variability of resource-grade marine ferromanganese nodules and crusts in the Pacific Ocean: A review of biogeochemical and physical controls. *Geochemistry* **2022**, *82*, 125741. [CrossRef]
12. Kolbe, H.; Siapno, B. Manganese nodules. Further resources of nickel and copper on the deep ocean floor. *Geoforum* **1974**, *5*, 63–82. [CrossRef]
13. Fritz, B.; Heidak, P.; Vasters, J.; Kuhn, T.; Franken, G.; Schmidt, M. Life cycle impact on climate change caused by metal production from deep sea manganese nodules versus land-based deposits, Resources. *Conserv. Recycl.* **2023**, *193*, 106976. [CrossRef]
14. Fox, S.J. 'Exploiting—Land, sea and space: Mineral superpower' In the name of peace: A critical race to protect the depths and heights. *Resour. Policy* **2022**, *79*, 103066. [CrossRef]
15. Heijlen, W.; Franceschi, G.; Duhayon, C.; Van Nijen, K. Assessing the adequacy of the global land-based mine development pipeline in the light of future high-demand scenarios: The case of the battery-metals nickel (Ni) and cobalt (Co). *Resour. Policy* **2021**, *73*, 102202. [CrossRef]
16. Kung, A.; Svobodova, K.; Lèbre, E.; Valenta, R.; Kemp, D.; Owen, J.R. Governing deep sea mining in the face of uncertainty. *J. Environ. Manag.* **2021**, *279*, 111593. [CrossRef]
17. Childs, J. Geographies of deep sea mining: A critical review. *Extr. Ind. Soc.* **2022**, *9*, 101044. [CrossRef]
18. McLellan, B.C. Sustainability Assessment of Deep Ocean Resources. *Procedia Environ. Sci.* **2015**, *28*, 502–508. [CrossRef]
19. Alhaddad, S.; Mehta, D.; Helmons, R. Mining of deep-seabed nodules using a Coandă-effect-based collector. *Results Eng.* **2023**, *17*, 100852. [CrossRef]
20. Jones, D.O.B.; Ardron, J.A.; Colaço, A.; Durden, J.M. Environmental considerations for impact and preservation reference zones for deep-sea polymetallic nodule mining. *Mar. Policy* **2020**, *118*, 103312. [CrossRef]
21. Kaikkonen, L.; Venesjärvi, R.; Nygård, H.; Kuikka, S. Assessing the impacts of seabed mineral extraction in the deep sea and coastal marine environments: Current methods and recommendations for environmental risk assessment. *Mar. Pollut. Bull.* **2018**, *135*, 1183–1197. [CrossRef]
22. Van Nijen, K.; Van Passel, S.; Squires, D. A stochastic techno-economic assessment of seabed mining of polymetallic nodules in the Clarion Clipperton Fracture Zone. *Mar. Policy* **2018**, *95*, 133–141. [CrossRef]

23. Cecchetto, M.M.; Moser, A.; Smith, C.R.; van Oevelen, D.; Sweetman, A.K. Abyssal seafloor response to fresh phytodetrital input in three areas of particular environmental interest (APEIs) in the western clarion-clipperton zone (CCZ). *Deep. Sea Res. Part I Oceanogr. Res. Pap.* **2023**, *195*, 103970. [CrossRef]
24. Hyman, J.; Stewart, R.A.; Sahin, O.; Clarke, M.; Clark, M.R. Visioning a framework for effective environmental management of deep-sea polymetallic nodule mining: Drivers, barriers, and enablers. *J. Clean. Prod.* **2022**, *337*, 130487. [CrossRef]
25. Toro, N.; Robles, P.; Jeldres, R.I. Seabed mineral resources, an alternative for the future of renewable energy: A critical review. *Ore Geol. Rev.* **2020**, *126*, 103699. [CrossRef]
26. Błażewicz, M.; Jóźwiak, P.; Menot, L.; Pabis, K. High species richness and unique composition of the tanaidacean communities associated with five areas in the Pacific polymetallic nodule fields. *Prog. Oceanogr.* **2019**, *176*, 102141. [CrossRef]
27. Gooday, A.J.; Holzmann, M.; Caulle, C.; Goineau, A.; Kamenskaya, O.; Weber, A.A.T.; Pawlowski, J. Giant protists (xenophyophores, Foraminifera) are exceptionally diverse in parts of the abyssal eastern Pacific licensed for polymetallic nodule exploration. *Biol. Conserv.* **2017**, *207*, 106–116. [CrossRef]
28. Hirt, W.C.; Rice, D.A.; Shirts, M.B. Flotation of cobalt-rich ferromanganese crust from the Pacific Ocean. *Miner. Eng.* **1991**, *4*, 535–551. [CrossRef]
29. von Heimendahl, M.; Hubred, G.L.; Fuerstenau, D.W.; Thomas, G. A transmission electron microscope study of deep-sea manganese nodules. *Deep. Sea Res. Oceanogr. Abstr.* **1976**, *23*, 69–79. [CrossRef]
30. Boughriet, A.; Ouddane, B.; Wartel, M.; Lalouj, C.; Cordier, C.; Gengembre, L.; Sanchez, J.P. On the oxidation states of Mn and Fe in polymetallic oxide/oxyhydroxide crusts from the Atlantic Ocean. *Deep. Sea Res. Part I Oceanogr. Res. Pap.* **1996**, *43*, 321–343. [CrossRef]
31. Blöthe, M.; Wegorzewski, A.; Müller, C.; Simon, F.; Kuhn, T.; Schippers, A. Manganese-Cycling Microbial Communities Inside Deep-Sea Manganese Nodules. *Environ. Sci. Technol.* **2015**, *49*, 7692–7700. [CrossRef] [PubMed]
32. Wegorzewski, A.V.; Grangeon, S.; Webb, S.M.; Heller, C.; Kuhn, T. Mineralogical transformations in polymetallic nodules and the change of Ni, Cu and Co crystal-chemistry upon burial in sediments. *Geochim. Cosmochim. Acta* **2020**, *282*, 19–37. [CrossRef]
33. Mewes, K.; Mogollón, J.M.; Picard, A.; Rühlemann, C.; Kuhn, T.; Nöthen, K.; Kasten, S. Impact of depositional and biogeochemical processes on small scale variations in nodule abundance in the Clarion-Clipperton Fracture Zone. *Deep. Sea Res. Part I Oceanogr. Res. Pap.* **2014**, *91*, 125–141. [CrossRef]
34. Aplin, A.C.; Cronan, D.S. Ferromanganese oxide deposits from the Central Pacific Ocean, II. Nodules and associated sediments. *Geochim. Cosmochim. Acta* **1985**, *49*, 437–451. [CrossRef]
35. Sharma, R. *Deep-Sea Mining: Resource Potential, Technical and Environmental Considerations*; Springer: Berlin/Heidelberg, Germany, 2017.
36. Senanayake, G. Acid leaching of metals from deep-sea manganese nodules—A critical review of fundamentals and applications. *Miner. Eng.* **2011**, *24*, 1379–1396. [CrossRef]
37. Fu, Y.; Wen, H. Variabilities and enrichment mechanisms of the dispersed elements in marine Fe–Mn deposits from the Pacific Ocean. *Ore Geol. Rev.* **2020**, *121*, 103470. [CrossRef]
38. Koschinsky, A.; Halbach, P. Sequential leaching of marine ferromanganese precipitates: Genetic implications. *Geochim. Cosmochim. Acta* **1995**, *59*, 5113–5132. [CrossRef]
39. Popoola, S.O.; Adegbie, A.T.; Akinnigbagbe, E.A.; Unyimadu, J.P. Geochemistry of ferromanganese micronodules recovered from sediment-core in the western Nigeria continental margin, Eastern Equatorial Atlantic: Implications on the genesis and depositional history. *J. Afr. Earth Sci.* **2021**, *184*, 104369. [CrossRef]
40. Zhang, Z.; Du, Y.; Gao, L.; Zhang, Y.; Shi, G.; Liu, C.; Zhang, P.; Duan, X. Enrichment of REEs in polymetallic nodules and crusts and its potential for exploitation. *J. Rare Earths* **2012**, *30*, 621–626. [CrossRef]
41. Randhawa, N.S.; Hait, J.; Jana, R.K. A brief overview on manganese nodules processing signifying the detail. *Hydrometallurgy* **2016**, *165*, 166–181. [CrossRef]
42. Parida, K.; Samal, A.; Das, D.; Chintalpudi, S.N. Effect of calcination temperature on Indian Ocean manganese nodules. Mössbauer, XRD, FT-IR and TG-DTA studies. *Thermochim. Acta* **1999**, *325*, 69–76. [CrossRef]
43. Meshram, P.; Pandey, B.D.; Abhilash. Perspective of availability and sustainable recycling prospects of metals in rechargeable batteries—A resource overview. *Resour. Policy* **2019**, *60*, 9–22. [CrossRef]
44. Mukherjee, A.; Raichur, A.M.; Natarajan, K.A.; Modak, J.M. Recent Developments in Processing Ocean Manganese Nodules—A Critical Review. *Miner. Process. Extr. Metall. Rev.* **2004**, *25*, 91–127. [CrossRef]
45. Sen, P.K. Metals and materials from deep sea nodules: An outlook for the future. *Int. Mater. Rev.* **2010**, *55*, 364–391. [CrossRef]
46. Toro, N.; Rodríguez, F.; Rojas, A.; Robles, P.; Ghorbani, Y. Leaching manganese nodules with iron-reducing agents—A critical review. *Miner. Eng.* **2021**, *163*, 106748. [CrossRef]
47. Sridhar, V.; Verma, J.K. Extraction of copper, nickel and cobalt from the leach liquor of manganese-bearing sea nodules using LIX 984N and ACORGA M5640. *Miner. Eng.* **2011**, *24*, 959–962. [CrossRef]
48. Pandey, B.D.; Kumar, V. Extraction of copper and nickel from ammoniacal leach liquor of Indian Ocean sea nodules. *Hydrometallurgy* **1991**, *26*, 35–45. [CrossRef]
49. Si, S.; Anand, S.; Nam, C.W.; Park, K.H.; Das, R.P. Dissolution Studies On Cu-Ni-Co-Fe Matte Obtained from Manganese Nodules. In Proceedings of the Fifth ISOPE Ocean Mining Symposium, Tsukuba, Japan, 15–19 September 2003.

50. Rokukawa, N. Development For Hydrometallurgical Process of Cobalt Rich Crusts. In Proceedings of the First ISOPE Ocean Mining Symposium, Tsukuba, Japan, 21–22 November 1995.
51. Sridhar, R.; Jones, W.E.; Warner, J.S. Extraction of copper, nickel and cobalt from sea nodules. *JOM* **1976**, *28*, 32–37. [CrossRef]
52. Sridhar, R. Thermal upgrading of sea nodules. *JOM* **1974**, *26*, 18–22. [CrossRef]
53. Xiang, Z.; Zequan, H.; Yujun, S.; Yongjun, M.; Shuguang, Q. The Smelting-Rusting-Solvent Extraction Process To Recover Valuable Matels From Polymetallic Nodules. In Proceedings of the Third ISOPE Ocean Mining Symposium, Goa, India, 8–10 November 1999.
54. Kotlinski, R.; Stoyanova, V.; Hamrak, H.; Avramov, A. An Overview of the Interoceanmetal (IOM) Deep-Sea Technology Development (Mining and Processing). In Proceedings of the Workshop on Polymetallic Nodule Mining Technology-Current Status and Challenges Ahead, Chennai, India, 18–22 February 2008; pp. 18–22.
55. Keber, S.; Brückner, L.; Elwert, T.; Kuhn, T. Concept for a Hydrometallurgical Processing of a Copper-Cobalt-Nickel Alloy Made from Manganese Nodules. *Chem. Ing. Tech.* **2020**, *92*, 379–386. [CrossRef]
56. Xue, Z.; Feng, Y.; Li, H.; Ma, R.; Ju, J.; Dong, L.; Bao, W.; Wang, J.; Fan, P.; Zhu, Z. Extraction of various valuable elements from oceanic manganese nodules using coal gasification slag via reduction roasting-acid leaching process. *J. Ind. Eng. Chem.* **2024**, *129*, 390–402. [CrossRef]
57. Li, J.; Hu, Y.; Yue, L.; Zeng, L.; Wu, S.; Wang, M.; Guan, W.; Cao, Z.; Li, Q.; Zhang, G. Recovery of critical battery metals from cobalt-rich deep-sea polymetallic nodules: Selective carbothermal reduction based on kinetics. *J. Environ. Chem. Eng.* **2024**, *12*, 111631. [CrossRef]
58. Han, K.N.; Hoover, M.; Fuerstenau, D.W. Ammonia-ammonium leaching of deep-sea manganese nodules. *Int. J. Miner. Process.* **1974**, *1*, 215–230. [CrossRef]
59. Kmetova, D.; Stofko, M.; Kmet, S. Ammoniacal leaching for extraction of non-ferrous metals from deep-sea nodules. *Int. J. Miner. Process.* **1985**, *15*, 145–153. [CrossRef]
60. Jana, R.K.; Murthy, D.S.R.; Nayak, A.K.; Mahanty, M.S.; Tiwary, S.K.; Akerkar, D.D. Leaching of roast-reduced polymetallic sea nodules to optimise the recoveries of copper, nickel and cobalt. *Int. J. Miner. Process.* **1990**, *30*, 127–141. [CrossRef]
61. Srikanth, S.; Alex, T.C.; Agrawal, A.; Premchand, P. Reduction Roasting Of Deep Sea Manganese Nodules Using Liquid And Gaseous Reductants. In Proceedings of the Second ISOPE Ocean Mining Symposium, Seoul, Republic of Korea, 24–26 November 1997.
62. Jana, R.K.; Pandey, B.D.; Premchand. Ammoniacal leaching of roast reduced deep-sea manganese nodules. *Hydrometallurgy* **1999**, *53*, 45–56. [CrossRef]
63. Mishra, D.; Srivastava, R.R.; Sahu, K.K.; Singh, T.B.; Jana, R.K. Leaching of roast-reduced manganese nodules in NH_3–$(NH_4)_2CO_3$ medium. *Hydrometallurgy* **2011**, *109*, 215–220. [CrossRef]
64. Jana, R.K.; Akerkar, D.D. Studies of the metal-ammonia-carbon dioxide-water system in extraction metallurgy of polymetallic sea nodules. *Hydrometallurgy* **1989**, *22*, 363–378. [CrossRef]
65. Agarwall, J.C.; Barner, H.E.; Beecher, N.; Davies, D.S.; Kust, R.N. Kennecott process for recovery of copper, nickel, cobalt and molybdenum from ocean nodules. *Miner. Eng.* **1979**, *12*, 1704–1709.
66. Agarwal, J.; Wilder, T. Recovery of Metal Values from Manganese Nodules. U.S. Patent 3,788,841, 29 January 1974.
67. Das, R.P.; Anand, S.; Das, S.C.; Jena, P.K. Leaching of manganese nodules in ammoniacal medium using glucose as reductant. *Hydrometallurgy* **1986**, *16*, 335–344. [CrossRef]
68. Acharya, S.; Das, R.P. Kinetics and mechanism of the reductive ammonia leaching of ocean nodules by manganous ion. *Hydrometallurgy* **1987**, *19*, 169–186. [CrossRef]
69. Anand, S.; Das, S.C.; Das, R.P.; Jena, P.K. Leaching of manganese nodule in ammoniacal medium using ferrous sulfate as the reductant. *Metall. Trans. B* **1988**, *19*, 331–334. [CrossRef]
70. Mohanty, P.S.; Ghosh, M.K.; Anand, S.; Das, R.P. Leaching of manganese nodules in ammoniacal medium with elemental sulphur as reductant. *Trans. Inst. Min. Metall. Sec. C* **1994**, *103*, 151–155.
71. Das, R.P.; Anand, S. Aqueous Reduction of Polymetallic Nodule For Metal Extraction. In Proceedings of the Second ISOPE Ocean Mining Symposium, Seoul, Republic of Korea, 24–26 November 1997.
72. Niinae, M.; Komatsu, N.; Nakahiro, Y.; Wakamatsu, T.; Shibata, J. Preferential leaching of cobalt, nickel and copper from cobalt-rich ferromanganese crusts with ammoniacal solutions using ammonium thiosulfate and ammonium sulfite as reducing agents. *Hydrometallurgy* **1996**, *40*, 111–121. [CrossRef]
73. Das, P.K.; Anand, S.; Das, R.P. Minimization of Ni(II) precipitation in the Ni(II) NH_3 SO_2 $(NH_4)_2SO_4$ MnO_2 system. *Int. J. Miner. Process.* **1997**, *50*, 77–86. [CrossRef]
74. Sanjay, K.; Subbaiah, T.; Anand, S.; Das, R.P. Manganese Recovery From Leach Liquors/Residues Generated During Hydrometallurgical Processing of Manganese Nodules. In Proceedings of the Third ISOPE Ocean Mining Symposium, Goa, India, 8–10 November 1999.
75. Rokukawa, N. Extraction of nickel, cobalt and copper from ocean manganese nodules with mixed solution of ammonium carbanate and ammonium sulphite. *J. Min. Metall. Inst. Jpn.* **1990**, *106*, 205–209.
76. Das, R.P. India's demonstration metallurgical plant to treat ocean nodule. In Proceedings of the Fourth ISOPE Ocean Mining Symposium, Szczecin, Poland, 23–27 September 2001; pp. 163–166.

77. Mittal, N.K.; Sen, P.K. India's first medium scale demonstration plant for treating poly-metallic nodules. *Miner. Eng.* **2003**, *16*, 865–868. [CrossRef]
78. Kanungo, S.B.; Jena, P.K. Reduction leaching of manganese nodules of Indian Ocean origin in dilute hydrochloric acid. *Hydrometallurgy* **1988**, *21*, 41–58. [CrossRef]
79. Jana, R.K.; Singh, D.D.N.; Roy, S.K. Alcohol-modified hydrochloric acid leaching of sea nodules. *Hydrometallurgy* **1995**, *38*, 289–298. [CrossRef]
80. Gupta, B.; Deep, A.; Singh, V.; Tandon, S.N. Recovery of cobalt, nickel, and copper from sea nodules by their extraction with alkylphosphines. *Hydrometallurgy* **2003**, *70*, 121–129. [CrossRef]
81. Rodriguez, M.P.; Aja, R.; Miyares, R.C. Optimization of the Existing Methods for Recovery of Basic Metals from Polymetallic Nodules. In Proceedings of the Tenth ISOPE Ocean Mining and Gas Hydrates Symposium, Szczecin, Poland, 22–26 September 2013.
82. Abramowski, T.; Stefanova, V.; Causse, R.; Romanchuk, A. Technologies for the processing of polymetallic nodules from Clarion Clipperton Zone in the Pacific Ocean. *J. Chem. Technol. Metall.* **2017**, *52*, 258–269.
83. Kanungo, S.B.; Das, R.P. Extraction of metals from manganese nodules of the Indian Ocean by leaching in aqueous solution of sulphur dioxide. *Hydrometallurgy* **1988**, *20*, 135–146. [CrossRef]
84. Anand, S.; Das, S.C.; Das, R.P.; Jena, P.K. Leaching of manganese nodules at elevated temperature and pressure in the presence of oxygen. *Hydrometallurgy* **1988**, *20*, 155–167. [CrossRef]
85. Barik, R.; Sanjay, K.; Mishra, B.K.; Mohapatra, M. Micellar mediated selective leaching of manganese nodule in high temperature sulfuric acid medium. *Hydrometallurgy* **2016**, *165*, 44–50. [CrossRef]
86. Torres, D.; Ayala, L.; Saldaña, M.; Cánovas, M.; Jeldres, R.I.; Nieto, S.; Castillo, J.; Robles, P.; Toro, N. Leaching Manganese Nodules in an Acid Medium and Room Temperature Comparing the Use of Different Fe Reducing Agents. *Metals* **2019**, *9*, 1316. [CrossRef]
87. Vu, H.; Jandová, J.; Lisá, K.; Vranka, F. Leaching of manganese deep ocean nodules in $FeSO_4$–H_2SO_4–H_2O solutions. *Hydrometallurgy* **2005**, *77*, 147–153. [CrossRef]
88. Jandová, J.; Lisá, K.; Vu, H.; Vranka, F. Separation of copper and cobalt–nickel sulphide concentrates during processing of manganese deep ocean nodules. *Hydrometallurgy* **2005**, *77*, 75–79. [CrossRef]
89. Shen, Y.-F.; Xue, W.-Y.; Li, W.; Li, S.-D.; Liu, X.-H. Recovery of Mn^{2+}, Co^{2+} and Ni^{2+} from manganese nodules by redox leaching and solvent extraction. *Trans. Nonferrous Met. Soc. China* **2007**, *17*, 1105–1111. [CrossRef]
90. Padhan, E.; Sarangi, K.; Subbaiah, T. Recovery of manganese and nickel from polymetallic manganese nodule using commercial extractants. *Int. J. Miner. Process.* **2014**, *126*, 55–61. [CrossRef]
91. Zhang, Y.; Liu, Q.; Sun, C. Sulfuric acid leaching of ocean manganese nodules using phenols as reducing agents. *Miner. Eng.* **2001**, *14*, 525–537. [CrossRef]
92. Zhang, Y.; Liu, Q.; Sun, C. Sulfuric acid leaching of ocean manganese nodules using aromatic amines as reducing agents. *Miner. Eng.* **2001**, *14*, 539–542. [CrossRef]
93. Hariprasad, D.; Mohapatra, M.; Anand, S. Reductive Leaching of Manganese Nodule Using Saw Dust in Sulphuric Acid Medium. *Trans. Indian Inst. Met.* **2018**, *71*, 2971–2983. [CrossRef]
94. Li, J.; Wu, X.; Zhou, X.; Zeng, L.; Wu, S.; Wang, M.; Guan, W.; Cao, Z.; Li, Q.; Zhang, G. A near-zero waste process for the full-component utilization of deep-sea polymetallic nodules based on reductive leaching with SO_2 followed by separation and recovery. *Hydrometallurgy* **2024**, *223*, 106207. [CrossRef]
95. Ghosh, M.K.; Barik, S.P.; Anand, S. Sulphuric acid leaching of polymetallic nodules using paper as a reductant. *Trans. Indian Inst. Met.* **2008**, *61*, 477–481. [CrossRef]
96. Venkataseetharaman, A.; Mishra, G.; Ghosh, M.K.; Das, G.K. Role of Glycerol Oxidation Pathways in the Reductive Acid Leaching Kinetics of Manganese Nodules Using Glycerol. *ACS Omega* **2021**, *6*, 14903–14910. [CrossRef] [PubMed]
97. Havlik, T.; Laubertova, M.; Miskufova, A.; Kondas, J.; Vranka, F. Extraction of copper, zinc, nickel and cobalt in acid oxidative leaching of chalcopyrite at the presence of deep-sea manganese nodules as oxidant. *Hydrometallurgy* **2005**, *77*, 51–59. [CrossRef]
98. Toro, N.; Pérez, K.; Saldaña, M.; Jeldres, R.I.; Jeldres, M.; Cánovas, M. Dissolution of pure chalcopyrite with manganese nodules and waste water. *J. Mater. Res. Technol.* **2020**, *9*, 798–805. [CrossRef]
99. Toro, N.; Saldaña, M.; Gálvez, E.; Cánovas, M.; Trigueros, E.; Castillo, J.; Hernández, P.C. Optimization of Parameters for the Dissolution of Mn from Manganese Nodules with the Use of Tailings in An Acid Medium. *Minerals* **2019**, *9*, 387. [CrossRef]
100. Toro, N.; Saldaña, M.; Castillo, J.; Higuera, F.; Acosta, R. Leaching of Manganese from Marine Nodules at Room Temperature with the Use of Sulfuric Acid and the Addition of Tailings. *Minerals* **2019**, *9*, 289. [CrossRef]
101. Zhao, F.; Jiang, X.; Wang, S.; Feng, L.; Li, D. The Recovery of Valuable Metals from Ocean Polymetallic Nodules Using Solid-State Metalized Reduction Technology. *Minerals* **2020**, *10*, 20. [CrossRef]
102. Li, F.; Xu, S.; Qiu, J.; Chen, Z.; Du, W.; Ju, D.; Xie, K. Optimization of Gas–Solid Co-Reduction Conditions for Deep-Sea Polymetallic Nodules. *JOM* **2023**, *75*, 5718–5728. [CrossRef]
103. van de Velde, S.; Callebaut, I.; Gao, Y.; Meysman, F.J.R. Impact of electrogenic sulfur oxidation on trace metal cycling in a coastal sediment. *Chem. Geol.* **2017**, *452*, 9–23. [CrossRef]
104. Kaksonen, A.H.; Deng, X.; Bohu, T.; Zea, L.; Khaleque, H.N.; Gumulya, Y.; Boxall, N.J.; Morris, C.; Cheng, K.Y. Prospective directions for biohydrometallurgy. *Hydrometallurgy* **2020**, *195*, 105376. [CrossRef]

105. Ehrlich, H.L. Ocean manganese nodules: Biogenesis and bioleaching possibilities. *Min. Metall. Explor.* **2000**, *17*, 121–128. [CrossRef]
106. Lee, E.Y.; Noh, S.-R.; Cho, K.-S.; Ryu, H.W. Leaching of Mn, Co, and Ni from manganese nodules using an anaerobic bioleaching method. *J. Biosci. Bioeng.* **2001**, *92*, 354–359. [CrossRef] [PubMed]
107. Mehta, K.D.; Pandey, B.D.; Mankhand, T.R. Studies on kinetics of biodissolution of metals from Indian Ocean nodules. *Miner. Eng.* **2003**, *16*, 523–527. [CrossRef]
108. Nakazawa, H.; Sato, H. Bacterial leaching of cobalt-rich ferromanganese crusts. *Int. J. Miner. Process.* **1995**, *43*, 255–265. [CrossRef]
109. Mehta, K.D.; Das, C.; Pandey, B.D. Leaching of copper, nickel and cobalt from Indian Ocean manganese nodules by *Aspergillus niger*. *Hydrometallurgy* **2010**, *105*, 89–95. [CrossRef]
110. Mehta, K.D.; Das, C.; Kumar, R.; Pandey, B.D.; Mehrotra, S.P. Effect of mechano-chemical activation on bioleaching of Indian Ocean nodules by a fungus. *Miner. Eng.* **2010**, *23*, 1207–1212. [CrossRef]
111. Kang, J.X.; Liu, Z.G.; Jiang, Z.X.; Wang, Y.Y.; Wang, X. The role of *Acidithiobacillus ferrooxidans* in the ocean polymetallic nodules reductive bioleaching: Electrochemical insights into interface processes. *Chem. Eng. Process. Process Intensif.* **2024**, *202*, 109869. [CrossRef]
112. Kumari, A.; Natarajan, K.A. Electroleaching of polymetallic ocean nodules to recover copper, nickel and cobalt. *Miner. Eng.* **2001**, *14*, 877–886. [CrossRef]
113. Kumari, A.; Natarajan, K.A. Electrochemical processing of ocean manganese nodules with microbial enhancement to recover valuable metals. *Min. Metall. Explor.* **2002**, *19*, 137–147. [CrossRef]
114. Kumari, A.; Natarajan, K.A. Electrobioleaching of polymetallic ocean nodules. *Hydrometallurgy* **2001**, *62*, 125–134. [CrossRef]
115. Kumari, A.; Natarajan, K.A. Cathodic reductive dissolution and surface adsorption behavior of ocean manganese nodules. *Hydrometallurgy* **2002**, *64*, 247–255. [CrossRef]
116. Matsuo, K.; Nitta, M.; Aomura, K. The activities of manganese nodules and manganese-iron mixed oxide for decomposition of isopropyl alcohol. *J. Catal.* **1978**, *54*, 446–449. [CrossRef]
117. Hai, J.; Liu, L.; Tan, W.; Hao, R.; Qiu, G. Catalytic oxidation and adsorption of Cr(III) on iron-manganese nodules under oxic conditions. *J. Hazard. Mater.* **2020**, *390*, 122166. [CrossRef] [PubMed]
118. Parida, K.M.; Mallick, S.; Dash, S.S. Studies on manganese nodule leached residues 2. Adsorption of aqueous phosphate on manganese nodule leached residues. *J. Colloid Interface Sci.* **2005**, *290*, 22–27. [CrossRef]
119. Randhawa, N.S.; Das, N.N.; Jana, R.K. Selenite adsorption using leached residues generated by reduction roasting–ammonia leaching of manganese nodules. *J. Hazard. Mater.* **2012**, *241–242*, 486–492. [CrossRef]
120. Parida, K.M.; Gorai, B.; Das, N.N. Studies on Indian ocean manganese nodules III: Adsorption of aqueous selenite on ferromanganese nodules. *J. Colloid Interface Sci.* **1997**, *187*, 375–380. [CrossRef]
121. Pan, L.; Zhang, A.-B.; Sun, J.; Ye, Y.; Chen, X.-G.; Xia, M.-S. Application of ocean manganese nodules for the adsorption of potassium ions from seawater. *Miner. Eng.* **2013**, *49*, 121–127. [CrossRef]
122. Mallick, S.; Dash, S.S.; Parida, K.M. Adsorption of hexavalent chromium on manganese nodule leached residue obtained from NH_3-SO_2 leaching. *J. Colloid Interface Sci.* **2006**, *297*, 419–425. [CrossRef] [PubMed]
123. Agrawal, A.; Sahu, K.K. Kinetic and isotherm studies of cadmium adsorption on manganese nodule residue. *J. Hazard. Mater.* **2006**, *137*, 915–924. [CrossRef] [PubMed]
124. Li, J.; Hu, Y.; Yue, L.; Cao, Z.; Li, Q.; Zeng, L.; Guan, W.; Wang, M.; Zhang, G.; Wu, S. Selective co-extraction of critical metals from deep ocean polymetallic nodule leachate and preparation of battery-grade Ni-Co-Mn solution. *Sep. Purif. Technol.* **2022**, *302*, 122096. [CrossRef]
125. Michalcová, A.; Orlíček, M.; Novák, P. Microstructure and Properties of Natural Alloy Prepared by Aluminothermic Reduction of Deep-Sea Nodules and Processed by Rapid Solidification. *Manuf. Technol.* **2024**, *24*, 235–240. [CrossRef]
126. Parhi, P.K.; Park, K.-H.; Kim, H.-I.; Park, J.-T. Recovery of molybdenum from the sea nodule leach liquor by solvent extraction using Alamine 304-I. *Hydrometallurgy* **2011**, *105*, 195–200. [CrossRef]
127. Parhi, P.K.; Park, K.H.; Nam, C.W.; Park, J.T.; Barik, S.P. Extraction of rare earth metals from deep sea nodule using H_2SO_4 solution. *Int. J. Miner. Process.* **2013**, *119*, 89–92. [CrossRef]

metals

MDPI

Review

Recent Advances in Indium Recovery

Francisco Jose Alguacil

Centro Nacional de Investigaciones Metalurgicas (CSIC), Avda, Gregorio del Amo 8, 28040 Madrid, Spain; fjalgua@cenim.csic.es

Abstract: Though indium has been removed from the fifth list (2023) of critical raw materials for the European Union list of critical metals, its recovery is still of paramount importance due to its wide use in a series of high-tech industries. As its recovery is closely associated with zinc mining, the recycling of In-bearing wastes is also of interest, for both profitable and environmental reasons. With unit operations (in hydrometallurgy and pyrometallurgy or extractive metallurgy) playing a key role in the recycling of indium, the present work reviewed the most recent innovations (2024) regarding the use of these operations in the recovery from this valuable metal from different solid or liquid wastes.

Keywords: indium; leaching; adsorption; liquid–liquid extraction; recovery; hydrometallurgy

1. Introduction

Though indium recently disappeared from the 2023-CRM list of the European Union [1], this metal is still considered to be of the utmost necessity for the development of a series of smart technologies for a number of reasons [2].

It is generally considered that indium has no independent deposits, though some minerals such as indite (indium–iron sulfide) contain it; thus, its production is associated to zinc mining, with China being the country which controls the vast majority of indium reserves. The majority of the global indium production comes from the by-products in the zinc refining process, where indium is enriched to varying degrees due to various smelting methods. This element is mainly concentrated in the furnace slag, flue dust, anode mud, leaching residues, and flotation tailings. To supplement or as an alternative to the above, waste indium tin oxide (ITO), solar cells, and alloys, among others, are secondary resources of indium, and its recycling is worth consideration. However, with only small amounts of the element in any given device, recovering indium from a post-consumer, i.e., LCD scrap, is considered cost prohibitive [3].

As in the case of other strategic metals, indium prices have seen relative growth over the past years (Table 1); on 25 October 2024, the price increased to $681.34/kg [4].

Table 1. Variation of the indium price in the last seven years (adapted from [4]).

Year	$/kg, on 1 January
2018	380.10
2019	428.61
2020	315.11
2021	315.10
2022	477.10
2023	442.30
2024	561.60

This increase in price together with the current global environmental spirit, and despite the first sentence of this introduction, makes the recovery of indium for waste material a consideration for many R + D + I (research, development and innovation) groups in academia, industry, and governmental scientific institutions.

Several recent reviews considered the recycling of indium, either from LCDs [5] to CIGS thin-film photovoltaic modules [6] and waste in general [7,8]. To help achieve positive results in recycling indium, molecular recognition technology (MRT) is also recognized for its usefulness and is used to recover critical metals (including indium) from minerals and process waste [9]. Also, eco-friendly living organisms have been reported to be useful in recycling metal [10].

The present review summarizes recent investigations (2024) on the various operations involved in the recovery of indium from various sources. These sources are solid wastes or aqueous solutions (leachates), resulting from the treatment of these solid wastes, or mimic solutions resembling real ones. Among the operations involved in this recovery, leaching, adsorption, liquid–liquid extraction, etc., are considered.

2. Research Methodology

References were taken from the Scopus "www.scopus.com" and Web of Science https://webofscience.clarivate.cn/wos/woscc/basic-search databases (both accessed on 1–31 September 2024) under the following terms: indium recovery, indium and leaching, indium and liquid–liquid extraction, indium and ion exchange resins, and indium, and adsorption, for the year 2024.

3. Leaching

Various electronic devices (smartphones, monitors, and television screens) were used to recover the indium contained within them [11]. After dismantling and milling, the leaching operation was conducted at 80 °C, using 1 M sulfuric acid as a leachant solution, two hours of reaction time, and 25% In-bearing solid. As a result, nearly 90% of the indium contained in the solid source was leached, though calcium, aluminum, and iron were dissolved in the process. No data were included in the publication about the separation of the various elements from the leachate.

The next reference [12] investigated the bioleaching of indium from discarded LCD panels, both on high pulp density shredded (Sh-LCDs) and powdered (P-LCDs) materials. Two acidophilic consortiums, the mixed sulfur-iron pathways and sulfur pathways, were used in the experimentation. Indium bioleaching efficiencies through the mixed sulfur-iron pathway were 60% and 100% for Sh-LCDs and P-LCDs, respectively, with three mechanisms involved in the extraction of indium from the samples: acidolysis (reaction of aromatic acetoxy group (or other ester-leaving group) with a carboxylic acid driven by the elimination of acetic acid (or other acid)), complexolysis (reaction between the dissolved metal and the acid or chelating agent), and redoxolysis (microorganisms produce catalytic compounds which regulate the oxidation potential of the solution). In the case of the sulfur pathway, the indium recovery rates were 10% for Sh-LCD and 55% for P-LCD simple. No data about the treatment of the solution for In recovery were included in the work.

Due to their properties (adjustable solubility, high chemical stability, tunability, low cost, etc.) [13], deep eutectic solvents (DESs) are considered a type of green solvent with multiple uses. One of these uses is to act as a substitute for mineral acids in the leaching operation of various metal-bearing materials. This work [14] utilized short-chain dicarboxylic acid-choline chloride DESs to dissolve indium tin oxide, though the type of the acid has some influence on the rate of dissolution (Table 2). The results showed that tin was leached as $[Sn^{4+}(OH^-)_3(H_2O)]^+$ (dominant species) and $[Sn^{4+}MA^{2-}(Cl^-)_3]^-$ anion. In the case of indium, the dissolved species were $[In^{3+}(Cl^-)_4]^-$ (dominant), $[In^{3+}MAH^-Cl^-)_4]^{2-}$, $[In^{3+(}MA^{2-})_3]^{3-}$,and $[In^{3+}MA^{2-}(Cl^-)_2]^-$. In the above formulations, MA stands for maleic acid. Tin was separated from indium by precipitation as SnO_2 (cassiterite) through dilution and hydrothermal treatment. From the resulting solution, indium was precipitated as $In(OH)_3$, then, the calcination of the solid at 600 °C transformed the hydroxide into In_2O_3 oxide.

Table 2. Influence of dicarboxylic acids on ITO leaching.

Acid	% In Leached	% Sn Leached
Oxalic	>99	>99
Maleic	>99	>99
Succinic	59.7	70.8

Temperature: 80 °C; time: 5 h. Adapted from [14].

The presence of lead in perovskite solar cells (PSCs) is a challenge with respect to the safe disposal of these materials [15]. In these cells, the commonly used hole transport layer is Spiro-OMeTAD, which was separated in the chlorobenzene medium. Once this layer was separated, the resulting solid was treated with aqua regia for ten minutes in order to yield a liquid phase and a solid ITO substrate, which served to fabricate new cells. Note the use of chlorobenzene: this organic solvent presented a high flammability, and though at a first instance it is considered a non-carcinogenic agent (group D) [16], other documents alleged the carcinogenic assessment of the solvent [17]. Its handling can be the cause of the development of several illnesses in humans.

The ultrasonic microwave leaching of indium from a zinc oxide dust was investigated [18]. The best conditions for the leaching operation were summarized as follows: ultrasonic power 200 W, microwave power 200 W, leaching time 30 min, sulfuric acid concentration 180 g/L, and L:S ratio 8:1. Under these conditions, the leaching rates of indium, zinc (97.69%), aluminum, (74.57%), and iron (48.62%) were reached, these results being better than those derived from the sole use of ultrasound or microwave methodologies. There were no data regarding the separation of the metals dissolved in the process.

In the treatment of zinc residues, a practical dissolution problem is often found when indium-bearing zinc ferrite (IBZF) is present in the starting material. The next reference tried to solve this problem by using a deep eutectic solvent formed by oxalic acid and choline chloride [19], with water utilized as a co-solvent and diluent. Under specific leaching conditions (temperature 90° C, liquid:solid ratio 20:1, molar ratio of oxalic acid:choline chloride 1:1, molar ratio of DES:H_2O 1:3, and reaction time 3.5 h), the leaching efficiencies of indium, zinc, and iron were 96.27%, 95.55%, and 96.33%, respectively. By the dilution of the leachate with water (leachate:water 1:4), zinc was precipitated (93.06% efficiency), and after, the ultraviolet irradiation of the supernatant produced the precipitation of iron (98.07%). Indium in the final solution was adsorbed onto activated carbon. No data were given in the manuscript about the indium-desorption step.

In the next investigation [20], a procedure was proposed for the treatment of a silver concentrate containing 3010 g/t Ag, 1.22% Pb, and 649 g/t In. The process consists of pre-oxidation followed by oxygen pressure sulfuric acid leaching. Under the optimal conditions of a sulfuric acid pre-oxidation step, concentrated sulfuric acid, temperature 60 °C, acid-to-ore mass ratio of 0.6, and 15 h, the P-O and-OH polar groups of flotation reagents on the surface of silver concentrate particles are neutralized through sulfuric acid oxidation. In the case of the oxygen pressure leaching step, the best conditions were as follows: temperature 155 °C, 160 g/L sulfuric acid, three hours of reaction time, oxygen partial pressure of 0.8 MPa, and a liquid/solid ratio of 7 mL/g. The dissolution rates were 94.8%, 98.4% and 91.1% for zinc, copper and indium, respectively, while the lead and silver grades in the leach residue increased to 2.88% and 7754 g/t, respectively. The use of sodium lignosulfonate in the oxygen pressure leaching step was mentioned, but the manuscript did not give conditions for the separation of the solubilized metals (95.1 g/L zinc, 1.23 g/L copper, 15.7 g/L total iron (0.30 g/L Fe(II)), 98.9 mg/L In in 57.8 g/L sulfuric acid medium).

The next study also considered the utilization of the oxidative pressure leaching process, but in this case a model was built on the treatment of a ZnS concentrate [21]. Using various input data derived from the literature, the model demonstrated the next efficiencies in the recovery of zinc (95.62%), sulfur (86.62%), indium (82.42%), gallium (82.55%), cadmium (85.90%), cobalt (85.57%), germanium (68.98%), and copper (62.53%), with sulfuric acid, oxygen, and lime as the main chemical inputs, while zinc electrowinning constituted over 90% of the total electricity consumption.

A thermal–biological hybrid method was utilized in the treatment of waste LCDs [22]. In the thermal treatment (20 wt% B_2O_3, 1100 °C and 70 min), metals were transferred from the glass phase to the B_2O_3 phase. After, the biometabolites of *Aspergillus niger* were used for the bioleaching of In, Sr, Al, and As from the thermally treated product. The best bioleaching conditions were as follows: pulp density 10 g/L, temperature 70° C, and a leaching time of two days, and the recovery rates of 82.6% Al, 70.8% As, 64.5% In, and 36.2% Sr were reached. No data were included about the separation recovery of the elements from the resulting solution.

The next reference [23] seemed to be a consequence of the results obtained from the previous reference [22]. Following a thermal treatment in very similar conditions as in [22], the resulting material was bioleached with *Aspergillus niger* utilizing various rate models. Kinetic evaluation indicated an overall mixed control mechanism between chemical reactivity and diffusion. Response surface methodology identified the optimal conditions as follows: temperature 70 °C and leaching time of 29 h. Under these conditions, the leaching efficiencies were as follows: 81.4% for Al, 69.1% for As, 60.0% for In, and 33.3% for Sr. As the results were very similar to those derived from [22], as in the previous reference, there was no information about the separation of the metals from the leachate.

At work, a process for recycling waste ITO targets has been investigated to recover both indium and tin [24]. Indium oxide from waste was leached using 2 M sulfuric acid and H_2O_2, at 80 °C for five hours using a L:S ratio of 4 mL:g. Under these conditions, tin remained in the same residue as SnO_2. The In-bearing solution was processed using oxalic acid, with the best conditions summarized as follows: oxalate/indium mol ratio of 1.8, 30 °C and 1 h. The precipitate ($In_2(C_2O_4)_3{\cdot}6H_2O$) from this operation was calcined (400 °C for four hours) to yield indium oxide.

It was found [25] that in the leaching (HCl + H_2O_2 medium) of spent surface mounted device (SMD) light emitting diodes (LEDs) considerable concentrations (Table 3) of accompanying metals dissolved together with indium (and gallium), thus, a replacement to the above medium is of the utmost necessity. Thus, desferrioxamine E is considered as the replacement to treat these spent SMD-LEDs. Best conditions for indium dissolution (24%) is reached at pH 7.3, 416 µM desferrioxamine E, solid:liquid ratio 5 g:L, "room temperature" (literally form the published manuscript) and one day.

Table 3. Metal concentrations in the leachate of spent LCDs.

In	Ga	Cu	Sn	Pb	Fe	Al	Si
6.4 mg/L	28.1 mg/L	29.3 g/L	3.2 g/L	1.8 g/L	0.22 g/L	0.32 g/L	0.23 g/L

Leaching solution: 3 M HCl + H_2O_2; temperature: 75 °C; time: 4 h. L:S ratio: 10. Adapted from [25].

This work [26] seemed to be a variation of the previous reference [25]. In this case, the source for indium was spent LCD screens, which were leached with the same desferrioxamine E chelating agent. The best conditions for optimal indium recovery (32%) were established as follows: pH: 5.5, S:L ratio 1 g:L, 25 °C, 5 mM chelating agent, and for 90 days. A comparison of these results with those of [25] showed important differences; in [25], indium recovery dropped after two days and, in [26], 90 days were needed to leach indium. Another important difference between both references is the optimal pH to leach indium; in [25], at a pH of 5.5 indium was poorly leached (<5%) and, in [26], this 5.5 value was the optimum for indium recovery.

4. Adsorption

In [25,26], the use of desferrioxamine E to recover indium from waste materials was mentioned. Since there are about 250 known siderophores, the selection of a suitable complexant for a given metal seemed somewhat unpracticable. Thus, the next reference [27] used a density functional theory-based approach to efficiently identify siderophores with increased selectivity towards target metals, i.e., indium. Thus, the binding properties of fimsbactin A towards indium are checked and compared with those of the reference siderophore desferrioxamine B (DFOB). The calculations showed that fimsbactin A presented 128% higher selectivity for indium than DFOB, whereas the experimental validation of the method indicated that fimsbactin A has up to 40% more selective indium binding than DFOB.

In the next reference [28], and in order to separate indium from iron (III) using electrospun fibers, the presence of fluorine-containing groups in the fibers were used to regulate the Lewis basicity of the N-containing fibers adsorption sites. The results demonstrated that in single metal solutions the adsorption capacity of both elements is the same, whereas in mixed In–Fe solutions, indium is loaded onto the fibers preferentially to iron. Desorption was performed in a sulfuric acid medium, the concentration of In(III) in the solution 73.9 being times higher than that of Fe(III).

The synthesis of a metal–organic framework adsorbent, MOF-808-TA (TA, tartaric acid) was reported [29]. Indium was better removed from the solution as the pH increased from 1 (1% efficiency) to 4 (near 40% efficiency). This element was removed (>80%) from solutions preferably to a series of other elements (<5%, Na, K, Ca, Mg, Co, Cu, Fe, Mo, Zn, Al, Sn) but Eu (near 75%) or Dy (near 80%). Indium can be desorbed (24 h) using 0.01 M HCl medium.

Two adsorbents, mesoporous silica (SBA-15) and ligand-modified (pyromellitic dianhydride) mesoporous silica SiO_2/PMDA, were used to investigate its performance on In removal from solutions [30]. SBA-15 material did not perform well in indium removal. Instead, the use of the modified silica allowed for the selective separation (pH 3) of indium in the presence of gallium, zinc, and tin. Desorption was investigated using 0.1 M nitric acid solutions.

The use of P507@MAC adsorbent on indium removal was investigated in the next reference [31]. This adsorbent resulted from the impregnation of P507 extractant (2-ethylhexyl 2-ethylhexyl phosphate, $C_{16}H_{35}O_3P$, (R(RO)POOH)) on coconut shell-derived mesoporous activated carbon. Indium was removed well from the solution at pH 1, allowing for the selective separation of this element from others (Ga, Cu, Cd, Zn). Indium desorption is affordable at moderate (3–5 M) nitric acid concentrations. The adsorbent was utilized in the separation of gallium and indium from CIGS photovoltaic modules leach solution (pH 1). After two consecutive steps, indium was completely removed from the solution, leaving gallium and other elements in it. Gallium was then removed from the solution (pH 2).

Since the recovery of indium from tail liquid zinc ore has difficulties due to the large concentrations of zinc and acid present in the solution, an adsorbent derived from the introduction of a network of zirconium glyphosine on carbon nanotubes was formed [32]. The composite (xZrGP-CNT) (x denoted the mass ratio of glyphosine to $ZrCl_4$) was used in the adsorption of indium from synthetic solutions. The maximum metal adsorption (around 80%) was derived from experiments carried out at pH values 3.5–4, increasing the removal percentage to around 90% when x = 2 or 3 and widening the pH range until 2. It was concluded that the mechanism of indium adsorption is due to electrostatic attraction, ion exchange reaction, and coordination interaction. Indium(III) was preferentially adsorbed (separation factors in the 1.7 value order) in the presence of Fe(II):Al(III):Zn(II):Mg(II):Ca(II) using a solution (unknown pH) with a concentration ratio of 1(In):1.5(Fe):5(Al):30(Zn):0.6(Mg):0.7(Ca). About 92% of indium was desorbed using 1 M nitric acid solution.

In the next reference [33], a nanofiber (PCA-Nanofiber) was fabricated by introducing an electronegative fluorine-containing group to modulate the electronic structure of adsorption sites used on indium recovery. In the present case, maximum indium adsorption occurred at pH 2, whereas desorption was performed using 1 M sulfuric acid.

Nanoscale metal–organic framework UiO-66-$(COOH)_2$ particles were embedded in sodium alginate (SA)-calcium structure to fabricate UiO-66-$(COOH)_2$/SA composite beads (approximate size of 3 mm) [34]. This adsorbent removes indium from solutions of pH 4, whereas 0.05 M HCl solutions were used to desorb indium from metal-loaded beads.

Though it is complex to compare the performance of the various adsorbents in terms of indium uptake, mainly due to the very different experimental conditions used in the investigations, Table 4 summarizes the different maximum indium loadings on selected adsorbents. With greater discrepancies between the results presented in [30] and the other references, it is worth noting here the close similarity in the adsorbents used in [28] (published online 21 December 2023) and [33] (published online 26 November 2023) and also in the maximum indium uptakes.

Table 4. Maximum indium uptakes onto different adsorbents.

Adsorbent	$^{a}[In]_{max}$, mg/g	Reference
PCA-nanofibers	135.7 (single In solution)	[28]
PCA-nanofibers	111.6 (mixed In-Fe(III) sol.)	[28]
MOF-808-TA	172.4	[29]
SBA-15	5.2	[30]
SiO_2-PMDA	38.0	[30]
2ZrGP-CNT	193.7	[32]
PCA-nanofibers	137.9	[33]
AM-nanofiber	43.1	[33]

[a] Values from the Langmuir isotherm model.

5. Liquid–Liquid Extraction

Crown ethers (CEs) with varied cavity sizes, benzo-18-crown-6 ($B_{18}C_6$), benzo-15-crown-5 ($B_{15}C_5$), and benzo-12-crown-4 ($B_{12}C_4$), were synthesized [35] and used in the extraction of indium from a KCl medium [36]. With $InCl_4^-$ being the predominant metal-species in the aqueous phase, the extraction order followed the sequence $B_{18}C_6 > B_{15}C_5 > B_{12}C_4$. The metal extraction was attributed to the coordination of the crown ether with K^+ to form $[CE\text{-}K]^+$ at the two-phase interface, which further associates with $InCl_4^-$ to create the complex of CE-$KInCl_4$. Moreover, $B_{18}C_6$ showed excellent selectivity toward In^{3+} over competing ions such as Fe^{3+}, Al^{3+}, Zn^{2+}, Sn^{2+} and Ca^{2+} in a mixed system. Indium was recovered (98.1% efficiency) from the loaded organic phase by using 1 M HCl as a strippant. Dichloromethane was used to dissolve the extractants and this misuse or poor choice of the diluent clearly precluded the practical use of these systems due to the harmful characteristics of this chemical [16,17].

Due to their properties, ionic liquids in general and particularly phosphonium-based ionic liquids are being considered as extractants for many metals. Cyphos IL104 (tri-hexyl(tetradecyl)phosphonium bis(2,4,4-trimethylpentyl)phosphinate dissolved in toluene was used to extract indium from a feed solution derived from the leaching of discarded LCD screens [37,38]. From the In-Sn-loaded organic solution, indium was selectively stripped with 4 M nitric acid. In a leaching-extraction procedure, the ionic liquid $[P_{44410}][Br_3]$ (tributyldecylphosphonium bromide) was used to recover indium (and gallium) from GaAs and InAs semiconductors [38,39]. After this first step, arsenic was first recovered by stripping with 4 M NaBr, then gallium was stripped using water, and finally indium was recovered from the organic phase using 0.25–1.3 M NaOH solutions as strippants.

Organic phases of D2EHPA (di(2-ethylhexyl)phosphoric acid) and 2-ethylhexanol in kerosene or petroleum paraffins were used to extract indium [40]. The optimum organic phase composition to extract this metal contained 15.8% extractant and 4.7% alcohol. In this extraction stage, indium recovery was in the 97.5–99.5 range. Mixtures of sulfuric acid and sodium chloride were used as strippants with efficiencies of around 99%. Though in the Abstract it is mentioned that the introduction of the alcohol into the organic phase helps to improve the indium "re-extraction" compared to D2EHPA alone, it is more than possible that the use of the alcohol prevented the formation of a third phase (or second organic phase) after the disengagement phase either in the extraction and/or stripping stages. The main problem with using D2EHPA to extract indium is the co-extraction of iron(III) [41]

The extraction of different technologically critical elements from leach solutions of multiple sources using ionic liquids was reviewed [42]. Generally, these extractants have low selectivity toward Ga, Ge, and In compared to Fe(III), Al(III), and Zn(II), which are typically present at high concentrations in the multielemental solutions.

Spent liquid crystal display panels were used as a source for indium recovery. The treatment of these panels [43] led to a solution which was processed using an ethanol-mediated dodecylamine (DDA) phase to which n-hexane was added prior to the separation phase [44]. Indium extraction (about 90%) occurred at pH values in the 7.5–9 range; however, tin, aluminum, and iron were extracted within the same efficiency, as well as copper (80%). In(III) loaded in the organic phase was then converted into InOOH nanoparticles with "abundant" oxygen vacancies through a low-temperature hydrothermal process. The nanoparticles (polyhedral morphology with average size 15.4 nm) were used to electrocatalyze CO_2 reduction.

6. Miscellaneous Operations

Besides the most used unit operations in the recovery of indium and reviewed above. Scientists try to resolve the problem using different methodologies. The results are reviewed below.

Ultrasound-enhanced calcium carbonate precipitation was used to increase indium recovery on a zinc oxide dust leachate (68 g/L Zn(II), 0.56 g/L In(III), 1.5 g/L Fe(II), 0.4 g/L Al(III)) [45]. The results of this investigation showed that the best conditions were as follows: a precipitation endpoint pH of 4.0 and an ultrasound power of 200 W. The precipitation yield of indium under these conditions was 99.79%. Compared with conventional precipitation, the utilization of ultrasounds reduced both the time to reach equilibrium, and the amount of calcium carbonate used in the process. It should be noted that together with indium, aluminum was precipitated within the same rate order (96.75%), whereas the rate of zinc precipitation reached 29.36%. Thus, the precipitated $In(OH)_3$ may be contaminated to some extent by these two elements.

Membranes, and particularly supported liquid membranes, are a widely proposed methodology in the removal of metals from liquid solutions or effluents [46]. In this methodology, an extractant or an organic phase (extractant and diluent as used in liquid–liquid extraction) are immobilized in the micropores of a hydrophobic solid support, forming the supported liquid membrane. Indium was recovered from an aqueous solution using one of these liquid membranes, acting Cyphos IL104 diluted in kerosene as the organic phase for the transport process [47]. The following five process parameters were evaluated: indium concentration (10–100 mg/L), extractant concentration (0.05–0.2 M), feed phase acidity (0.01–3 M nitric acid), chloride ion concentration (0.5–4 M), and the stripping agent concentration (0.1–5 M nitric acid). Experimental results were compared with the models resulting in the utilization of the response surface methodology (RSM with $r^2 = 0.9589$) and artificial neural network (ANN with $r^2 = 0.9860$)). The best-optimized data were as follows: 73.92 mg/L (In), 0.157 M (Cyphos IL104), 1.386 M (nitric acid in the feed phase), 2.99 M (chloride ion), and 3.06 M (nitric acid in the stripping phase).

Copper indium gallium selenide solar cells were used as the source phase for indium recovery using microwave pyrolysis, thermal oxidation, and thermal chlorination technologies [48]. Under microwave pyrolysis conditions (332–430 °C and 200–300 W), the cells lost about 90% of their weight. Selenium was then recovered using thermal oxidation at 800 °C for 30 min; this time was increased to 40 min if the temperature decreased to 600 °C. Further, a first thermal chlorination step (two hours at 260 °C) allowed the recovery of gallium (56% efficiency with 100% pure gallium), and a second step (two hours at 340 °C) allowed the recovery of gallium (40% efficiency and 20% purity) and indium (46% efficiency and 55% purity). Gallium can be completely recovered in the first stage if longer reaction times were used. If this option was chosen, the purity of indium would rate near

68%, this being an unsatisfactory result. To increase this rate, the usage of one or more additional technologies is necessary to achieve the complete separation of indium and other metals.

Gallium-based liquid metal (G-LM) waste is produced during the preparation of liquid metals and related products, this being a waste source of gallium and indium. In the next investigation [49], a two-stage vacuum distillation method was used to recover Ga–In alloy from these wastes. Firstly, low-temperature vacuum distillation was used to recover 98.23% of Zn under optimal conditions (1–5 Pa, 500 °C, one hour). Secondly, the Ga–In alloy was recovered through high-temperature vacuum distillation. Under the optimal conditions (1–5 Pa; 1250 °C, two hours), the recovery rates of gallium and indium were 85.14 and 95.23%, and the content of the Ga–In alloy in the product was 99.46%.

The next study [50] presents evaporation parameters and the characteristics of indium in the process of vacuum distillation to direct the refining of this metal using vacuum distillation. The investigation was carried out on a 92In–8Sn alloy. The following was concluded: (i) the measured volatilization rate of indium is temperature-dependent in the 900–1100 °C range (5 Pa and 25 mm crucible depth); (ii) the volatilization rate of indium in the alloy was found to be proportional to the crucible depth (25–65 mm at 1050 °C and 5 Pa); (iii) the determined mass transfer coefficient for indium showed positive and negative correlations with temperature and crucible depth, respectively; and (iv) a kinetic model for indium volatilization was derived and showed that the volatilization of indium in the alloy is mainly constrained by the transfer of mass in the vapor phase.

As is widely known, end-of-life LEDs are a source for the recovery of indium or rare earth elements. The next investigation [51] used these LEDs to recover indium by a prior thermochemical step with hydrochloric acid-containing vapor (650 °C) from polyvinyl chloride as a chlorine source. The results from this operation showed that indium is mainly concentrated (42–70 ppm) in liquid products, especially in the water phase. Indium can be recovered from the aqueous phase by electrolytic extraction after its concentration has been increased by reduction or by the removal of water [52]. Further, the LED residue is leached with sulfuric acid (32%) and hydrogen peroxide in a L:S ratio of 10. From the leachate, the heavy rare earth element lutetium can be extracted tributyl phosphate/1-octanol.

The infamous copper indium gallium selenide solar cells were used for the recovery of silver and indium tin oxide particles [53]. The process to achieve this used two-step ultrasonic leaching with 0.1 M nitric acid solutions. In the first step, indium tin oxide particles were liberated through the selective dissolution of the zinc-rich layer underneath, using low ultrasound power for 3 min. In the second step, the same conditions as in the first step were applied, but now a high-power ultrasound was used for 15 min to recover the silver grid lines. Both the indium tin oxide (70.5 wt% purity) and the silver grid (95.0 wt% purity) particles were recovered by filtration, and there was no loss of silver in the leachates, though some indium was dissolved, together with zinc, in the first step. Copper indium gallium selenide material remained in the substrate of the second step and some losses of this material were produced along the various steps of the process.

Scraps from indium phosphide (InP) semiconductors are another source for indium. However, the recycling of these INP scraps (78.2% In, 18.3% P) presented some challenges mainly due to phosphorus flammability and the release of toxic and irritating smokes during combustion. To resolve these challenges, the next investigation proposed a "controlled-pressure pyrolysis–spray condensation" process [54]. Firstly, InP is decomposed (1050 °C, micronegative pressure for five hours) into phosphorus and indium; phosphorus then evaporates into the vapor phase and is sprayed with warm water (55 °C) for condensation in this same medium. Industrial α-yellow phosphorus with a purity greater than 98 wt% (recovery rates exceeding 90%) was obtained and stored in water. From the first step, an indium ingot (99.59 wt% purity) was recovered (98.2% efficiency rate).

7. Conclusions

Though nowadays the EU does not consider indium as a critical metal, the recovery of this metal from different wastes is still conducted. Apparently, there is not a consolidated market for recycling some (if not all) of these wastes with a specific emphasis on indium recovery.

In the leaching step, future research should focus on enhancing the amplification technology of ultrasonic microwave leaching equipment. Also, the utilization of siderophores in eco-friendly technologies should be taken into consideration for the recovery of indium (and many different critical metals) from various industrial waters and leachates or bioremediation approaches. The potential uses of ionic liquids and deep eutectic solvents to substitute more aggressive acidic compounds in the leaching operation cannot be forgotten. As substitutes to these mineral acids, the use of organic acids may be considered.

Adsorption processing is a technology to be considered, even though a series of adsorbents tend to lose their initial adsorption properties under continuous use; thus, one of the future hot spots in this field should be the obtention of more stable adsorbents under more extreme working conditions, i.e., acidities of the feed or desorption solutions.

In the case of liquid–liquid extraction, the rational design of ionic liquids and task-specific ionic liquids, their recycling, process scaling-up, and detailed economic viability studies need to be adequately investigated to accelerate the potential commercialization of these chemicals in the processing of indium.

Of course, the development of alternative technologies to the established ones is always of interest.

In any case, the real value of all the above technologies must be proven in the recovery of indium from real waste solid materials, or in the treatment of real leach solutions, and under continuous use (pilot plant); thus, efficiency in the recovery and selectivity of indium must be areas to consider.

Funding: This research received no external funding.

Data Availability Statement: No new data were created or analyzed in this study. Data sharing is not applicable to this article.

Acknowledgments: To the CSIC (Spain) and the Editorial Office for waiving the APC.

Conflicts of Interest: The author declares no conflicts of interest.

References

1. Grohol:, M.; Veeh, C. *Study on the Critical Raw Materials for the EU 2023-Final Report*; Publications Office of the European Union: Brussels, Belgium, 2023. [CrossRef]
2. Muller, B.; Van de Velde, K. Indium as Critical and Strategical Raw Material. Imat e.V. 2023. Available online: https://imat-ev.eu/ (accessed on 25 September 2024).
3. American Chemical Society Green Chemistry Institute. Indium: An Illuminating Element. ACS. 2024. Available online: https://www.acs.org/greenchemistry/research-innovation/endangered-elements/indium.html (accessed on 20 September 2024).
4. Available online: https://strategicmetalsinvest.com/indium-prices/ (accessed on 16 September 2024).
5. Meng, J.; Chen, G.; Xu, L.; Xia, D.; Zhao, Z. Recycling of indium from waste LCD. *Xiyou Jinshu/Chin. J. Rare Met.* **2024**, *48*, 883–894. [CrossRef]
6. Wang, J.; Feng, Y.; He, Y. Advancements in recycling technologies for waste CIGS photovoltaic modules. *Nano Energy* **2024**, *128*, 109847. [CrossRef]
7. Huang, Y.; Wang, M.; Liu, B.; Su, S.; Sun, H.; Yang, S.; Han, G. The extraction and separation of scarce critical metals: A review of gallium, indium and germanium extraction and separation from solid wastes. *Separations* **2024**, *11*, 91. [CrossRef]
8. Kluczka, J. A review on the recovery and separation of gallium and indium from waste. *Resources* **2024**, *13*, 35. [CrossRef]
9. Huang, Y.; Qiu, Y.; Zhang, Z.; Wang, W.; Peng, W.; Cao, Y. A perspective on molecular recognition technology for recovering critical metals from minerals and processing wastes. *Sep. Purif. Technol.* **2024**, *347*, 127734. [CrossRef]
10. Golzar-Ahmadi, M.; Bahaloo-Horeh, N.; Pourhossein, F.; Norouzi, F.; Schoenberger, N.; Hintersatz, C.; Chakankar, M.; Holuszko, M.; Kaksonen, A.H. Pathway to industrial application of heterotrophic organisms in critical metals recycling from e-waste. *Biotechnol. Adv.* **2024**, *77*, 108438. [CrossRef]

11. Castro, J.P.; Garcia, J.A.; Souza, R.G.; Pereira-Filho, E.R. Indium recovery from end-of-life e-waste: Important details related to spectroanalytical determination and recycling viability. *J. Brazilian Chem. Soc.* **2024**, *35*, e-20230201. [CrossRef]
12. Constantin, A.; Pourhossein, F.; Ray, D.; Farnaud, S. Investigating the acidophilic microbial community's adaptation for enhancement indium bioleaching from high pulp density shredded discarded LCD panels. *J. Environ. Manag.* **2024**, *365*, 121521. [CrossRef]
13. Gao, Y.; Fan, M.; Cheng, X.; Liu, X.; Yang, H.; Ma, W.; Guo, M.; Li, L. Deep eutectic solvent: Synthesis, classification, properties and application in macromolecular substances. *Int. J. Biol. Macromol.* **2024**, *278*, 134593. [CrossRef]
14. Jin, X.; Jin, B.; Rao, L.; Cao, K.; Huang, Z.; Chen, F.; Huang, Q. Separation of indium and tin from ITO powders with short-chain dicarboxylic acid-ChCl deep eutectic solvents: Indium tin leaching and splitting mechanism. *Proc. Saf. Environ.* **2024**, *185*, 1268–1276. [CrossRef]
15. Le Khac, D.; Chowdhury, S.; Soheil Najm, A.; Luengchavanon, M.; mebdir Holi, A.; Shah Jamal, M.; Hua Chia, C.; Techato, K.; Selvanathan, V. Efficient laboratory perovskite solar cell recycling with a one-step chemical treatment and recovery of ITO-coated glass substrates. *Sol. Energy* **2024**, *267*, 112214. [CrossRef]
16. Grinstead, K. U.S. Environmental Protection Agency. Available online: https://www.epa.gov/sites/default/files/2015-04/documents/chlorobenzene.pdf (accessed on 16 September 2024).
17. Unnamed. IRIS Advanced Search. U.S. Environmental Protection Agency. Available online: https://cfpub.epa.gov/ncea/iris/search/index.cfm (accessed on 16 September 2024).
18. Li, S.; Wang, H.; Wang, S.; Xie, F. Study on the kinetics and mechanism of ultrasonic-microwave synergistic enhancement for leaching indium from zinc oxide dust. *Chem. Pap.* **2024**, *78*, 3667–3685. [CrossRef]
19. Liu, J.; Chen, B.; Huang, Y.; Cao, Y.; Chen, J.; Wang, L.; Liu, Y.; Fan, Y. Efficient and clean treatment of indium-bearing zinc ferrite: A new approach using a water-regulated deep eutectic solvent. *Sep. Purif. Technol.* **2024**, *347*, 127576. [CrossRef]
20. Liu, Q.; Li, C.X.; Gu, Z.H.; Li, C.W.; Wang, Q.L. Study on pre-oxidation of silver concentrate and leaching behaviour of Zn, Cu and In during oxygen-pressure leaching. *Hydrometallurgy* **2024**, *228*, 106358. [CrossRef]
21. Nan, T.; Yang, J.; Aromaa-Stubb, R.; Zhu, Q.; He, H.; Lundström, M. Extracting valuable metals from zinc sulfide concentrate: A comprehensive simulation-based life cycle assessment study of oxidative pressure leaching. *Miner. Eng.* **2024**, *216*, 108888. [CrossRef]
22. Parsa, A.; Bahaloo Horeh, N.; Mousavi, S.M. A hybrid thermal-biological recycling route for efficient extraction of metals and metalloids from end-of-life liquid crystal displays (LCDs). *Chemosphere* **2024**, *352*, 141408. [CrossRef]
23. Parsa, A.; Bahaloo-Horeh, N.; Mousavi, S.M. A kinetic study of indium, aluminum, arsenic, and strontium extraction from LCDs using biometabolites produced by *Aspergillus niger*. *Miner. Eng.* **2024**, *205*, 108441. [CrossRef]
24. Xu, L.; Chen, G.; Zhang, X.; Yang, Y.; Leng, C.; Yang, C.; Tian, Y.; Zhao, Z. Waste ITO target recycling for efficient indium recovery through a closed-loop process. *J. Environ. Chem. Eng.* **2024**, *12*, 112136. [CrossRef]
25. Zheng, K.; Benedetti, M.F.; Jain, R.; Pollmann, K.; van Hullebusch, E.D. Recovery of gallium (and indium) from spent LEDs: Strong acids leaching versus selective leaching by siderophore desferrioxamine E. *Sep. Purif. Technol.* **2024**, *338*, 126566. [CrossRef]
26. Zheng, K.; Benedetti, M.F.; Jain, R.; Guy, B.M.; Pollmann, K.; van Hullebusch, E.D. Selective leaching of indium from spent LCD screens by siderophore desferrioxamine E. *J. Hazard. Mater.* **2024**, *469*, 134013. [CrossRef]
27. Hintersatz, C.; Tsushima, S.; Kaufer, T.; Kretzschmar, J.; Thewes, A.; Pollmann, K.; Jain, R. Efficient density functional theory directed identification of siderophores with increased selectivity towards indium and germanium. *J. Hazard. Mater.* **2024**, *478*, 135523. [CrossRef] [PubMed]
28. Li, M.; Zhang, L.; Xu, D.; Cheng, N.; Wang, M.; Shao, P.; Yang, L.; Yao, Z.; Zhao, C.; Feng, J.; et al. Deep separation between In(III) and Fe(III) ions by regulating the Lewis basicity of adsorption sites on electrospun fibers. *Adv. Funct. Mater.* **2024**, *34*, 2313443. [CrossRef]
29. Li, P.; Liu, S.; Tian, H.; Li, K.; Zhao, X. An anionic metal-organic framework decorated with tartaric acid for enhanced adsorption of indium ions. *Micropor. Mesopor. Mat.* **2024**, *370*, 113059. [CrossRef]
30. Protsak, I.; Stockhausen, M.; Brewer, A.; Owton, M.; Hofmann, T.; Kleitz, F. Enhanced selective extraction of indium and gallium using mesoporous sorbents. *Chem. Eng. J.* **2024**, *498*, 154468. [CrossRef]
31. Wang, W.; Xu, X.; Lai, D.; Xu, Q.; Li, J.; Wang, Y. Selective isolation of gallium and indium from waste photovoltaic modules facilitated by extractant-mesoporous activated carbon composites. *Sep. Purif. Technol.* **2024**, *330*, 125510. [CrossRef]
32. Xiong, Y.; Rao, T.; Lou, Z.; Cui, J.; Yu, H.; Feng, X. A novel zirconium glyphosine carbon nanotube composite for fast and ultraeffective In(III) capture. *Sep. Purif. Technol.* **2024**, *350*, 128004. [CrossRef]
33. Zhang, L.; Zhao, B.; Wang, L.; Shao, P.; Yang, L.; Zhao, C.; Wang, L.; Yang, J.; Yang, H.; Li, M. Regulating the electronic microenvironment of adsorption sites in nanofiber for promoting In(III) capture performance. *Sep. Purif. Technol.* **2024**, *332*, 125880. [CrossRef]
34. Zhao, X.; Zhang, X.; Gao, X.; Liu, B. Facile preparation of metal-organic framework beads for effective capture of indium ion. *J. Solid State Chem.* **2024**, *329*, 124452. [CrossRef]
35. Yao, D.; Ge, T.; Xu, L.; Chen, G.; Yao, C.; Yang, C.; Tian, Y.; Zhao, Z. Complexation mechanism of crown ether with indium in the presence of KI: Toward efficient recovery of indium from secondary resources. *Sep. Purif. Technol.* **2023**, *308*, 122936. [CrossRef]
36. Chen, G.; Xiong, Y.; Xu, L.; Yao, C.; Zhang, X.; Yang, C.; Tian, Y. Recovery of indium by solvent extraction with crown ether in the presence of KCl and stripping with HCl: A mechanistic study. *Hydrometallurgy* **2024**, *229*, 106378. [CrossRef]

37. Dhiman, S.; Gupta, B. Cyphos IL 104 assisted extraction of Indium and recycling of indium, tin and zinc from discarded LCD screen. *Sep. Purif. Technol.* **2020**, *237*, 116407. [CrossRef]
38. Dhiman, S.; Agarwal, S.; Gupta, H. Application of phosphonium ionic liquids to separate Ga, Ge and In utilizing solvent extraction: A review. *J. Ion. Liq.* **2024**, *4*, 100080. [CrossRef]
39. Van den Bossche, A.; Vereycken, W.; Vander Hoogerstraete, T.; Dehaen, W.; Binnemans, K. Recovery of gallium, indium, and arsenic from semiconductors using tribromide ionic liquids. *ACS Sustain. Chem. Eng.* **2019**, *7*, 11451–11459. [CrossRef]
40. Grigorieva, N.A.; Fleitlikh, I.Y.; Zagrebin, S.A.; Kozlov, K.M. Indium extraction from sulfuric acid solutions of zinc production. *Tsvet. Metall.* **2024**, *2024*, 40–45. [CrossRef]
41. Zheng, K.; Benedetti, M.F.; van Hullebusch, E.D. Recovery technologies for indium, gallium, and germanium from end-of-life products (electronic waste)—A review. *J. Environ. Manag.* **2023**, *347*, 119043. [CrossRef]
42. Silwamba, M.; Chibesa, M.; Oteng, E.V.; Alagha, L. Extraction of critical elements-gallium, germanium, and indium-using ionic Liquids: A review. *Miner. Process. Extr. Metall. Rev.* **2024**, 1–25. [CrossRef]
43. Zhang, K.; Li, B.; Wu, Y.; Wang, W.; Li, R.; Zhang, Y.; Zuo, T. Recycling of indium from waste LCD: A promising non-crushing leaching with the aid of ultrasonic wave. *Waste Manag.* **2017**, *64*, 236–243. [CrossRef]
44. Wu, Y.; Ren, Z.; Liu, H.; Guo, F.; Tian, S.; Yang, J. Phase transfer-assisted indium recovery from spent liquid crystal display panels and its extension in preparing indium-based electrocatalysts for CO_2 reduction. *Chem. Eng. J.* **2024**, *492*, 152119. [CrossRef]
45. Li, S.; Wang, H.; Wang, S.; Xie, F.; Sun, X. Mechanism and kinetics of ultrasound-enhanced $CaCO_3$ precipitation for indium enrichment in zinc oxide dust leaching solution. *Ultrason. Sonochem.* **2024**, *110*, 107046. [CrossRef]
46. Kaczorowska, M.A. The latest achievements of liquid membranes for rare earth elements recovery from aqueous solutions—A mini review. *Membranes* **2023**, *13*, 839. [CrossRef]
47. Hemmati, A.; Asadollahzadeh, M.; Torkaman, R. Assessment of metal extraction from e-waste using supported IL membrane with reliable comparison between RSM regression and ANN framework. *Sci. Rep.* **2024**, *14*, 3882. [CrossRef] [PubMed]
48. Huang, Y.-F.; Chen, Y.; Chiueh, P.-T.; Lo, S.-L. Metal recovery from copper indium gallium selenide solar cells by using microwave pyrolysis, thermal oxidation and thermal chlorination. *Process Saf. Environ.* **2024**, *190*, 226–232. [CrossRef]
49. Li, Z.; Chen, Z.; Ma, W.; Cai, C.; Li, S. Clean and efficient recovery of gallium–indium alloy from gallium-based liquid metal waste using a two-stage vacuum distillation method. *Vacuum* **2024**, *221*, 112884. [CrossRef]
50. Pang, J.; Dong, C.; Xu, B.; Wu, H.; Kong, L.; Yang, B. Kinetics of indium separation and recovery from In–Sn alloys in ITO waste target products via vacuum evaporation. *Vacuum* **2024**, *222*, 113076. [CrossRef]
51. Peer, M.; Fehn, T.; Hofmann, A.; Berninger, B.; Kunz, W. Recycling of critical metals from light-emitting diodes. *Chem. Ing. Tech.* **2024**, *00*, 1–6. [CrossRef]
52. Illes, B.I.; Kekesi, T. The relative efficiency of electrowinning indium from chloride electrolytes. *J. Appl. Electrochem.* **2022**, *53*, 271–284. [CrossRef]
53. Teknetzi, I.; Click, N.; Holgersson, S.; Ebin, B. An environmentally friendly method for selective recovery of silver and ITO particles from flexible CIGS solar cells. *Sust. Mater. Technol.* **2024**, *39*, e00844. [CrossRef]
54. Zhang, J.; Kong, L.; Yang, B.; Xu, B. Clean and sustainable recovery of valuable materials from InP scrap via controlled-pressure pyrolysis–spray condensation. *J. Clean. Prod.* **2024**, *475*, 143613. [CrossRef]

MDPI AG
Grosspeteranlage 5
4052 Basel
Switzerland
Tel.: +41 61 683 77 34

Metals Editorial Office
E-mail: metals@mdpi.com
www.mdpi.com/journal/metals

www.ingramcontent.com/pod-product-compliance
Lightning Source LLC
Chambersburg PA
CBHW061436210926
51850CB00008B/1526
9783725864423